Proceedings of the Lebedev Physics Institute
Academy of Sciences of the USSR
Series Editor: N.G. Basov

The Proceedings of the Lebedev Physics Institute of the Academy of Sciences of the USSR

Series Editor: Academician N.G. Basov

Volume 165 **Solitons and Instantons, Operator Quantization,** Edited by V.L. Ginzburg.

Volume 166 **The Nonlinear Optics of Semiconductor Lasers,** Edited by N.G. Basov.

Volume 167 **Group Theory and the Gravitation and Physics of Elementary Particles,** Edited by A.A. Komar.

Volume 168 **Issues in Intense-Field Quantum Electrodynamics,** Edited by V.L. Ginzburg.

Volume 169 **The Physical Effects in the Gravitational Field of Black Holes,** Edited by M.A. Markov.

Volume 170 **The Theory of Target Compression by Longwave Laser Emission,** Edited by G.V. Sklizkov.

Volume 171 **Research on Laser Theory,** Edited by A.N. Orayevskiy.

Volume 172 **Phase Conjugation of Laser Emission,** Edited by N.G. Basov.

Volume 173 **Quantum Mechanics and Statistical Methods,** Edited by M.M. Sushchinskiy.

Volume 174 **Nonequilibrium Superconductivity,** Edited by V.L. Ginzburg.

Volume 175 **Luminescence Centers of Rare Earth Ions in Crystal Phosphors,** Edited by M.D. Galanin.

Volume 176 **Classical and Quantum Effects in Electrodynamics,** Edited by A.A. Komar.

Proceedings of the Lebedev Physics Institute
Academy of Sciences of the USSR
Series Editor: N.G. Basov
Volume 174
Supplemental Volume 2

ELECTRON LIQUID THEORY OF NORMAL METALS

Edited by V.P. Silin

Translated by Kevin S. Hendzel

NOVA SCIENCE PUBLISHERS
COMMACK

NOVA SCIENCE PUBLISHERS
283 Commack Road
Suite 300
Commack, New York 11725

This book is being published under exclusive English Language rights granted to Nova Science Publishers, Inc. by the All-Union Copyright Agency of the USSR (VAAP).

Library of Congress Cataloging-in-Publication Data

Teorii͡a ėlektronnoĭ zhidkosti normalʹnykh metallov.
English.
Electron fluid theory of normal metals.

Translation of: Teorii͡a ėlektronnoĭ zhidkosti normalʹnykh metallov.
"Proceedings of the Lebedev Physics Institute of the Academy of Sciences of the USSR ... v. 174. Supplemental volume 2."
1. Fermi surfaces. 2. Quantum field theory. 3. Free electron theory of metals. 4. Cyclotron waves. I. Silin, V. P. (Viktor Pavlovich) II. Trudy Fizicheskogo instituta. English ; 174 (Supplement 2) III. Title.
QC176.8.F4T4613 1988 530.4'1 88-1619
ISBN 0-941743-19-5

The original Russian-language version of this book was published by Nauka Publishing House in 1985.

Printed in the United States of America

CONTENTS

Quantum Spin-Acoustic Waves in Metals

V.I. Okulov, V.P. Silin

Abstract: Results are given from the theory of quantum spin-acoustic waves: longitudinal waves that may propagate in a metal when quantization of conduction electron motion in a magnetic field is significant. Initial equations describing such waves are discussed in detail and an analysis of their spectrum is provided. The interaction between quantum waves and oscillations in the ion lattice of the metal is examined. The influence of Fermi-liquid interaction of the electrons on the spectrum of quantum waves is analyzed. Equations are given that describe the interaction of quantum spin-acoustic waves with helicons when propagating at an angle with respect to the magnetic field. The existence conditions and possibilities for observing quantum waves are discussed.

INTRODUCTION

Unique properties appear in metals in strong magnetic fields and at low temperatures due to the fact that the conduction electrons are on discrete quantum levels [1, 2]. The energy of electron motion transverse to the magnetic field is a multiple of the cyclotron quantum energy $\hbar\Omega$, and this has been detected in a variety of effects where the quantity $\hbar\Omega$ exceeds the thermal energy T. Among the interesting phenomena in metals determined by such powerful (or commonly called quantizing) fields are a number of phenomena under the heading of "quantum waves." Quantum wave theory was developed rather long ago, although there has been no experimental investigations of such waves. Among other reasons this is due, in our view, to insufficient knowledge of the predictions and fundamental physical concepts of quantum wave theory. This survey chapter will be devoted to this gap.

In introducing the reader to the many issues associated with quantum waves we will briefly outline the simplest concepts of quantum wave physics, as well as the history of this area. We should focus on the primary reason for the existence of quantum waves which is related to the fact that the velocity of electrons with Fermi energy will adopt only discrete values due to both the discreteness of the quantum energy levels of an electron in a magnetic field and due to the fact that the sharp boundary on the Fermi electron distribution is equal in the limit to zero temperature. Fig. 1 shows the quantum energy levels as a function of the value of a continuous parameter: the projection of electron velocity v_z onto the magnetic field direction. The intersection of the energy terms with the Fermi level (the energy is equal to the Fermi energy ζ) determines the discrete set of Fermi velocities of the conduction electrons. This property of the conduction electrons in a quantizing magnetic field is also responsible for the existence of quantum waves which was noted in the first studies devoted to quantum wave theory [3, 4].

The discrete values of the electron Fermi velocities also results in a picture where Cherenkov absorption of waves by electrons which produces Landau damping is in the longwave limit allowed only with corresponding and discrete values of the wave phase velocities. Even before quantum wave research the restriction of Cherenkov electron interaction in a quantizing magnetic field was employed in study [5] to predict quantum oscillations in the acoustic absorption coefficient. In addition to forbidden Cherenkov absorption of waves by electrons in specific frequency ranges, the discrete values of the Fermi velocities makes it possible to represent the system of electrons as a set of different electron groups belonging to different Landau levels in a magnetic field. It is precisely the suppression of Landau damping and the grouping of electrons into groups with different Fermi velocities that has made it possible to predict the generation of new (quantum) waves. A certain analogy of quantum waves may be seen in the phenomenon of plasma acoustic waves discussed earlier for metals with several types of current carriers [6].

The initial concepts used to describe quantum waves in studies [3, 4] that are based on a model of noninteracting conduction electrons neglecting their spin, were repeated and developed independently in studies [7-10]. The coupled quantum waves and helicons were investigated using the same physical model in study [11] where the term "quantum waves" was first introduced. Transverse electromagnetic quantum waves were also investigated in metals [12] and semimetals [13]. Demikhovskiy and Protogenov provided a survey of the material from the research outlined above as well as subsequent investigations of quantum waves limited to the scope of a simple electron gas model in 1976 [14]. As a result of this limitation study [14] did not address a wide range of issues that have been investigated in rather extensive detail related to accounting for electron spin, interelectron interaction and the interaction of electron oscillations with the ion lattice: factors that are critical in real metals.

In the late 1960's we carried out a number of studies in conjunction with P.S. Zyryanov [15-20] that revealed a number of new phenomena associated with quantum waves. Quantum spin oscillations were predicted that even exist in the simple free electron model that does, however, account for the nearly constant differential in the cyclotron quantum energy $\hbar\Omega$ and the spin splitting energy $\hbar\Omega_0$. Qualitative variations in the regularities characterizing quantum waves occur, as noted in studies [15-20], due to interelectron interaction. This is manifest in both the shape and configuration of the dispersion curves as well as in the manifestation of new types of waves. In this regard we should note that only in rare exceptions will quantum waves be described by the model responsible for the results provided in survey [14]. A number of subsequent studies [21-24] provided a generalization of quantum wave theory in the electron liquid of a metal.

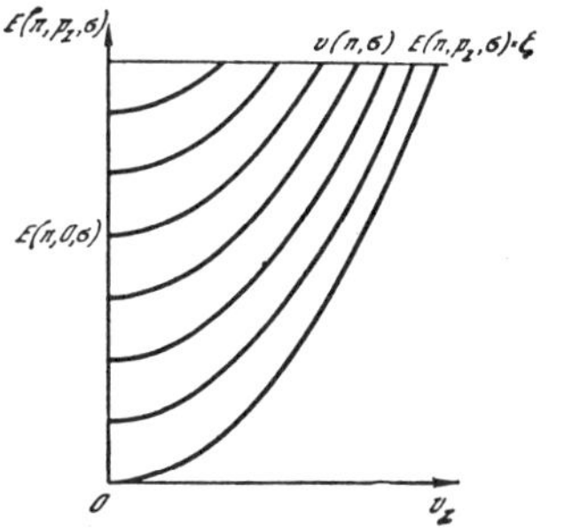

Fig. 1. Illustration of the discreteness of longitudinal electron velocities in a quantizing magnetic field.

Another qualitative result of several studies [15-20] and developed later in [24-26] is the prediction of quantum electron wave interaction with acoustic waves and the existence of coupled quantum spin-acoustic waves. Having mentioned the unique aspect of this phenomenon that allows investigation of the various partial contributions to the elastic modulii, we emphasize that investigating coupled acoustic and quantum waves may be the most effective method for detecting quantum waves by acoustic excitation. Quantum waves may also be related to optical phonons as demonstrated in study [27].

Today the interest in quantum waves in metals and a quantizing magnetic field may also be traced to the fact that a number of their theoretically-investigated properties have also been discovered in other quantum excitations. This specifically relates to waves in thin films whose investigation also requires dealing with discrete electron energy levels produced by dimensional quantization effects [28].

Overall quantum wave theory is a rapidly developing field of solid state physics requiring adequate models of the electron structure of metals in order to successfully investigate resulting problems. Hence, an entire set of surveys would be required to provide a comprehensive coverage of accumulated material. However, in our view it is most advantageous to cover the material accumulated to date for one unique type of quantum wave; specifically, electron spin-acoustic waves accompanied by an electrical field of longitudinal polarization, spin excitations of like polarization and metallic lattice vibrations. Hence our goal in this chapter will be to discuss in sufficient detail the modern concepts and results of quantum spin-acoustic wave theory at once simplifying the necessary physical models without loss of rigor.

1. The physical nature and fundamental properties of quantum spin-acoustic waves

A consistent treatment of results from quantum wave theory will include an examination of the formulation of the initial equations and an analysis of the corresponding dispersion equation for the various specific levels. In this section we will first provide a brief qualitative discussion of the nature and fundamental properties of quantum spin-acoustic waves based on theoretical conclusions that will be formulated in subsequent sections.

We wish to emphasize the simple facts relating to the description of the equilibrium state of the conduction electrons in a metal located in a quantizing magnetic field. In order to understand the essence of the phenomena in this case a simple model of a metal is sufficient in which the electron energy is taken as isotropic and a quadratic function of their momentum. Thus the electron energy in a magnetic field takes the form [1]:

$$E(n, p_z, \sigma) = \hbar\Omega\left(n + \frac{1}{2}\right) + \frac{p_z^2}{2m} + \frac{\sigma}{2}\hbar\Omega_0. \tag{1.1}$$

Here m is the effective electron mass; $\Omega = |e|B/mc$ is the cyclotron frequency; n is the number of the transverse energy level, p_z is the momentum projection in the direction of the field (the z axis); $\sigma = \pm 1$ is the spin index; $\hbar\Omega_0$ is the spin splitting energy.

As a rule the frequency Ω_0 of electrons in a metal is quite different from the cyclotron frequency Ω. One reason is the difference in the effective (cyclotron) mass and the free electron mass. Another reason is that the exchange electron interaction and the (generally speaking, less significant) spin-orbital interaction with the lattice field change the spin splitting energy. The influence of exchange interaction on Ω_0 is described by the formula [29]: $\Omega_0 = \omega_s/(1 + B_0)$, where B_0 is the parameter characterizing the intensity of exchange interaction, while ω_s ignoring spin-orbital effects coincides with the spin-flip frequency of the free electron. Here we should note that both Ω_0 and ω_s may appear in different experimental situations. Specifically, paramagnetic resonance of conduction electrons is observed at ω_s, while spin splitting of quantum oscillation peaks in the de Haas-van Alphen effect is determined by ω_0. We emphasize that the observed differential between Ω_0 and Ω is clearly reflected in the properties of quantum spin-acoustic waves discussed below.

Quantum waves will exist only at sufficiently low temperatures corresponding to virtual complete degeneration of the electron system. Hence the only occupied states will be those with energies $e(n, p_z, \sigma)$ less than ζ: the chemical potential (the Fermi energy) in a magnetic field. It follows that the longitudinal velocity $v_z = p_z/m$ of electrons at the transverse energy level labeled n with a given spin projection value may adopt any values in the range

$$-v(n,\sigma) \leqslant v_z \leqslant v(n,\sigma), \tag{1.2}$$

where

$$v(n,\sigma) = \sqrt{2\left[\zeta - \hbar\Omega\left(n+\frac{1}{2}\right) - \frac{\sigma}{2}\hbar\Omega_0\right]/m} \tag{1.3}$$

is the critical longitudinal velocity.

Our waves are accompanied by perturbations in the electron system that do not change the transverse electron energy, i.e., do not cause transitions between quantum states with different values of n. Under such perturbations the electron system behaves as a group of one-dimensional quasi particles whose possible velocities lie within the range of (1.2). The systems of one-dimensional quasi particles are labeled by the quantum numbers n, σ and the total number of such systems is equal to the number of occupied transverse energy levels, i.e., the sum of the integers of the ratios $(\zeta - \hbar\Omega/2 - \alpha\hbar\Omega_0/2)/\hbar\Omega$ for both values of σ.

The chemical potential ζ is determined by the normalization condition (see [2, formula (58.10]):

$$\frac{|e|B}{(2\pi\hbar)^2 c}\sum_{n,\sigma}\int_{-mv(n,\sigma)}^{mv(n,\sigma)} dp_z = \frac{2|e|B}{(2\pi\hbar)^2 c}\sum_{n,\sigma} mv(n,\sigma) = n_e \equiv \frac{k_F^3}{3\pi^2}. \tag{1.4}$$

Here n_e is the electron concentration; k_F is the Fermi wave vector, while summation over n, σ is taken with respect to the occupied states, i.e., over all n, σ for which the radicand in (1.3) is positive. The magnetic field dependence of the chemical potential is quite significant with only a few occupied levels. Thus, in the ultraquantum limit, when only a single transverse energy level is occupied with the magnetic moment projected along the direction of the field, i.e., the transverse energy of all electrons is identical and is equal to $(1/2)\hbar(\Omega - \Omega_0)$, we have, from (1.4):

$$\zeta = \left[\frac{2}{9}\lambda^6 k_F^6 + \frac{1}{2}\left(1 - \frac{\Omega_0}{\Omega}\right)\right]\hbar\Omega, \tag{1.5}$$

where $\lambda = (c\hbar/|e|b)^{1/2}$ is the magnetic quantum length. The occupation condition of a single level is reduced to the inequality $2(\lambda k_f)^6 < 9$. The quantity $\lambda k_F \equiv \gamma$ is determined by the electron concentration n_e and the magnetic field strength: $\delta \approx 0.02 n_e^{1/3} B^{-1/2}$ CGSE units. In typical conditions of moderate field strengths $\gamma \gg 1$ and a chemical potential close to the Fermi energy in the absence of a magnetic field:

$$\zeta \approx \varepsilon_F = {}^1/_2\gamma^2\hbar\Omega, \tag{1.6}$$

where the electrons occupy a large number $(\sim\gamma^2)$ of transverse energy levels. The dependence of ζ on the field B (the parameter γ) becomes comparatively weak even with ten occupied levels.

We will now proceed to a discussion of waves where quantization of electron motion in the magnetic field is significant. We will consider a wave of electron density oscillations:

$$n(\mathbf{r},\ t) = n_e + n(k,\ \omega) \exp(ikz - i\omega t),$$

propagating along the direction of a constant magnetic field. Since the perturbation to the electron system is independent of the coordinates x, y, no transitions between transverse energy levels will occur in the propagation of this wave, and the electrons will behave as one-dimensional quasi particles. In the ultraquantum limit, where all electrons belong to the same group of one-dimensional quasi particles, the only possible waves are longitudinal waves accompanied by an uncompensated perturbation of electron charge that are analogous to Langmuir plasma waves. If two transverse quantization levels are occupied, this means that there are two groups of one-dimensional quasi particles with different Fermi velocities. In this case in addition to plasma waves there may also be waves in which the perturbation to the charge density of the electron group with a lower Fermi velocity is virtually compensated by perturbation of an electron group with a higher velocity. The electroneutrality causes such waves to have an acoustic dispersion relation where the frequency is linearly dependent on the wave vector with small k, while the phase velocity has a value lying between the Fermi velocities of the electrons of the two groups. It is these acoustic waves that are called quantum waves, since their primary generation mechanism involves quantization of the electron energy levels in a magnetic field.

When the electrons occupy a large number of transverse quantization levels, the number of one-dimensional quasi particle groups becomes equally large. Such "multiple-grading" of the electron system will produce many types (branches) of quantum waves, while the phase velocities of these waves adopt values lying between all possible values of the boundary maximum velocities of the various groups of electrons $\upsilon(n,\ \sigma)$. In the transition to the quasi-classical case of so many occupied levels that the influence of thermal motion and collisions make it possible to consider the distribution of boundary values of the longitudinal velocities to be continuous there are virtually no gaps for the phase velocities of the quantum waves, and the quantum waves themselves become impossible.

It follows from the determination of Fermi values of the longitudinal velocities $\upsilon(n,\ \sigma)$ (1.3) that with unequal frequencies Ω_0 and Ω each of the electron groups at n, σ corresponds to one of the values of spin projection in the direction of the field. In the propagation of the quantum wave this causes there to be oscillations in the x-component of the spin density or a local change in the population of the spin electron energy sublevels. Exchange electron interaction enhances such spin oscillations; this makes possible spin waves in the particular case of equal frequencies of Ω_0 and Ω as well. We emphasize that the quantum spin waves that exist in this case have longitudinal polarization, which qualitatively differentiates these from the familiar transversely-polarized spin waves in normal and ferromagnetic metals [29] and from transverse quantum spin waves [15].

We note the similarity between quantum spin-acoustic waves and plasma acoustic waves that may exist without a magnetic field in metals with two or more current carriers [6]. It is, however, important to emphasize their significant differential related to the collisionless Landau damping effect. Specifically, in the case of regular plasma acoustic waves there exists, generally speaking, a significant quantity of current carriers with velocities near the phase velocity of the wave. This satisfies the Čherenkov electron/wave interaction condition and produces Landau damping. A qualitatively different situation exists in the case of the quantum spin-acoustic waves we are discussing. Indeed, with small values of the wave vector and ignoring the thermal spread of the electron energy, Landau damping may be experienced by waves for which, consistent with the Čherenkov interaction condition

$$\omega - kv(n, \sigma) = 0 \tag{1.7}$$

the phase velocity of the quantum waves coincides with the boundary Fermi value of electron velocity $v(n, \sigma)$. Since the phase velocity does not coincide with this value, condition (1.7) is not satisfied, and there is no damping. This argument requires some refinement when applied to high values of k, where we use a general conservation law in place of classical condition (1.7):

$$\hbar\omega - E(n, p_z, \sigma) + E(n, p_z - \hbar k, \sigma) = 0, \tag{1.8}$$

which includes the electron energies before ($E(n, p_z - \hbar k, \sigma) \leqslant \zeta$) and after ($E(n, p_z, \sigma) \geqslant \zeta$) absorption of the wave energy quantum. The values of ω and k for which equality (1.8) is satisfied with the noted energy constraints form a range of frequencies and wave vectors in which collisionless damping is nonzero. Consistent with equalities (1.8) and (1.1) this region, unlike the continuous Landau damping region in a metal in the absence of a quantizing magnetic field (between the curves $\omega = \hbar k(k_F + k/2)/m$ and $\omega = \hbar k(k/2 - k_F)/m$ (Fig. 2a) is more complex in structure. Transparency windows appear in the continuous quasi-classical damping range; within these windows conservation law (1.8) will not be satisfied. The detailed regularities behind the transparency windows will be discussed below in section 3. Their structure is illustrated in Fig. 2b, c. Here we simply note that with small values of k the transparency appear as "lobes" formed by parabola segments. With large wave vectors, when $\gamma k > 1$, we have a set of triangular windows, two sides of which are also formed by parabolas. Finally, the third type includes a transparency window bounded by the parabola $\omega = kv_{\min} - \hbar k^2/2m$, where $v_{\min}$ is the longitudinal velocity $v(n, \sigma)$ related to the last occupied transverse quantization level. The dispersion curves of quantum waves may be distributed within the transparency windows. In the interelectron interaction approximation, i.e., in the electron gas model, the dispersion curves pass only through the lobe transparency windows (Fig. 3a). With small values of k the dispersion law is limited, and with growth of k the phase velocity of the quantum wave becomes dependent on k, and the dispersion curves terminate at the vertices of the lobe windows. One qualitative manifestation of interelectron interaction, as noted in study [21] (see also [23]), is quantum wave production in triangular transparency windows (Fig. 3b).

The shape of the transparency windows was discussed in detail in survey [14] for the particular case $\Omega_0 = 0$ or $\Omega_0 = \Omega$ where the spin splitting of the electron energy levels does not play a significant role. It is clear from Fig. 2 that in actual conditions $\Omega_0 \neq \Omega$ spin splitting of the electron energy levels produces a difference in the size of windows of an identical type. Such a differential may be sufficiently large and depend on the ratio of frequencies Ω_0 and Ω. Various resulting possibilities in this regard will be discussed in detail in section 3.

The number of transparency windows is determined by the number of occupied quantization levels. The number of occupied levels drops with growth of magnetic field strength; in other words, each electron group with a corresponding minimum velocity vanishes one after the other, and the number of windows and branches of the quantum waves drop. With growth of the magnetic field strength the velocities $\upsilon(n, \sigma)$ also drop and the slope of the lobe windows is reduced correspondingly (Fig. 2). This causes the phase velocities of the corresponding quantum waves to drop. They reach their minimal values when the corresponding electron group vanishes. During the reduction in the phase velocities of the quantum waves there exists strong coupling of the slow quantum waves to regular acoustic waves in the metal when their frequencies coincide.

When the magnetic field strength changes the lobe transparency windows intersect the dispersion curve of sound represented by the straight line $\Omega = ku_s$ in the ω, k plane. Giant quantum oscillations in the acoustic absorption coefficient arise as a result of this intersection [5]: there is no collisionless damping in the transparency windows, and outside the windows it is comparatively large. Moreover, sound interaction with quantum spin-acoustic waves produces an anomalous change in sound velocity with a change in the magnetic field when the dispersion curve of sound falls within the transparency window and is close to the dispersion curve of the quantum wave (Fig. 4). This phenomenon has been predicted and investigated in studies [17, 25, 26] (see also study [24]).

The existence of sharp boundaries between the transparency and damping regions for acoustic waves in a quantizing magnetic field has a significant influence on the dependence of sound frequency on the wave vector. Within a lobe transparency window the dispersion curve of the sound wave approaches the boundary exponentially and makes contact (Fig. 4c). Outside the window in the damping region this curve reaches the boundary at a different point. Thus, anomalously strong dispersion of sound velocity exists near the boundary of the transparency window together with a jump in the absorption coefficient. This effect also occurs at the boundary of other transparency windows (triangular windows) intersected by the acoustic dispersion curve. This effect is analogous to the Kohn effect in the absence of a magnetic field, although there are significantly more powerful anomalies in the dispersion relation due to the one-dimensional nature of motion of the electrons from each of the subsystems interacting with sound on the quantization levels.

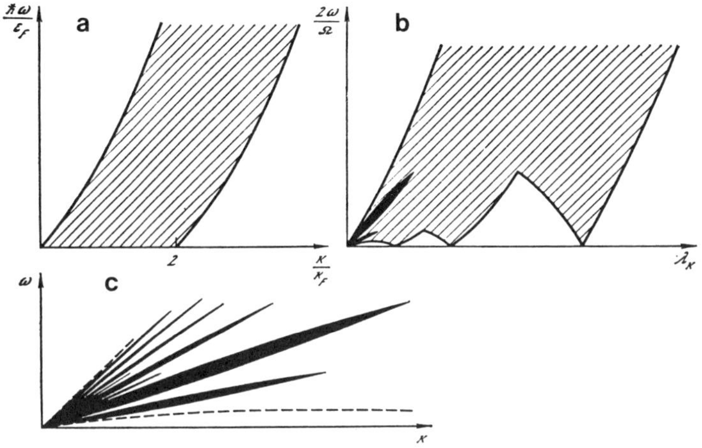

Fig. 2. Landau damping region and transparency windows
a - zero magnetic field, hatched damping region; b - quantizing magnetic field with two occupied levels; the lobe windows are the dark areas and the parabolic and triangular windows are the light areas; c - the lobe transparency windows (dark) for the case of ten occupied levels

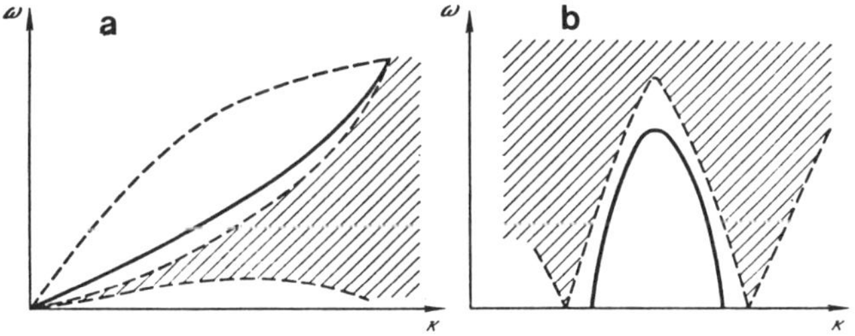

Fig. 3. Dispersion curves of a quantum wave (schematic representation)
a - in a lobe transparency window; b - in a triangular transparency window with interelectron exchange interaction

The regularities characteristic of quantum waves are largely determined by the Fermi-liquid interaction of conduction electrons. The new effect identified above results from interelectron interaction: the appearance of quantum waves in the triangular transparency windows. This same type of effect will occur for waves in lobe

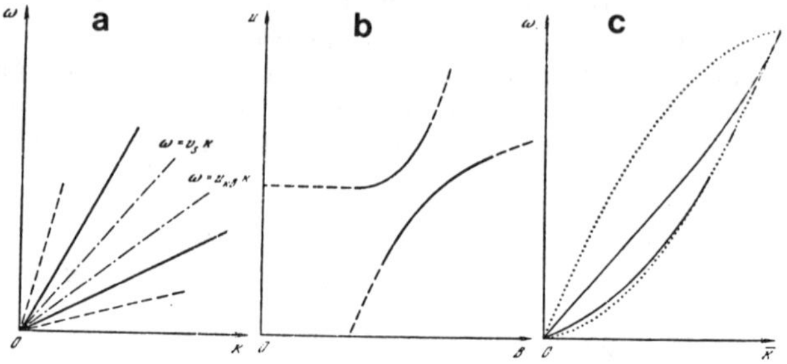

Fig. 4. Illustration of quantum wave interaction with sound
a - "spreading" of the dispersion curves. The dot and dash lines represent the dispersion curves of noninteracting waves, while the dotted lines represent the boundaries of a lobe transparency window; b - the field dependence of the phase velocities of coupled waves; c- the dispersion curves of coupled acoustic and quantum waves in a lobe transparency window

windows when the energies of electrons with opposite spin projections are sufficiently proximate. This effect is manifest as the existence of spin density waves: quantum longitudinal spin waves resulting entirely from electron exchange interaction. This effect was predicted in studies [16, 19]. In other cases where the energies for the various spin projections are significantly isolated, exchange interaction will influence the quantum wave spectrum and may qualitatively alter the entire progression of the dispersion curves if the interaction constant is sufficiently large. Specifically, with a negative and sufficiently large exchange interaction constant waves may appear whose dispersion curves pass through a parabolic transparency window or the higher frequency far boundary of the damping region.

The phenomena described above occur from longitudinal wave propagation in the direction of field *B*. With inclined propagation quantum spin-acoustic waves will also exist, although they are coupled to the transverse electromagnetic field oscillations. In this case we have interaction between the longitudinal quantum waves and electromagnetic waves having spiral polarization in the plane perpendicular to the field *B*: helicons. This effect was investigated in studies [11, 20]. It produces discontinuities in the dispersion relation of the helicon and distorts the dispersion curves of the quantum waves. The relationship between quantum waves and the transverse electromagnetic field enhances the possibility for their observation.

Transverse electromagnetic field oscillations in quantum wave propagation may also arise from the anisotropy of the electron dispersion law. Although this issue has not been investigated in detail, it is nonetheless quite clear that this fact may be related to the increased complexity of the dispersion relation of the quantum waves with sufficiently large k. However, the acoustic nature of the dispersion of the quantum wave frequencies in the limit of small k is conserved with an arbitrary Fermi surface shape if, of course, the energy spectrum of transverse electron motion in the magnetic field is discrete (i.e., the cross-sections of the Fermi surface by planes perpendicular to the field B are closed), otherwise the quantum waves could not exist at all. Anisotropy of the Fermi surface also produces spherical transparency windows with large k (in the order of the inverse cyclotron orbital radius or higher), since with such values of k the form of the windows is highly dependent on the form of the electron dispersion law.

This then is a qualitative picture of phenomena associated with the existence of quantum spin-acoustic waves in metals. In summary we may state that these waves appear as collective acoustic excitations in the electron system and the metal lattice in a quantizing magnetic field. The existence of longitudinal oscillations in spin density are characteristic of these waves.

2. Dispersion equation for longitudinal quantum waves. Discussion of theoretical principles

Quantum waves correspond to solid states slightly out of equilibrium. In a quantizing magnetic field the equilibrium electron density matrix f has in the energy representation only diagonal elements $f(n, p_z, \sigma)$ equal to the energy-dependent $E(n, p_z, \sigma)$ Fermi distribution function. In the nonequilibrium state corresponding to wave propagation the density matrix has a nonequilibrium addition ρ satisfying the following kinetic equation:

$$\frac{\partial}{\partial t}\rho+\frac{1}{\tau}\rho+\frac{1}{i\hbar}[\rho, \mathcal{H}_0]=\frac{1}{i\hbar}[W(t), f]. \tag{2.1}$$

Here $\mathcal{H}_0$ is the Hamiltonian of an electron in a constant magnetic field whose eigenvalues $E(n, p_z, \sigma)$ are determined by formula (1.1); $W(t)$ is the interaction energy operator of electron interaction with the wave-perturbing field and, finally, the component ρ/τ describes the collisional influence which will henceforth be considered small and will be required only to determine the existence conditions of the quantum waves.

For longitudinal waves with frequency ω and the wave vector k propagating along the magnetic field, the matrix elements of perturbation energy $W(t)$ are nonzero only for electron states with momentum projections p_z and $p_z-\hbar k$ having identical quantum numbers n, σ. Introducing the convention $W(n, p_z, \sigma, \omega, k)\exp(-i\omega t)$ for such matrix elements as well as analogous conventions for the matrix elements of the nonequilibrium density matrix $\rho(n, p_z, \sigma, \omega, k)\exp(-i\omega t)$ we have consistent with equation (2.1):

$$\rho(n, p_z, \sigma, \omega, k) = \frac{[f(n, p_z - \hbar k, \sigma) - f(n, p_z, \sigma)]\, W(n, p_z, \sigma, \omega, k)}{\hbar\omega + E(n, p_z - \hbar k, \sigma) - E(n, p_z, \sigma) + i\hbar/\tau} \equiv$$
$$\equiv \Gamma(n, p_z, \sigma, \omega, k)\, W(n, p_z, \sigma, \omega, k). \tag{2.2}$$

The W operator is the linear functional of the nonequilibrium density matrix ρ. The simplest approximation for it is obtained in the self-consistent field approximation (the "chaotic phase approximation") where it is assumed that the electrons are under the influence of a mean electrical field with the potential $\Phi(z, t) = \Phi(k, \omega) \exp(-i\omega t)$. In this approximation the interaction energy $W(n, p_z, \sigma, \omega, k)$ coincides with the electron energy in the self-consistent field of the wave

$$W_e(\omega, k) = e\Phi(k, \omega). \tag{2.3}$$

The self-consistent field potential is determined by the Poisson equation

$$k^2\Phi(k, \omega) = 4\pi\, [en(k, \omega) + q(k, \omega)], \tag{2.4}$$

where the amplitudes of the charge densities of the electrons $en(k, \omega)$ and the ion lattice $q(k, \omega)$ enter into the right half of this equation. The nonequilibrium electron density is determined by the density matrix (2.2):

$$n(k, \omega) = \frac{|e|B}{(2\pi\hbar)^2 c} \sum_{n, \sigma} \int dp_z \rho(n, p_z, \sigma, \omega, k) = n^+(k, \omega) + n^-(k, \omega), \tag{2.5}$$

where $n^+(k, \omega)$ and $n^-(k, \omega)$ are the contributions of electrons with two opposite spin projection directions ($\sigma = \pm 1$), respectively. In order to find the amplitude of oscillations in the ion charge density $q(k, \omega)$ we must consider lattice vibrations. We will introduce a local perturbation that drives the lattice $u(z, t) = u(k, \omega) \exp(ikz - i\omega t)$ from equilibrium. Then

$$q(k, \omega) = -Q\left[\frac{\partial u(z, t)}{\partial z}\right]_{k, \omega} = -Qiku(k, \omega), \tag{2.6}$$

where Q is the charge per unit of volume of the lattice. Oscillations in lattice displacement are described by the equation of motion

$$-\omega^2\rho_m u(k, \omega) = F(k, \omega). \tag{2.7}$$

Here ρ_m is the mass density while $F(k, \omega)$ is the amplitude of the force acting on the lattice whose explicit form may appear differently in different models. In the simplest model of a uniform distribution of equilibrium ion charge (the "jelly" model) the force acting on the lattice is determined by the self-consistent field

$$F^e(k, \omega) = -Qik\Phi(k, \omega). \tag{2.8}$$

In such a self-consistent field model

$$n(k,\omega) = -\chi_0(\omega,k)\,e\Phi(k,\omega) = -\sum_\sigma X^\sigma(\omega,k)\,e\Phi(k,\omega), \tag{2.9}$$

where $X^\sigma(k, \omega)$ infers polarizability of electrons with a given spin projection and consistent with (2.2), (2.3), and (2.5) takes the form

$$X^\sigma(\omega,k) = -\frac{|e|B}{(2\pi\hbar)^2 c}\sum_n \int dp_z \Gamma(n, p_z, \sigma, \omega, k). \tag{2.10}$$

In the self-consistent field approximation equations (2.2)-(2.9) represent a closed system of equations whose solvability condition produces the following dispersion equation:

$$1 - \frac{\Omega_p^2}{\omega^2} + \frac{4\pi e^2}{k^2}\chi_0(\omega,k) = 0, \tag{2.11}$$

where $\Omega_p = (4\pi Q^2/\rho_m)^{1/2}$ is the ion plasma frequency. In this equation obtained in study [30] quantization of electron motion is manifest in the expression for polarizability $\chi_0(\omega, k)$ and its general form is the same as in the case of no magnetic field [31].

A number of important quantum wave properties may be understood even in the self-consistent field approximation. We will discuss certain general concepts deriving from equation (2.11). We note that in the absence of a magnetic field equation (2.11) describes two types of longitudinal waves: high-frequency plasma waves and low-frequency acoustic waves. We will be interested in low-frequency waves with a wavelength significantly exceeding the screening radius of the Coulomb field. In these conditions we may ignore unity in the left half of equation (2.11), corresponding to the insignificance of the left half of Poisson equation (2.4) for both the quantum waves and the acoustic waves. In phenomena related to quantum waves an important role is played by the properties of acoustic waves and hence we will examine several properties separately characterized by dispersion equation (2.11). Since the phase velocity ω/k of the acoustic waves is small compared to the Fermi electron velocity v_F, in the absence of a magnetic field the frequency dispersion of the electron polarizability $\chi_0(\omega, k)$ will only result in small corrections to the acoustic frequency. In this particular case the adiabatic approximation is suitable. In this case we may write the following simple expression for the polarizability for wave vectors small compared to the Fermi vector k_F:

$$\chi_0^{\text{кл}}(\omega,k) = \eta\left(1 + \frac{i\pi\omega}{2kv_F}\right), \tag{2.12}$$

where $\eta = mk_F/\pi^2\hbar^3$ is the electron state density on the Fermi surface. The imaginary part of $\chi_0^{\text{кл}}(l, k)$ describes collisionless Landau damping and is written for the case where the acoustic wavelength is small compared to the free path length of the electron $l_e (kl_e \gg 1)$. Formula

(2.12) consistent with (2.11) produces the following dispersion equation:

$$\omega^2 = k^2 u_0^2 \left(1 - \frac{i\pi u_0}{2v_F}\right), \tag{2.13}$$

that determines the frequency and damping decrement of sound consistent with studies [32, 33]. Here we have for sonic velocity in the "jelly" model:

$$u_0 = (Q^2/e^2 \rho_m \eta)^{1/2}. \tag{2.14}$$

Quantization of electron motion in a strong magnetic field qualitatively changes the properties of the acoustic waves. For example, in the case of frequencies less than the collision frequency ($\omega\tau \ll 1$), when the frequency polarizability dispersion, as in the case of no magnetic field is insignificant, the following two quantum effects may arise. First, the state density η entering into formula (2.14) is replaced by the quantum quantity η_{KB} having an oscillating dependence on the magnetic field strength, which produces quantum oscillations in acoustic velocity, similar to those that appear in the de Haas-van Alphen effect (see study [34]). Second, the quantization-generated transparency windows in the Landau damping region produce giant quantum oscillations in sound absorption [5] propagating in the direction of the magnetic field. Such giant oscillations have been examined based on dispersion equation (2.11) in studies [35, 36]. In the opposite high frequency case ($\omega\tau \gg 1$) the frequency polarizability dispersion $\chi_0(\omega, k)$ related to the discreteness of the electron levels in the magnetic field becomes significant. In this case the polarizability in the quantizing magnetic field is equal to the sum of the polarizabilities of the one-dimensional electron subsystems having critical Fermi velocities $v(n, \sigma)$ (see section 1) whose magnitude may be approached by the phase velocity ω/k. As a result the dispersion law of acoustic frequency not only changes, but in certain cases new solutions of dispersion equation (2.11) corresponding to quantum waves arise. We note that the spectrum of quantum waves based on equation (2.11) has been investigated in studies [7-10, 25]. However in these studies, with the exception of study [10], the spin splitting of the electron energy levels was entirely ignored; this corresponds to $\Omega_0 = 0$ in formula (1.1), and hence it was not possible to describe a number of important phenomena. In study [10] spin splitting was incorporated by numerical calculation of the quantum wave frequencies in the case of few occupied quantization levels. Moreover, studies [7-12] ignored the contribution of the lattice ($\Omega_p \to 0$) and as a result the interaction of acoustic and quantum waves was neglected. The properties of quantum waves will be examined in detail below and at this point we will make a number of generalizations of the physical model employed in this case. This will allow formulation of a more realistic approach that accounts for the important properties of interelectron interaction and electron/lattice interaction. This approach will also allow prediction of new effects related to the qualitative difference between a real metal and a simple self-consistent field and "jelly" model and, second, will make it possible to understand the significance and possibility for using a simple model.

As we know the self-consistent field approximation does not account for the correlation interaction that differentiates a real degenerate electron liquid from an electron gas. It is important for quantum wave theory to account for the part of the interaction related to the spin dependence and which is often referred to as the exchange interaction. In this connection we will now introduce representations of a simple exchange interaction model. We note that oscillations in the mean electron density in a quantizing magnetic field are virtually always accompanied by oscillations in the longitudinal spin density component $s^z(z, t) = s^z(k, \omega) \exp(ikz - i\omega t)$. Here, in accordance with formula (2.5) we have:

$$s^z(k, \omega) = {}^1/_2 [n^+(k, \omega) - n^-(k, \omega)]. \tag{2.15}$$

In accordance with equality (2.15) the longitudinal spin density of oscillations will always be nonzero when electrons with opposite spins make different contributions to their total density. Oscillations in spin density (2.15) generate an alternating effective exchange force field that acts on the electron spin in the same manner as the Weiss field acts on the spin in a ferromagnetic. The amplitude of such a field is proportional to $s^z(k, \omega)$, and therefore we may write the following expression for the electron/spin oscillation exchange interaction energy:

$$W_{ex}(\sigma, \omega, k) = 4\psi s^z_{\sigma\sigma} s^z(k, \omega) = 2\sigma \frac{B_0}{\eta} s^z(k, \omega). \tag{2.16}$$

Here ψ is the dimensional energy constant that corresponds to the exchange integral in the Hartree-Fock approximation, while B_0 is the exchange interaction parameter corresponding to this constant. Subject to (2.16) the interaction energy in equation (2.2) is equal to the sum $W_e + W_{ex}$. Therefore we may now find from equation (2.2) the following relation between the $n(k, \omega)$ and $s^z(k, \omega)$ amplitudes and the amplitude of the self-consistent field potential:

$$s^z(k, \omega) = -\frac{1}{2} e\Phi(k, \omega) \frac{\chi_1(\omega, k)}{1 + \psi\chi_0(\omega, k)}, \tag{2.17}$$

$$n(k, \omega) = -e\Phi(k, \omega)\chi(\omega, k) \equiv -e\Phi(k, \omega)\left\{\chi_0(\omega, k) - \frac{\psi[\chi_1(\omega, k)]^2}{1 + \psi\chi_0(\omega, k)}\right\}, \tag{2.18}$$

where $\chi_1(\omega, k) = X^+(\omega, k) - X^-(\omega, k)$. The dispersion equation also changes form now in accordance with formula (2.18). Specifically, in place of (2.11) in order to account for smallness of the Coulomb field screening radius compared to wavelengths of interest to us $k^2 \ll k_D^2 = 4\pi e^2\eta$ we obtain

$$-\frac{\Omega_p^2}{\omega^2} + \frac{4\pi e^2}{k^2}\chi(\omega, k) = 0. \tag{2.19}$$

Now the wave properties are determined by the electron polarizability in the quantizing magnetic field $\chi(\omega, k)$ accounting for exchange interaction.

The approximation of effective exchange field (2.16) is very useful, although it cannot claim to be exhaustive in its capacity to describe interelectron interaction. A comprehensive description may be achieved by incorporating correlation electron interaction where in place of formula (2.16) we utilize the following expression

$$W_c(n, p_z, \sigma, \omega, k) = \frac{|e|B}{(2\pi\hbar)^2 c} \sum_{n', \sigma'} \int dp_z \{\varphi(n, p_z, n', p_z') + \\ + \sigma'\sigma\psi(n, p_z, n', p_z')\} \rho(n', p_z'\sigma', \omega, k).$$

The functions $\varphi(n, p_z, n', p_z')$ and $\psi(n, p_z', n', p_z')$ are analogous to Fermi-liquid interaction functions from quasi-classical electron liquid theory [37]. Quantum wave theory was developed in study [21] fully accounting for such interaction (see also [23, 24]). In this study we will not consider the properties of quantum waves entirely accounting for the regularities of interelectron correlations, and below we will use a simple approximation of correlation interaction corresponding to approximation of the constants φ and ψ, when

$$W_0(\sigma, \omega, k) = \varphi n(k, \omega) + 2\sigma\psi s^z(k, \omega) = 1/\eta \{A_0 n(k, \omega) + 2\sigma B_0 s^z(k, \omega)\}.$$

The role of the first spin-independent component containing the constant φ or the dimensionless interaction parameter A_0 is manifest, as will be demonstrated below, in the generalization of the description of electron/lattice interaction (see Appendix 1).

The simplicity of the uniform lattice charge distribution model employed above (the "jelly" model) often engenders certain doubts as to its suitability. Bearing in mind the need to establish an interrelationship between quantum waves and the longwave phonons in the metal, we will now refine the description of electron/lattice interaction after fitting the theoretical model for a more consistent description of the actual structure of the ion density. The fact that by virtue of lattice action on the electrons the conduction electron energy, in addition to the dependence on the self-consistent field and the electron distribution, will also be dependent on the ion density plays an important role in this refinement. Hence, with acoustic wave-induced deformation of the ions, the electron energy changes, which introduces one additional so-called deformation contribution to the interaction energy $W(n, p_z, \sigma, \omega, k)$

$$W_d(n, p_z, \omega, k) = \Lambda(n, p_z)\left[\frac{\partial u(z, t)}{\partial z}\right]_{k,\omega} = iku(k, \omega)\Lambda(n, p_z). \tag{2.22}$$

The quantity $\Lambda(n, p_z)$ is the deformation potential introduced in this case analogous to the process used in the quasi-classical theory of acoustic propagation [38, 39] (another method of introducing deformation potential is used in studies [40, 41]). For simplicity we will consider Λ to be constant in subsequent calculations.

The deformation action of the lattice on the electrons (2.22) produces corresponding counteraction described by the force:

$$F^d(k, \omega) = \frac{|e|B}{(2\pi\hbar)^2 c_i} \sum_{n,\sigma} \int dp_z \rho(n, p_z, \sigma, \omega, k)\, ik\Lambda, \tag{2.23}$$

acting on the lattice from the electrons. In addition to this contribution which serves as a supplement to force (2.8) determined by the self-consistent field, it is also necessary to account for the intrinsic ion elasticity (independent of the conduction electrons) corresponding to the force

$$F^i(k, \omega) = \lambda^0 \left[\frac{\partial^2 u(z, t)}{\partial z^2}\right]_{k,\omega} = -\lambda^0 k^2 u(k, \omega), \tag{2.24}$$

where λ^0 is the modulus of intrinsic elasticity of the ion lattice. Finally the force entering into the right half of equation (2.7) determining lattice oscillations now consists of three components:

$$F(k, \omega) = F^e(k, \omega) + F^d(k, \omega) + F^i(k, \omega). \tag{2.25}$$

Subject to expression (2.25) in the equation of motion of the ions and using the sum of expressions (2.3), (2.21) and (2.22) for the electron interaction energy, we find the relations relating the perturbation amplitudes:

$$-e\Phi(k, \omega) - ik\Lambda u(k, \omega) = \left[\varphi + \frac{1}{\chi(\omega, k)}\right] n(k, \omega), \tag{2.26}$$

$$s^z(k, \omega) = \frac{n(k, \omega)\chi_1(\omega, k)}{2\chi(\omega, k)[1 + \psi\chi_0(\omega, k)]} = \frac{Q}{e}\, \frac{ik\chi_1(\omega, k)\, u(k, \omega)}{2\chi(\omega, k)[1 + \psi\chi_0(\omega, k)]} \tag{2.27}$$

and the following dispersion equation:

$$-\frac{\Omega_p^2}{\omega^2 - v_s^2 k^2} + \frac{4\pi e^2}{k^2}\chi(\omega, k) = 0. \tag{2.28}$$

Here

$$v_s^2 = \frac{\lambda^0}{\rho_m} + \frac{2Q\Lambda}{e\rho_m} + \frac{Q^2\varphi}{e^2\rho_m} = \frac{\lambda^0}{\rho_m} - \frac{2Z\Lambda}{M} + A_0 u_0^2, \tag{2.29}$$

where $\rho_m = n_i M$; $Q = -Zen_i$; M is the ion mass; $(-Z_e)$ is the ion charge, while u_0 is determined by formula (2.14).

Equation (2.28) served as the basis for analyzing the quantum wave spectrum in studies [17, 19], while the general formulation of acoustic and quantum wave theory producing equation (2.28) is discussed in article [24] (see also [42]). We should emphasize that accounting for the ion elasticity, and the deformation interaction independent of the spins of the Fermi-liquid interaction produces "bare" velocity of sound so that, for example, in the absence of a quantizing magnetic field we have in place of (2.4) for the acoustic velocity

$$\frac{\omega}{k} = \sqrt{v_s^2 + u_0^2} = \left[\frac{\lambda^0}{\rho_m} - \frac{2Z\Lambda}{M} + \frac{Z^2 n_i}{M\eta}(1 + A_0)\right]^{1/2}. \tag{2.30}$$

We should emphasize that selection of an adequate model of electron-lattice interaction for correct formulation of the principles of quantum wave theory is quite critical, since, for example, the Fröhlich model [43] produces incorrect results that may be found in study [44] devoted to an investigation of anomalies in acoustic velocity. A critique of these results may be found in study [24]. Appendix 1 contains a comparison of our approach to the dielectric constant formalism used in the microscopic approach to the problem of electron/lattice interaction, and the drawbacks of the Fröhlich model are also noted. We should also make mention of the fact that the principles of quantum wave theory outlined above require some generalization in the case of wave propagation at an angle to the magnetic field. Such a generalization is necessary for the theory of interaction between quantum and helicon waves presented in Appendix 2. Finally, the problems of investigating quantum waves in metals with an anisotropic Fermi surface are discussed in Section 7. Sections 4, 5, and 6 of this chapter are devoted to a discussion of results deriving from dispersion equation (2.28).

3. Electron polarizability in a quantizing magnetic field. Transparency windows

Before providing results that derive from the dispersion equation for quantum waves we will discuss the characteristic properties of electron polarizability in conditions of quantization of electron motion, which is necessary to understand the phenomena under consideration. Consistent with relations (2.9) and (2.18) the $\chi(\omega, k)$ polarizability is expressed by the following formula:

$$\chi(\omega, k) = X^{+}(\omega, k) + X^{-}(\omega, k) - \frac{\psi[X^{+}(\omega, k) - X^{-}(\omega, k)]^2}{1 + \psi[X^{+}(\omega, k) + X^{-}(\omega, k)]}, \tag{3.1}$$

where the functions $X^{\sigma}(\omega, k)$ are the electron gas polarizabilities with a given spin projection that are in turn determined by equalities (2.10), (2.2). We will also take the temperature to be zero temperature. Then after integration with respect to p_z we have from formula (2.10)

$$X^{\sigma}(\omega, k) = \frac{\Omega m \eta}{4k_F \hbar k} \sum_{n=0}^{N(\sigma)} \int_{w(n,\sigma,-k)}^{w(n,\sigma,k)} dv \left\{ \frac{1}{v + u + i/\tau k} + \frac{1}{v - u - i/\tau k} \right\}, \tag{3.2}$$

where $N(\sigma) = [(\zeta/\hbar\Omega) - 1/2 - \sigma(\Omega_0/2\Omega)]$ is the number of the last occupied level for a given σ (here [...] infers taking of the entire part); $u = \omega/k$ is the phase velocity of the waves, while $w(n, \sigma, k) = v(n, \sigma) = \hbar k/2m$. It is clear from expression (3.2) that the function $X^{\sigma}(\omega, k)$ is equal to the sum $N(\sigma) + 1$ of the polarizabilities of one-dimensional subsystems with corresponding Fermi velocities $v(n, \sigma)$.

The quantum waves are a high-frequency process corresponding to satisfaction of the condition $\omega\tau \gg 1$. Hence in investigating these waves we may take the collision frequency $1/\tau$ in formula (3.2) equal to zero with good accuracy. In this limit the imaginary part of the function $X^{\sigma}(\omega, k)$ describes Landau damping which forbids wave existence in the electron system. In this regard we must above all determine on the ω, k plane the region within which the imaginary part of the polarizability when $1/\tau \to 0$ is nonzero and its auxiliary region of possible wave existence in a quantizing magnetic field.

First we remember that if we ignore the influence of quantization of electron motion the region of Landau damping, i.e., the region in which the imaginary part of the quasi-classical probability is nonzero is bounded on the ω, k plane by the parabolas $\omega = kv_F + \hbar k^2/2m$; $\omega = \hbar k^2/2m - kv_F$ and the axis $\omega = 0$ (see Fig. 2a). In other words in the quasi-classical limit collisionless wave damping is significant for all values of ω and k satisfying the inequalities:

$$\omega \leqslant kv_F + \hbar k^2/2m, \qquad \omega \geqslant \hbar k^2/2m - kv_F. \tag{3.3}$$

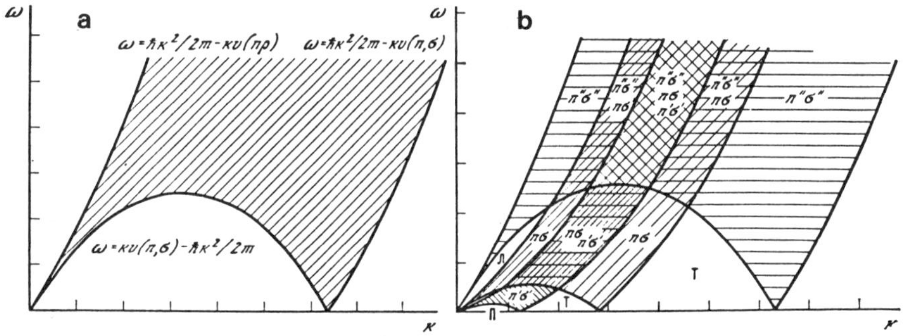

Fig. 5. Formation of transparency windows
a - the Landau damping region from electrons on the n, σ level (hatched); b - superposition of damping regions from three neighboring levels n'', σ'', n, σ, n', σ', for which $v(n', \sigma') < v(n, \sigma) < v(n'', \sigma'')$. Л.П.Т. - lobe, parabolic, and triangular transparency windows. The diagram also indicates the level of origin of damping in each part

The Landau damping region assumes a different form in a quantizing field. Section 1 has already accounted for the fact that by virtue of the quantum discreteness of the electron transverse motion energy levels transparency windows arise within the quasi-classical damping region (3.3). We will demonstrate this and will describe in detail the structure of such windows.

Proceeding from relation (3.1) we may easily see that in order to understand the properties of the imaginary part of the polarizability $\chi(\omega, k)$ it is sufficient to consider the "gas" polarizability $\chi_0(\omega, k) = X^{+}(\omega, k) + X^{-}(\omega, k)$. Consistent with formula (3.2) when $1/\tau \to 0$ we have

$$\operatorname{Im}\chi_0(\omega, k) = \frac{\Omega m\eta}{4k_F\hbar k}\sum_{\sigma}\sum_{n=(}^{N(\sigma)}\left\{\int_{w(n,\sigma,-k)}^{w(n,\sigma,k)} dv\,\delta(v-u) - \int_{0}^{-w(n,\sigma,-k)} dv\,\delta(v-u)\right\}. \tag{3.4}$$

The brackets under the integral sign in (3.4) represent the difference of the two integrals over υ, each of which is equal to either unity or zero. The first integral is equal to unity when $k\upsilon(n, \sigma) - \hbar k^2/2m < \omega < k\upsilon(n, \sigma) + \hbar k^2/2m$, while the second is equal to unity when $\omega < \hbar k^2/2m - k\upsilon(n, \sigma)$. Hence each component of the sum over n, σ in the right half of (3.4) is nonzero in the region on the ω, k plane bounded by the parabolas $\omega = k\upsilon(n, \sigma) + \hbar k^2/2m$, $\omega = k\upsilon(n, \sigma) - \hbar k^2/2m$ and $\omega = \hbar k^2/2m - k\upsilon(n, \sigma)$ (Fig. 5a). This region includes values of ω and k satisfying the relation:

$$\left| kv(n, \sigma) - \frac{\hbar k^2}{2m} \right| \leqslant \omega \leqslant kv(n, \sigma) + \frac{\hbar k^2}{2m}. \tag{3.5}$$

The entire Landau damping region is formed by superposition of such regions for all occupied n, σ levels. Thus, for example, by superimposing some selected region from (3.5) and two analogous regions originating from electrons on neighboring levels n', σ' and n'', σ'' with $\upsilon(n', \sigma') < \upsilon(n, \sigma) < \upsilon(n'', \sigma'')$ satisfied we obtain the damping region shown in Fig. 5b. We clearly see from the diagram that sections exist on the ω, k plane that are not covered by any of the damping regions from (3.5). It is precisely these regions that form the transparency windows. In Fig. 5b the two transparency windows form "lobes" and one window is bounded by parabolas:

$$\omega = kv(n', \sigma') + \frac{\hbar k^2}{2m}, \qquad \omega = kv(n, \sigma) - \frac{\hbar k}{2m}. \tag{3.6}$$

The two far right windows in Fig. 5b are triangular in shape. The boundaries of the left window are formed by the parabolas

$$\omega = kv(n, \sigma) - \frac{\hbar k^2}{2m}, \qquad \omega = \frac{\hbar k^2}{2m} - kv(n', \sigma') \tag{3.7}$$

and the case $\omega = 0$. Finally the last parabolic transparency window separates the lobe and triangular window groups (in Fig. 5b bounded on the top by the parabola $\omega = k\upsilon(n', \sigma') - \hbar k^2/2m$ and on the bottom by the axis $\omega = 0$).

In the general case after superposition of the damping regions from (3.5) generated by electrons with all occupied levels n, σ, both a set of lobe transparency windows and a set of a like number of rectangular windows are formed. Moreover, as a rule, a single parabolic window bounded by the axis $\omega = 0$ and the parabola $\omega = k\upsilon_{min} - \hbar k^2/2m$ is formed; here υ_{min} is the minimal velocity from $\upsilon(n, \sigma)$ corresponding to the last occupied quantization level. The parabolic window is absent only for magnetic field strength values for which $\upsilon_{min} = 0$.

The entire Landau damping region in the quantizing magnetic field together with the transparency windows fall between the parabolas $\omega = k\upsilon_{max} + \hbar k^2/2m$ and $\omega = \hbar k^2/2m - k\upsilon_{max}$, where υ_{max} is the critical velocity from $\upsilon(n, \sigma)$. The velocity υ_{max} with many occupied levels is close to the Fermi velocity without a field and drops with an increase in field strength. It follows that all transparency windows fall within quasi-classical damping region (3.3) where the imaginary part of polarizability is hardly small in the absence of quantum effects.

Having established the existence of transparency windows and their shape, we will discuss in greater detail the distribution and length of the lobe and triangular windows. As the primary window characteristic we will employ the position of its upper termination point, i.e., the points of intersection with the parabolas serving as its boundaries. The frequency $\omega_{\Gamma p}$ corresponding to this point characterizes the height of the window, while the corresponding wave vector $k_{\Gamma p}$ may be used as the characteristic of the horizontal length for lobe windows and as the position for triangular windows.

We will first determine possible values of the maximal frequency in the given window $\omega_{\Gamma p}$. As we see from Fig. 5b lobe and triangular windows are formed from each pair of neighboring levels n, σ with different values of $\upsilon(n, \sigma)$, and both such windows terminate at an identical frequency $\omega_{\Gamma p}$. We find from the formulae for the window boundaries (3.6), (3.7):

$$\omega_{\Gamma p} = \frac{m}{2\hbar}\left[v^2(n,\sigma) - v^2(n',\sigma')\right] = \Omega(n'-n) + {}^1/_2(\sigma'-\sigma)\,\Omega_0. \qquad (3.8)$$

This expression shows that each of the frequencies $\omega_{\Gamma p}$ is equal to the difference or sum of the frequency Ω_0 and a frequency that is a multiple of Ω. As a rule there exists a certain set of transparency windows with different values of $\omega_{\Gamma p}$, i.e., with significantly varying heights. Only in the idealized particular case when $\Omega = \Omega_0$ and the energies of levels with numbers n, + and $n + 1$ coincide pairwise do all windows terminate at $\omega_{\Gamma p} = \Omega$. In this particular case the general picture of the windows is the same as when $\Omega_0 = 0$, and this case is discussed in article [14].

Let the frequencies Ω_0 and Ω, as usual, be different. We will distribute the velocities $\upsilon(n, \sigma)$ in diminishing order, beginning at the maximal value $\upsilon_{max} = \upsilon(0, -)$. Then the value closest to $\upsilon(0, -)$ will be either $\upsilon(1, -)$ (when $\Omega_0 > \Omega$), or $\upsilon(0, +)$ (when $\Omega_0 < 0$). The transparency window corresponding to these highest values of the critical velocities (lobe or triangular) terminate in the first case when $\omega_{\Gamma p} = \Omega$ and in the second case when $\omega_{\Gamma p} = \Omega_0$. The subsequent windows may terminate at any frequencies $\omega_{\Gamma p}$.

In order to determine possible values of frequencies $\omega_{\Gamma p}$ in the case $\Omega_0 > \Omega$ we will take $l\Omega < \Omega_0 < (l + 1)\Omega$, l is an integer. Then the window corresponding to the gap between $\upsilon(n, +)$ and the neighboring lower value of the critical velocity terminates at the lowest frequency from Ω and $(l + 1)\Omega - \Omega_0$, and termination at $(l + 1)\Omega - \Omega_0$ will occur only when the level labeled $l + n + 1$ is occupied. Analogously the window terminating at the maximal frequency from Ω and Ω_0, $-l\Omega$ corresponds to the gap between $\upsilon(n, -)$ and the next velocity value, and termination at $\omega_{\Gamma p} = \Omega_0 - l\Omega$ will appear if $n > l$. In the limit of large Ω_0 ($\Omega_0 > N_{max}\Omega$, N_{max} is the higher number of $N(\sigma)$) only levels with $\sigma = -1$ are occupied, and all transparency windows terminate at $\omega_{\Gamma p} = \Omega$.

In the other characteristic case when $\Omega_0 < \Omega$ two types of windows arise: specifically, the gaps between $\upsilon(n, +)$ and $\upsilon(n + 1, +)$ correspond to windows terminating at $\omega_{\Gamma p} = \Omega - \Omega_0$ while the gaps between $\upsilon(n, -)$ and $\upsilon(n, +)$ correspond to windows with $\omega_{\Gamma p} = \Omega_0$. If $\Omega_0 \ll \Omega$, or $\Omega - \Omega_0 \ll \Omega$, the frequencies $\omega_{\Gamma p}$ are significantly different for the corresponding two types of windows.

Having examined the distribution of maximal frequencies in the windows $\omega_{\Gamma p}$ for all achievable relations between Ω_0 and Ω (i.e., the height distribution of the windows), we will proceed to a discussion of possible values of the wave vector $k_{\Gamma p}$ at the upper termination points. After again using formulae (3.6) and (3.7) for the boundaries of the windows originating from the neighboring levels n, σ and n', σ', we obtain

$$k_{\Gamma p} = m/\hbar[\upsilon(n, \sigma) - \upsilon(n', \sigma')] \quad \text{for a lobe window} \tag{3.9}$$

$$k_{\Gamma p} = m/\hbar[\upsilon(n, \sigma) + \upsilon(n', \sigma')] \quad \text{for a rectangular window} \tag{3.10}$$

If only a few quantization levels are occupied, all critical velocities $\upsilon(n, \sigma)$ are of the order $\sqrt{\hbar\Omega/m}$ and are separated by a like order. Consequently with small numbers $N(\sigma)$ the wave vectors $k_{\Gamma p}$ for different windows and differentials between them are in the order of the inverse quantum length λ^{-1}. In this case if there is no significant difference in the $\omega_{\Gamma p}$ frequencies, the dimensions of the, for example, lobe windows will be within an order of magnitude.

The situation is different with a large number of occupied levels ($N(\sigma) \gg 1$). In this situation the lobe windows corresponding to comparatively low velocities $\upsilon(n, \sigma)$ terminate as before at wave vectors in the order of λ^{-1}. At the same time for lobe windows having a comparatively high angle of inclination with respect to the axis $\omega = 0$ (i.e., $\upsilon(n', \sigma')$, $\upsilon(n, \sigma) \sim \upsilon F$) the wave vector at the termination point is significantly lower (in the order of the inverse cyclotron orbital radius R^{-1}). Thus, lobe windows with a small angle of inclination have a significantly greater longitudinal size; such windows are also larger in the transverse cross-section determined by the difference $\upsilon(n, \sigma) - \upsilon(n', \sigma')$. For triangular windows when $N(\sigma) \gg 1$ the values of the wave vector $k_{\Gamma p}$ are within an order of magnitude between λ^{-1} and k_F, while the transverse dimensions of such windows in the transition from comparatively low $k_{\Gamma p} \sim \lambda^{-1}$ to comparatively high $k_{\Gamma p} \sim k_F$ values drops significantly.

These conclusions are illustrated by Fig. 6 which shows sets of transparency windows in the case of five occupied levels for various relations between Ω and Ω_0. It is clear from Fig. 6 that in an actual situation windows of significantly different dimensions may exist simultaneously. Such is the structure of transparency windows/regions in which the imaginary part of the longitudinal electron polarizability in the quantizing field vanishes.

The role of quantum effects is also manifest in the fact that the function Im $\chi(\omega, k)$ in the region where it is not equal to zero differs significantly from the corresponding quasi-classical quantity. Indeed, according to formula (3.4), Im $\chi(\omega, k)$ is a staircase function whose magnitude is determined by the number of nonzero components in the sum over n, σ in the right half of (3.4). This number is determined in turn by how many regions in (3.5) (Fig. 5b) from various n, σ and n' and σ' are superimposed on one another for the values of ω, k considered. When $\omega < \Omega$ only two such regions will be superimposed or there will be no superposition. Specifically, if the frequency ω is less than the minimal value of $\omega_{\Gamma p}$, the regions in (3.5) will not be superimposed. This means that when $\omega < \Omega$ the quantity $\chi_0(\omega, k)$ is equal to either $\pi\Omega m\eta/4k_F\hbar k$, or $\pi\Omega m\eta/2k_F\hbar k$. On the other hand in the quasi-classical limit (see formula (2.12)) Im $\chi_0^{\text{кл}}\ \omega, k) = \pi\omega m\eta/2k_F\hbar k$, i.e., the imaginary part of the quasi-classical polarizability differs from its quantum values by a factor equal to either $\omega/2\Omega$ or ω/Ω which is in the order of or less than unity for the frequencies under consideration.

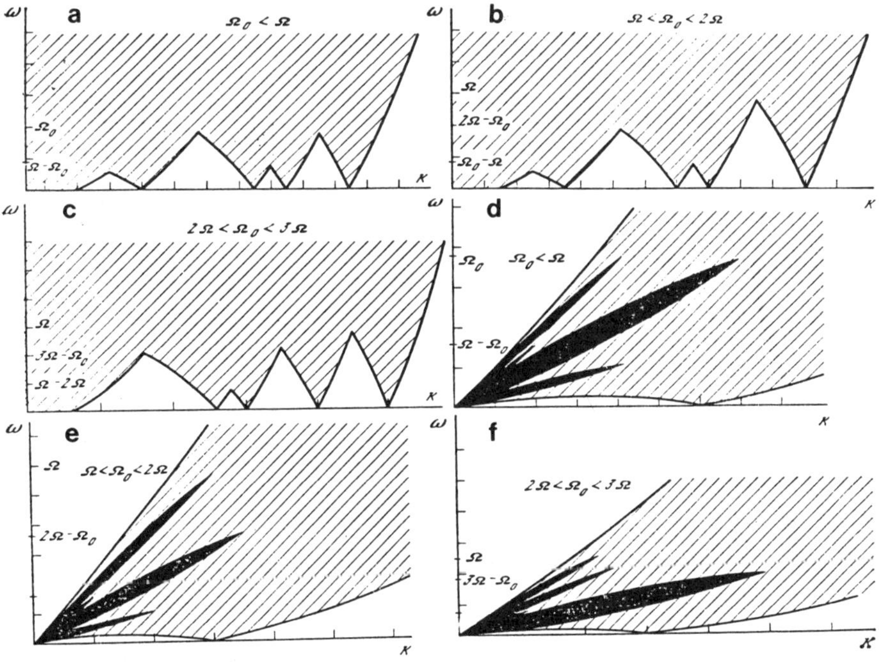

Fig. 6. Lobe and triangular transparency windows in the case of five occupied quantization levels for various relations between the frequencies Ω_0 and Ω
The lobe windows are represented by the dark regions, while the parabolic and triangular windows are represented by the clear regions, with the damping region hatched. In the portions of the diagram containing the triangular windows, the lobe windows are not shown. Frequency. ω is measured in units of $\Omega/2$, while the wave vector k is measured in units of λ^{-1}

Thus, the imaginary part of the polarizability $\chi_0(\omega, k)$ in quantization conditions is equal to zero in the transparency windows and is in the order of or much greater than its classical value in the damping region. The same case applies to the complete polarizability. The higher value of the imaginary part of the polarizability due to alternating windows and damping regions produces giant quantum oscillations of Landau damping.

In concluding this section we will consider the real part of the function $X^{\sigma}(\omega, k)$. Expression (3.2) for $1/\tau \to 0$ yields:

$$\operatorname{Re} X^{\sigma}(\omega, k) = \frac{\Omega m \eta}{4k_F \hbar k} \sum_{n=0}^{N(\sigma)} \ln \left| \frac{[v(n, \sigma) + \hbar k/2m]^2 - u^2}{[v(n, \sigma) - \hbar k/2m]^2 - u^2} \right|. \tag{3.11}$$

It is clear from this formula that the function Re $X^{\sigma}(\omega, k)$ is equal to the sum of the logarithmic components, each of which goes to plus or minus infinity at the boundaries of the transparency windows. This property of the real part of $X^{\sigma}(\omega, k)$ is closely related to the jumps of its imaginary part at the boundaries of the windows which is often referred to as a threshold effect, making possible the existence of solutions of the dispersion equation corresponding to quantum waves within the transparency windows.

Formula (3.6) has a simple form with small values of the wave vector in the vicinity of lobe transparency windows. Assuming the quantity $\hbar k/m|v(n, \sigma) - u|$ to be small, we may expand the logarithm in powers of the wave vector, thereby obtaining the following expression:

$$\operatorname{Re} X^{\sigma}(\omega, k) = \frac{\Omega \eta}{2k_F} \sum_{n=0}^{N(\sigma)} \frac{v(n, \sigma)}{v^2(n, \sigma) - u^2}. \tag{3.12}$$

When this expression is valid the damping regions are very narrow and are located along the lines $u = v(n, \sigma)$. The $\chi(\omega, k)$ polarizability in accordance with relation (3.1) is expressed through the functions $X^{\sigma}(\omega, k)$. Hence formulae (3.11) and (3.12) provide the basis for investigating quantum waves: the solutions of dispersion equation (2.28) in the transparency windows.

4. The spectrum of quantum waves in an electron gas

In this section we will provide the results from quantum wave theory in an electron gas, i.e., in a system of quasi particles whose interaction is negligible. As a result we may ignore exchange interaction and take the electron polarizability to be equal to $\chi_0(\omega, k)$. The other constraint in this section involves ignoring interaction between the electron quantum waves and regular acoustic waves. Such interaction is not significant if the phase velocities of these waves are significantly different from the acoustic velocity. If, on the other hand, the phase velocities are close to the acoustic velocities, a wave coupling effect arises that will be discussed in the following section.

The following dispersion equation derives from equation (2.28) subject to expression (3.11); this equation determines the spectrum of quantum waves in an electron gas:

$$\sum_{\sigma}\sum_{n=0}^{N(\sigma)} \ln\left|\frac{[v(n,\sigma)+\hbar k/2m]^2-u^2}{[v(n,\sigma)-\hbar k/2m]^2-u^2}\right|=0. \tag{4.1}$$

Here we drop the imaginary part of $\chi_0(\omega, k)$ since we will only consider waves in the transparency windows; $u = \omega/k$ is the phase velocity. Equation (4.1) describes waves that do not interact with sound and hence they do not account for the contribution of lattice vibrations. Such an equation ignoring spin splitting ($\Omega_0 = 0$) was examined in studies [7-10]. The influence of spin splitting on the quantum wave spectrum was investigated by P.S. Zyryanovand the authors of the present chapter [16, 19] (see also [10]).

For longwaves corresponding to the condition $k^{-1} \gg r = v_F/\Omega$ equation (4.1) takes the form

$$\sum_{\sigma}\sum_{n=0}^{N(\sigma)} \frac{v(n,\sigma)}{v^2(n,\sigma)-u^2}=0. \tag{4.2}$$

This relation no longer contains the wave vector k and it follows that in the limit $k \to 0$ the dispersion law of the waves is acoustic ($\omega \sim k$). Each of the roots of equation (4.2) corresponds to one of the branches of the quantum wave spectrum resulting from the dependence of electron polarizability on the phase velocity and caused by oscillations of the one-dimensional electron subsystems. We will discuss the primary properties of the phase velocities of quantum waves in the longwave limit.

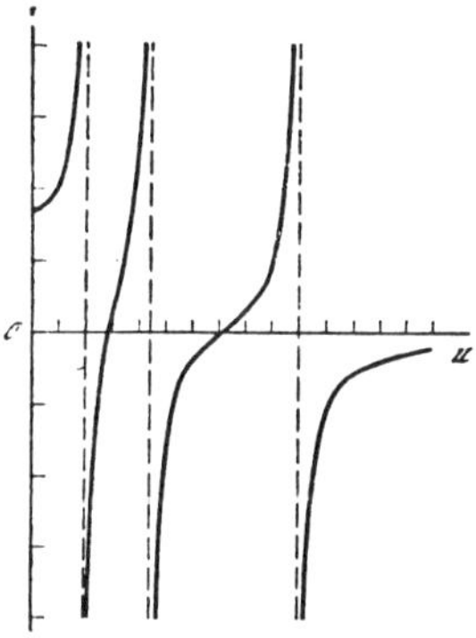

Fig. 7. The left half of equation (4.2) in the case of three occupied quantization levels plotted against u

The left half of equation (4.2) is shown in Fig. 7) (for the case of three occupied levels). The points of intcr section of the curves with the X-axis represent the phase velocities that are roots of equation (4.2). The number of roots of the equations is equal to the number of gaps between the various critical velocities $v(n, \sigma)$ and hence is one less than the number of occupied quantization levels. Obviously in the ultraquantum limit when only a single level is occupied, there are no roots of equation (4.2) and hence there are no quantum waves either. This follows from the nature of the quantum waves and corresponds to the impossibility of the existence of acoustic waves in an electron system with a single critical velocity.

In order to discuss qualitative effects the limiting cases are of interest; here we may obtain from equation (4.2) simple expressions for the phase velocities of the quantum waves. In the simplest case of two occupied quantum levels (for example, 0, - and 1, -, corresponding to $\Omega_0 > \Omega$) a quantum wave with the phase velocity

$$u = \sqrt{v(0, -)\, v(1, -)}. \tag{4.3}$$

is the only possible wave. Since $v(0, -) > u > v(1, -)$, it is obvious that the dispersion curve of this wave lies in a lobe transparency window. If three levels are occupied (for definiteness we speak of levels with critical velocities $v(0, -)$, $v(1, -)$, $v(0,+)$), we may easily see that equation (4.2) reduces to a biquadratic equation for the phase velocity determining its explicit form for two possible quantum waves. Of particular interest here is the case where the lowest of the critical velocities $v(0, +)$ is much less than the remaining velocities: $v(0, +) \ll v(0, -), v(1, -)$. This case is implemented with magnetic field values where the number of electrons on the (0, +) level is comparatively low, i.e., in the case of weak occupation of the level with the highest transverse motion energy. Here we may speak of two oscillation branches where one branch, as in the case when only two levels are occupied, corresponds to relative oscillations of two one-dimensional electron subsystems (0, -), (1, -) and a quantum wave with a phase velocity u_1 taking the form of (4.3). The second branch of the quantum waves is generated by electron oscillations on the weakly occupied level (0, +) with respect to the subsystems (0, -), (1, -) and this corresponds to the phase velocity determined by the formula:

$$u_2^2 = v(0, +)\, [1/v(0, -) + 1/v(1,-)]^{-1}. \tag{4.4}$$

Since $v(0, -) \sim \sqrt{\hbar|\Omega - \Omega_0|m}$ consistent with (1.3) while $v(1, -) \sim \sqrt{\hbar\Omega/m}$, in order to realize a slow quantum wave with velocity $u_2 \ll u_1$ the inequality $v(0, +) \ll \sqrt{\hbar|\Omega - \Omega_0|/m}, \sqrt{\hbar\Omega/m}$ must be satisfied. With an increase in the magnetic field strength the velocity $v(0, +)$ drops together with the degree of occupation of the level (0, +). The phase velocity u_2 drops simultaneously, remaining within the gap between $v(0, +)$ and $v(1, -)$.

The conclusion that a quantum wave exists with a low phase velocity with weak occupation of the level with the highest transverse energy may easily be generalized to the case of occupation of an arbitrary number of quantization levels. Indeed, let v_{min} be the smallest critical velocity $v(n, \sigma)$ corresponding to the weakly-occupied level. The velocity v_{min} is small compared to other velocities $v(n, \sigma)$ if $v_{min} \ll \sqrt{\hbar/\Omega - \Omega_0|/m}, \sqrt{\hbar\Omega/m}$. Then the corresponding minimal phase velocity of the quantum waves u_{min} is small compared to all $v(n, \sigma)$ except v_{min}, which makes it possible to ignore the quantity u^2 in all components in (4.2) except that containing v_{min}. Then we have:

$$u_{\min} = \sqrt{v_{\min}(V + v_{\min})}. \tag{4.5}$$

Here

$$V = \Big[\sum_{n,\sigma}{}' \frac{1}{v(n,\sigma)} \Big]^{-1}, \tag{4.6}$$

where the prime indicates that summation is taken over all occupied levels aside from the weakly-occupied level which corresponds to the minimal critical frequency $v_{\min}$. When the greatest number of levels are occupied it turns out that $v_{\min} \ll V$, and formula (4.5) takes the form of (4.4), when $v_{\min} \ll u_{\min} \ll V$.

When many quantization levels are occupied ($N(\sigma) \gg 1$) in addition to a quantum wave with minimal velocity (4.5) there exist several branches with phase velocities that are small compared to the Fermi velocity v_F. In order to explain this we note that when $N(\sigma) \gg 1$ the highest critical velocity $v(n, \sigma)$ is within an order of magnitude of v_F (as is the case with neighboring velocities), while the gaps between these velocities are of the order Ω/k_F. On the other hand there are small values of $v(n, \sigma)$ near $v_{\min}$ that are, like the gaps between them, of the order $\sqrt{\hbar\Omega/m}$. The phase velocities of the slow quantum waves lie in the gaps between the small values of $v(n, \sigma)$. These velocities ($u \ll v_F$) are close to the corresponding values of $v(n, \sigma)$, i.e., the difference $u - v(n, \sigma)$ for these is small compared to $\sqrt{\hbar\Omega/m}$. In order to explain this statement we will find an explicit expression for small values of u which will also confirm the inequality $|u - v(n, \sigma)| \ll v(n, \sigma)$.

We will assume that the velocity u is close to some value of $v(n, \sigma)$ that is small compared to v_F, for example, $v(n_0, +)$. Here it may turn out that the difference between the velocity $v(n_0, +)$ and the closest neighboring velocity $v(n, \sigma)$ (for example, $v(n_0 +1, -)$) will be small compared to the gap between $v(n_0, +)$ and $v(n_0 +1, -)$. Bearing in mind such a general case we will account for the smallness of both the difference $u - v(n_0, +)$ and the difference $u - v(n_0 + 1, -)$. We will isolate the two "resonance" terms containing the velocities $v(n_0, +)$ and $v(n_0 +1, -)$ in the sum in equation (4.2). The remaining part of the sum for these values of u will have a continuous dependence on the phase velocity and since the number of quantization levels is high it may be replaced approximately by the n integral which is equal to

$$\frac{2k_F}{\Omega}\left(1 + \frac{u}{2v_F} \ln \frac{v_F - u}{v_F + u}\right) \approx \frac{2k_F}{\Omega}. \tag{4.7}$$

Thus, equation (4.2) takes the form

$$\frac{v(n_0, +)}{v^2(n_0, +) - u^2} + \frac{v(n_0 + 1, -)}{v^2(n_0 + 1, -) - u^2} + \frac{2k_F}{\Omega} = 0. \tag{4.8}$$

Since $2k_F/\Omega \gg \upsilon^{-1}(n_0, +), \upsilon^{-1}(n_0 +1, -) \sim \sqrt{m/\hbar\Omega}$, the roots of equation (4.8) are the phase velocities of the slow quantum waves, which are indeed close to the values $\upsilon(n_0, +), \upsilon(n_0 +1, -)$. For the particular case $\upsilon(n_0, +) = \upsilon(n_0 +1, -)$ this fact was established in study [8], while the general form was examined in study [16]. Equation (4.8) is also valid for determining the minimal phase velocities, when ?"$n_0 = N(+)$. If here $\upsilon(n, +) \ll \upsilon(N +1, -)$ then in the left half of (4.8) we may ignore the second component, which yields expression (4.5) in which $V = \Omega/2k_F$. In the general case equation (4.8) determines the phase velocities of the two coupled electron quantum waves. The coupling of these waves is significant if the difference of the velocities $\upsilon(n_0, +)$ and $\upsilon(n_0 +1, -)$ is of the order Ω/k_F. If these critical velocities are significantly different, so $|\upsilon(n_0 +1) - \upsilon(n_0 + 1, -)| \gg \Omega/k_F$, we have for the phase velocities of the waves [16]:

$$u_1 = v(n_0, +) + \frac{\Omega}{4k_F}, \quad u_2 = v(n_0 + 1, -) + \frac{\Omega}{4k_F}. \tag{4.9}$$

It is obvious that waves (4.9) with velocities u_1, u_2 apply to different lobe transparency windows. In the limit opposite case (4.9) when $|\upsilon(n_0, +) - \upsilon(n_0 +1, -)| \ll \Omega/k_F$, we obtain [8]:

$$u_1 \approx u_2 = v(n_0, +) + \frac{\Omega}{2k_F}. \tag{4.10}$$

It follows from the examination of slow quantum waves that with a large number of occupied quantization levels there exist two types of such waves. First, the slowest wave whose velocity is significantly different from the critical electron velocity in the weakly-occupied quantization level υ_{min} exists in the case of weak occupation of the level with the greatest transverse energy. Second, there are many branches with phase velocities close to the critical velocities $\upsilon(n, \sigma)$. We note that the observation conditions for observing waves with velocities close to $\upsilon(n, \sigma)$ are more rigid than for the slowest wave since the lines $u = \upsilon(n, \sigma)$ on the ω, k plane with small values of the wave vector correspond to the edges of the damping regions (see Fig. 3) that effectively shift due to the influence of collisions and temperature, thereby causing strong damping of waves with velocities close to $\upsilon(n, \sigma)$.

Some of the quantum wave branches have phase velocities in the order of the Fermi velocity. We cannot write a simple analytic expression for these velocities. However, one possible qualitative conclusion is that the phase velocities of the fast waves lie in the central regions of the gaps between the critical velocities $\upsilon(n, \sigma)$. In other words, the phase velocities of the fast waves, as in the case of slow waves, have a differential in the order of Ω/k_F from the critical velocities $\upsilon(n, \sigma)$.

We will now discuss theoretical results relating to the short-wave portion of the spectrum of quantum waves in an electron gas based on dispersion equation (4.1). As noted above with small values of the wave vector the dispersion curves of the quantum waves lie within the lobe transparency windows. This means that

$$v(n', \sigma') + \hbar k/2m \leqslant u \leqslant v(n, \sigma) - \hbar k/2m, \tag{4.11}$$

where $\upsilon(n', \sigma')$ and $\upsilon(n, \sigma)$ are the closest values of the critical electron velocities. Above all we will establish the point of intersection between the dispersion curve and the boundary of the lobe transparency window. We note that with condition (4.11) all components of the sum in (4.1) containing the velocities $\upsilon(n'', \sigma'') \geqslant \upsilon(n, \sigma)$ are finite and positive, while those containing the velocities $\upsilon(n'', \sigma'') \leqslant \upsilon(n', \sigma')$ are finite and negative. As a result the left half of equation (4.1) goes to $+\infty$ at the upper bound of the window, when $u = \upsilon(n, \sigma) - \hbar k/2m$, while on the lower bound ($u = \upsilon(n', \sigma') + \hbar k/2m$) it goes to $-\infty$. It follows that in this case the dispersion curve cannot intersect either the upper or lower bounds of the lobe window separately, since then the sum in (4.1) cannot vanish. The only possibility for the dispersion curve of the quantum wave to intersect the boundary of the transparency windows remains at the point of intersection of the upper and lower bounds of the window, i.e., when $k = k_{\Gamma p} = m[\upsilon(n, \sigma) - \upsilon(n', \sigma')]/\hbar$ and $\omega = \omega_{\Gamma p}$ (see formulae (3.8) and (3.9)). It is precisely this point that is the termination point of the dispersion curve of the quantum wave within the lobe transparency window. Overall the dispersion dependence of the quantum wave frequency in the transparency window is described by a monotonic and continuous curve that becomes a straight line with small k, while with large values of the wave vector it terminates at the point with a maximal value of $k = k_{\Gamma p}$ (see Fig. 3a).

A simple dispersion law exists for a quantum wave in the case of occupation of two levels n, σ. If $\Omega_0 > \Omega$, then in this case it follows from (4.1) that

$$u^2 = v(0, -)\, v(1, -) + (\hbar k/2m)^2. \tag{4.12}$$

This formula describes the dispersion curve of a quantum wave across the entire transparency window and generalizes the longwave limit (4.3).

In the case of a large number of occupied quantization levels the characteristic properties of the dispersion curves are different for slow and fast waves, due to the fact that the transparency windows for slow waves extend up through values of $k_{\Gamma p} \sim \lambda^{-1}$ at the same time that for slow waves the range of the windows is significantly shorter: the maximal values of $k_{\Gamma p}$ is of the order R^{-1} (see section 3). The phase velocities of the fast waves have a weak dependence on the wave vector, while the dispersion curves of the slow waves when $kR \geqslant 1$ nearly repeat the shape of the lower boundaries of the transparency windows.

It is comparatively simple to obtain a formula describing the entire dispersion curve for the slowest quantum wave in the case of weak occupation of the level with the highest transverse energy. We will consider the case of many occupied levels and will isolate in the sum of equation (4.1) the component containing velocity v_{min} and one additional component which includes the next velocity $v(n, \sigma)$ after v_{min} (we will label this v_1). The dispersion curve of the slowest quantum wave lies in a lobe transparency window formed by the curves $\omega = kv_{min} + \hbar k^2/2m$ and $\omega = kv_1 - \hbar k^2/2m$. In this window the dependence on the phase velocity and the wave vector of the remaining components of the sum in (4.1) aside from the two isolated components is rather weak. We will ignore this dependency and will assume that in these remaining components $u^2 \ll [v(n, \sigma) \pm \hbar k^2/2m]^2$, $\hbar k/2m \ll v(n, \sigma)$ and will replace summation over n with integration (the case $N(\sigma) \gg 1$). Then we arrive at the following dispersion equation:

$$\ln \frac{u^2 - (v_{min} + \hbar k/2m)^2}{u^2 - (v_{min} - \hbar k/2m)^2} + \ln \frac{(v_1 + \hbar k/2m)^2 - u^2}{(v_1 - \hbar k/2m)^2 - u^2} + 4kR = 0. \qquad (4.13)$$

In the region of low values of k ($kR \ll 1$) this equation takes the form of (4.8) and when $v_{min} \ll v_1$ it produces formula (4.5) for the phase velocity of the slowest wave in the longwave limit. For arbitrary k equation (4.13) rewritten in the form

$$\frac{u^2 - (v_{min} + \hbar k/2m)^2}{u^2 - (v_{min} - \hbar k/2m)^2} \, \frac{(v_1 + \hbar k/2m)^2 - u^2}{(v_1 - \hbar k/2m)^2 - u^2} = e^{-4kR} \qquad (4.14)$$

is easily solved, thereby obtaining the $u(k)$ relation. If $kR \gtrsim 1$, the dispersion curve of the slowest quantum wave is close to the lower boundary of the transparency window ($u \approx v_{min} + \hbar k/2m$). In this range of values of the wave vector the following approximate expression derives from equation (4.14):

$$u = v_{min} + \hbar k/2m + v_{min} \frac{k_{rp} - k}{k_{rp} + k} \, \frac{2k}{k + 2mv_{min}/\hbar} \, e^{-4kR}. \qquad (4.15)$$

This expression, specifically, shows that the dispersion curve near the termination point $k = k_{rp} \gg R^{-1}$ virtually coincides with the lower boundary of the window. Such behavior near the termination points is characteristic of slow quantum waves with a large number of occupied levels. Fast waves with few occupied levels (see formula (4.12)) have dispersion curves that are not as close to the window boundaries in the vicinity of the termination points.

In concluding our discussion of the spectrum of quantum waves in an electron gas, we note an important fact relating to the physical nature of such waves. Specifically, it turns out that quantum wave propagation, as a rule, is accompanied by significant oscillations in the spin electron density. This was first demonstrated in study [16]. Specific expressions for the spin density amplitude $s^z(k, \omega)$ may be obtained by using general formula (2.17) and the approximate expressions given above for the phase velocities of the waves. For example, when two quantization levels with spin polarization $\sigma = -1$ are occupied and the quantum wave has the phase velocity (4.12) the polarizability of electrons with a positive spin is equal to zero ($X^+ = 0$), and hence according to formulae (2.17) and (2.18) the amplitude of the spin density $s^z(k, \omega)$ differs from the amplitude of the electron density only in the factor $-1/2\ n(k,\ \omega)$: $s^z(k,\ \omega) = -n(k,\ \omega)/2$. During the propagation of the slowest quantum wave there exist oscillations of the electrons in the state with velocity υ_{min} and with a single spin direction (for example, corresponding to $\sigma = +1$) that make a contribution to the corresponding polarizability X^+ comparable to approximately an identical contribution of all other electrons to X^+ and X^-. Hence it turns out that $X^+ - X^- \sim X^+ \sim X^+ + X^-$ and the $s^z(k,\ \omega)$ amplitude is again within an order of magnitude of $n(k,\ \omega)$. In the general case we may state that if some quantum wave corresponds to the resonance contribution to polarizability with only a single spin direction, this wave is related to significant oscillations in spin density. Specifically, waves with phase velocities (4.9) are accompanied by significant oscillations in spin density, at the same time that such oscillations are small in wave (4.10).

Above we examined the quantum waves in lobe transparency windows. We will now show that in triangular transparency windows equation (4.1) has no roots. A triangular window is determined by the inequalities $u \leqslant \upsilon(n,\ \sigma) - \hbar k/2m$, $u \leqslant \hbar k/2m - \upsilon(n,\ \sigma')$, $u \geqslant 0$, $\upsilon(n',\ \sigma') \leqslant \hbar k/2m \leqslant \upsilon(n,\ \sigma)$, where $\upsilon(n,\ \sigma)$, $\upsilon(n',\ \sigma')$ are the neighboring values of the critical velocities. We will determine that all components of the sum in (4.1) in this triangular window are positive. It follows directly from the inequalities determining the window area that the components in equation (4.1) that contain the velocities $\upsilon(n,\ \sigma)$, $\upsilon(n',\ \sigma')$ are positive, since the numerator in the expression under the logarithm sign is greater than the denominator. We will consider a given third component containing the velocity $\upsilon(n'',\ \sigma'')$ and will show that $[\upsilon(n'',\ \sigma'') - \hbar k/2m]^2 > u^2$ if velocity υ is within the selected triangular window. For this it is sufficient to substitute the maximum value of u in the selected window which is equal to $u_{max}(k) = \hbar k/2m - \upsilon(n',\ \sigma')$ when $\upsilon(n',\ \sigma') \leqslant \hbar k/2m \leqslant [\upsilon(n,\ \sigma) + \upsilon(n',\ \sigma')]/2$ or $u_{max}(k) = \upsilon(n\sigma) - \hbar k/2m$, when $[\upsilon(n,\ \sigma) + \upsilon(n',\ \sigma')]/2 \leqslant \hbar k/2m \leqslant \upsilon(n,\ \sigma)$ into the right half of the inequality. A simple test shows that $[\upsilon(n'',\ \sigma'') - \hbar k/2m]^2 > u^2_{max}(k)$ both in the case $[\upsilon(n'',\ \sigma'') - \hbar k/2m]^2 > u^2$ and when $\upsilon(n'',\ \sigma'') < \upsilon(n',\ \sigma')$ and as a result we have the inequality $[\upsilon(n'',\ \sigma'') - \hbar k/2m]^2 > u^2$ in the entire window. Thus, the left half of equation (4.1) in a triangular transparency window is greater than zero and there are no quantum waves in such windows. The same case is valid for a parabolic transparency window. We note, however, that this conclusion was obtained ignoring the influence of exchange interaction between electrons that may produce quantum waves in the triangular transparency windows, as discussed in section 6.

5. Coupled acoustic and quantum waves

The theory of quantum waves in an electron gas outlined above predict the existence of slow electron density waves in a quantizing magnetic field whose phase velocities are small compared to the Fermi velocity υ_F. When the velocities of the slow waves are close to the acoustic velocity, the oscillations in the ion charge density become coupled to the electrons, which produces coupled quantum waves accompanied by oscillations of both the electron system and the ion lattice. We will now present results from coupled wave theory developed in studies [17, 24-26].

In this section the entire examination will relate to a model of electron gas in an ion lattice, so that the initial equation for subsequent analysis is dispersion equation (2.28) with electron gas polarizability $\chi_0(\omega, k)$. Substituting expression (3.11) for the real part of $\chi_0(\omega, k)$ we write the dispersion equation in the transparency windows in the following manner:

$$-\frac{u_0^2}{u^2 - v_s^2} + \frac{1}{4kR}\sum_\sigma \sum_{n=0}^{N(\sigma)} \ln\left|\frac{[v(n,\sigma) + \hbar k/2m]^2 - u^2}{[v(n,\sigma) - \hbar k/2m]^2 - u^2}\right| = 0. \tag{5.1}$$

Equation (5.1) has a different first component in the left half from (4.1); this component is proportional to the quantity u_0^2 (see formula (2.14)) and describes the contribution of the ion oscillations. In the longwave limit ($kR \ll 1$) we have from (5.1):

$$-\frac{u_0^2}{u^2 - v_s^2} + \frac{\Omega}{2k_F}\sum_\sigma \sum_{n=0}^{N(\sigma)} \frac{v(n,\sigma)}{v^2(n,\sigma) - u^2} = 0. \tag{5.2}$$

As noted above only slow quantum waves interact with sound, since the acoustic velocity u_s is small compared to v_F. Here two characteristic cases are possible depending on the relationship between the velocity u_s and $\sqrt{\hbar\Omega/m}$ within an order of magnitude. If the condition $u_s \ll \sqrt{\hbar\Omega/m} \sim v_s\sqrt{\hbar\Omega/e_F}$ is satisfied, the sound will interact primarily with the slowest quantum wave having a phase velocity u_{min} only. In the other case when $u_s \sim \sqrt{\hbar\Omega/m}$ sound may interact with other branches of the quantum wave spectrum. We will first consider in detail the first case $u_s \ll \sqrt{\hbar\Omega/m}$ investigated in study [26] which, as a rule, corresponds to actual comparatively large magnetic field strengths that satisfy the quantization conditions.

We will first take $v_{min} \sim \sqrt{\hbar\Omega/m}$. Then in order to find the velocity of longwave sound we may ignore the dependence on u^2 in equation (5.2) in the sum over n, σ and obtain

$$u^2 = v_s^2 + u_0^2\left(\frac{\Omega}{2k_F v_{min}} + \frac{\Omega}{2k_F V}\right)^{-1}. \tag{5.3}$$

Here if many quantization levels are occupied, $V \approx \Omega/2k_F \ll \sqrt{\hbar\Omega/m}$ and the right half of formula (5.3) is reduced to the squared acoustic velocity in the absence of a magnetic field $u_s^2 = v_s^2 + u_0^2$. If only a few levels are occupied the multiplier in front of u_0^2 in (5.3) is other than unity, although it does not exceed it by more than an order of magnitude.

We will now assume that the velocity v_{min} is reduced (the field strength is increased). Then the phase velocity of the slowest quantum wave u_{min} will also drop. As a result with sufficiently small values of υ_{min} two waves will exist (an acoustic wave and a quantum wave) that are coupled due to the proximity of the phase velocities. Incorporating the dependence on u in the sum over n, σ in equation (5.2) only in the component containing υ_{min}, we obtain the following dispersion equation for the coupled waves with small values of k:

$$-\frac{u_0^2}{u^2-v_s^2}+\frac{\Omega}{2k_F}\frac{v_{min}}{v_{min}^2-u^2}+\frac{\Omega}{2k_F V}=0. \tag{5.4}$$

The roots of this equation may be written as

$$u_{1,2}^2=\frac{1}{2}\{U_s^2+u_{min}^2\pm\sqrt{(U_s^2-u_{min}^2)^2+8u_0^2V^2k_Fv_{min}/\Omega}\}, \tag{5.5}$$

where $U_s^2=v_s^2+u_0^2(2k_FV/\Omega)$ is a quantity that does not differ significantly from u_s^2, and virtually coincides with u_2^2 with many occupied levels.

In the limit of small υ_{min} ($\upsilon_{min} \ll V$, U_s^2/V) the following expressions derive from formula (5.5):

$$u_1^2=U_s^2+\frac{u_0^2}{U_s^2}V^2\frac{2k_Fv_{min}}{\Omega}, \tag{5.6}$$

$$u_2^2=\frac{2k_Fv_{min}}{\Omega}V^2\frac{v_s^2}{U_s^2}. \tag{5.7}$$

We may easily see that the squared phase velocity of the first wave u_1^2 changes from u_{min}^2 to the value in (5.6) at the same time that u_2^2 jumps from the squared acoustic velocity (formula (5.3)) to the critical value (5.7). At the point $\upsilon_{min} = 0$ the velocity u_2 vanishes, while u_1^2 becomes (5.3) (for complete correlation it is necessary to isolate from V^{-1} the contribution of the state with the next lowest critical velocity after υ_{min}).

Wave interaction is most strongly manifest in the range of values of υ_{min} in which the velocities u_s and u_{min} are of the same order. We will assume U_s does not exceed V within an order of magnitude ($U_s \lesssim V$). This condition for the case of few occupied levels is satisfied by virtue of the inequality $U_s \ll \sqrt{\hbar\Omega/m}$. If the number of occupied levels is high, it may be written as $\hbar\Omega/e_F \gtrsim U_s/\upsilon_F$. Then in expression (5.5) the component $8u_0^2V^2k_F\upsilon_{min}/\Omega$ under the root when $U_s \sim u_{min}$ is of the order $U_s^4(u_0^2k_FV/\Omega \sim u_s^2;\ V\upsilon_{min} \sim u_{min}^2 \sim u_s^2)$, and, consequently, the velocities u_1 and u_2 are in the order of u_s from u_s. Thus, in sufficiently strong magnetic fields, when $u_s \lesssim V$ two acoustic waves may exist in the metal whose phase velocities are separated from the acoustic velocity by a quantity in the order of the acoustic velocity itself. This effect is predicted in study [26]. The dependence of the phase velocities u_1, u_2 on υ_{min} predicted by formula (5.5) determines their dependence on the field B, since υ_{min} is proportional to $\sqrt{B_m - B}$, where B_m is the magnetic field induction at the peak.

We should emphasize the characteristic feature of this effect where the elastic properties of the electrons and the lattice are manifest differently in the effect. In the absence of a magnetic field the two contributions to the squared acoustic velocity (u_0^2, v_s^2) are not differentiable. On the other hand in conditions of sound/quantum wave interaction the phase velocities of the coupled waves are highly dependent on the relation between u_0 and v_s. Wave interaction is quite strongly manifest in the case where metallic elastic is wholly related to the electrons ($v_s = 0$). In this case the phase velocity u_2 in the limit of small v_{min} has a quadratic rather than a linear dependence on v_{min}, and in place of formula (5.7) when $v_{min} \ll V$, $k_F u_0^2/\Omega$ we have:

$$u_2^2 = v_{\min}^2. \tag{5.8}$$

Formulae (5.3)-(5.8) relate to the longwave region. We will now consider the range of comparatively high values of k and will write an approximate dispersion equation describing the coupled waves across the entire transparency windows. As in obtaining equation (4.13) we will ignore the insignificant dependence on u and k of all components of the sum in equation (5.1) aside from those containing the velocity v_{min} and the next level velocity v_1. Then consistent with (5.1) we obtain a simpler approximate equation:

$$\frac{1}{4kR}\frac{2k_FV'}{\Omega}\left\{\ln\left|\frac{(v_{\min}+\hbar k/2m)^2-u^2}{(v_{\min}-\hbar k/2m)^2-u^2}\right|+\ln\left|\frac{(v_1+\hbar k/2m)^2-u^2}{(v_1-\hbar k/2m)^2-u^2}\right|\right\}=-\frac{u^2-U_s'^2}{u^2-v_s^2}, \tag{5.9}$$

where the velocities V' and U' are determined by the formulae

$$\frac{1}{V'}=\sum_{n,\sigma}{}'' \frac{1}{v(n,\sigma)}, \qquad U_s'^2=v_s^2+u_0^2\frac{2k_FV'}{\Omega}. \tag{5.10}$$

Unlike (4.13) the contribution of the ion oscillations is accounted for in equation (5.9) and it is not assumed that there are many occupied levels. If we set $u_0 = 0$ (zero ion oscillations), $V' = \Omega/2k_F$ (many occupied levels), (5.9) becomes (4.13). An analysis of the solutions of equations (5.9) yields the dispersion curves shown schematically in Fig. 8. We will discuss the form of these curves.

When $v_{min} > v_s$ one of the curves is located in a parabolic transparency window whose boundary is described by the equation $u = v_{min} - \hbar k/2m$. The corresponding branch of the spectrum may be called the natural acoustic branch, since only ion oscillations will produce waves in the parabolic window. We should bear in mind that with strong wave coupling such a name is rather arbitrary. Another curve is located in a lobe transparency window and corresponds to a natural quantum wave whose dispersion relation is altered by ion oscillation influence. We will consider the acoustic branch in greater detail. In the parabolic transparency window the left half of dispersion equation (5.9) is positive and therefore the right half will also be positive, thereby yielding the relation $v_s^2 < u^2 < U_s'^2$. Then we may take

$u^2 \ll (v_1 - \hbar k/2m)^2$, $\hbar k/2m \ll v_1$ in the parabolic window by virtue of the smallness of the velocity $v_{\min}$ ($v_{\min} \ll v_1$), and we may expand the logarithm in (5.9) containing v_1 in terms of $\hbar k/2m$. Then the dispersion equation is simplified:

$$\frac{mV}{2\hbar k}\ln\frac{(v_{\min}+\hbar k/2m)^2-u^2}{(v_{\min}-\hbar k/2m)^2-u^2}=\frac{U_s^2-u^2}{u^2-v_s^2}\,. \tag{5.11}$$

In the limit of small k this equation becomes (5.4) and we will now consider the root u_2 (the lower sign in (5.5)) satisfying the condition $v_s^2 < u^2 < U_s^2$. With an increase in the wave vector k the dispersion curve reaches the boundary of the window at the point $u = v_s$, $k = k_\Gamma = 2m\,(v_{\min} - v_s/\hbar)$. In the vicinity of this point when $u - v_s \ll v_s$, $k_\Gamma - k \ll k_\Gamma$ equation (5.11) is transformed to $\ln(t - y) = -a/y$, where $t = v_s(k_\Gamma - k)/2v_{\min} k_\Gamma$; $y = mv_s(u - v_s)/\hbar\, v_{\min}k_\Gamma$; $a = 2k_F u_0^2/\Delta v_{\min}$. When $t \to 0$ the solution of the last equation rewritten as $y = t - \exp(-a/y)$ is given by the formula $y \approx t - \exp(-a/t)$ so that the form of the dispersion curve near the boundary of the transparency window is described by the following expression:

$$u=v_{\min}-\frac{\hbar k}{2m}-\frac{2v_{\min}}{v_s}(v_{\min}-v_s)\exp\left\{-\frac{2k_F u_0^2 k_\Gamma}{\Omega v_s(k_\Gamma-k)}\right\}. \tag{5.12}$$

This expression is valid if the exponential component is small compared to $v_{\min} - v_s - \hbar k/2m$.

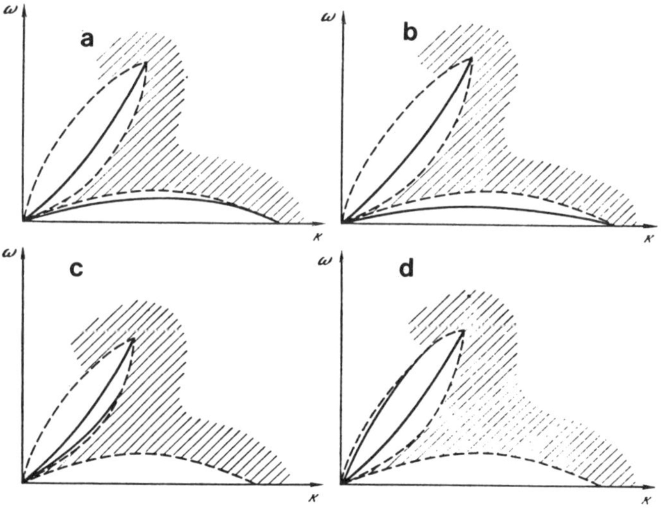

Fig. 8. The dispersion curves of a coupled acoustic and the slowest quantum wave in transparency windows
a - the acoustic branch in a parabolic window, $v_s \neq 0$; b - the acoustic branch in a parabolic window, $v_s = 0$; c - both branches in a lobe window intersecting with the lower boundary; d - both branches in a lobe window intersecting with the upper boundary

If we take $\upsilon_s \approx 0$ (the elasticity of the metal is determined primarily by conduction electrons), the acoustic branch will not fall within the lobe transparency window and will make contact with the boundary of the parabolic window only in the limit $\omega \to 0$ (Fig. 8b). In fact due to the condition $\omega\tau > 1$ the boundary of the wave in the parabolic window will be determined by the quantity $1/\tau$. If $\upsilon_s \neq 0$, then when $\upsilon_{min} < \upsilon_s$ both dispersion curves of the coupled waves will lie (with corresponding k) in a lobe transparency window (Fig. 8c). The dispersion curve of sound intersects the lower boundary of the window $u = \upsilon_{min} + \hbar k/2m$ at the point $k = k'_\Gamma = 2m(\upsilon_s - \upsilon_{min})/\hbar$ if the quantity k'_Γ determined by this equality is less than the maximal value of k in the lobe window $k_{\Gamma p} = m(\upsilon_1 - \upsilon_{min})/\hbar$, i.e., when $2\upsilon_s < \upsilon_1$. We remember that the dispersion curve of the quantum wave intersects the boundary of the window at precisely the point $k = k_{\Gamma p}$ and, consequently, in the case $k'_\Gamma < k_{\Gamma p}$ the natural quantum branch will have a higher frequency than the acoustic branch (Fig. 8c). A formula analogous to (5.12) may be obtained near the point of intersection with the lower boundary of the window for sound.

In the case when $2\upsilon_s > \upsilon_1$ (moderately strong magnetic fields) the inequality $k'_\Gamma < k_{\Gamma p}$ may be satisfied which reduces to the condition $2\upsilon_s - \upsilon_{min} > \upsilon_1$. In this condition the quantum wave (or, more precisely, the wave from the coupled waves whose dispersion curve terminates at $k = k_{\Gamma p}$) has a frequency lower than the acoustic wave in the lobe transparency window. As a result the acoustic branch intersects the upper boundary of the transparency window $u = \upsilon_1 - \hbar k/2m$ (Fig. 8d).

We have discussed the form of the dispersion curves of waves in transparency windows. In the damping region one of the two coupled waves (the natural quantum wave) is always significantly damped, i.e., it in fact does not propagate at the same time that the other wave (the natural sound wave) may propagate. In order to investigate the dispersion dependence of sound in the damping region we must include in dispersion equation (5.1) the imaginary part of the polarizability $\chi_0(\omega, k)$. For example, we will analyze the continuation of the dispersion curve of the sound wave from the parabolic transparency window when dispersion equation (5.11) and formula (5.12) are valid within the window. In the range $|y_{min} - \hbar k/2m| < u < \upsilon_{min} + \hbar k/2m$ and in the condition u, $\hbar k/2m \ll \upsilon_1$ we write the dispersion equation in the form

$$\frac{mV}{2\hbar k}\left[\ln\frac{(v_{min}+\hbar k/2m)^2 - u^2}{u^2-(v_{min}-\hbar k/2m)^2} + i\pi\right] = \frac{U_s^2-u^2}{u^2-v_s^2}\,. \tag{5.13}$$

We set $u^2 = u'^2(1 + i\gamma)$ and after isolating the real and imaginary parts of all functions in (5.13) we obtain two equations for determining the quantities u' and γ. One of these equations which includes the imaginary parts of the functions of u appears as follows:

$$\frac{\gamma}{(u'^2-v_s^2)^2+u'^4\gamma^2} = -\frac{\pi}{4kRu_0^2u'^2}\left\{1+\frac{1}{\pi}\times\right.$$

$$\times \operatorname{arctg} \frac{2(\hbar k/2m)\, v_{\min}\, u'^2\gamma}{u'^4\gamma^2 - [(v_{\min} + \hbar k/2m)^2 - u'^2]\,[u'^2 - (v_{\min} - \hbar k/2m)^2]} \Bigg\} . \qquad (5.14)$$

The second component in brackets in (5.14) does not exceed modulo 1/2. Hence bearing in mind the form of the left half of equation (5.14) we may easily understand that either when $kR \gg 1$ or if the velocity u' is close to v_s, the quantity $\gamma \sim (u'^2 - v_s^2)^2/kRu_0^2u'^2$ and is small compared to unity. In these conditions we may write the following formula:

$$\gamma = -\frac{\pi}{4kR}\,\frac{(u'^2 - v_s^2)^2}{u_0^2 - u'^2}\,b, \quad \frac{1}{2} \leqslant b \leqslant 1, \qquad (5.15)$$

where the quantity b is determined by equation (5.14) in which the component $u'^4\gamma^2$ in the denominator of the left half is ignored.

It was demonstrated above that the dispersion curve within the transparency window reaches the window boundary when $u = v_s$, $k = k_\Gamma = 2m(v_{\min} - v_s)/\hbar$. If the dispersion curve from the damping region were to intersect the boundary at this point, then damping at the boundary would vanish in accordance with (5.15), and would grow continuously when moving away from the boundary. However it turns out that the behavior of the dispersion curve is more complex. If we write the dispersion equation for u' near the point $u = v_s$, $k = k_\Gamma$ as done above in deriving formula (5.12) we obtain the equation $\ln[(y - t)^2 + \pi y^2/a^2] = -2ay$ (the conventions are the same as those used in deriving (5.12)) that has no solutions when $t \to 0$, $y \to 0$. It follows that the point of intersection with the boundary of the transparency window of the dispersion curve from the damping region is different from the case of a curve from the transparency window, and when intersecting the boundary both the damping and phase velocity of the wave jump.

We may obtain comparatively simple results for the phase velocity of an acoustic wave in the damping region when $kR \gg 1$. Near the boundary of the window the inequality $kR \gg 1$ takes the form $|v_{\min} - u| \gg \Omega/k_F$, and in order to satisfy the equality the velocity $v_{\min}$ must be sufficiently high. Taking advantage of the fact that when $kR \gg 1$, γ is small, we obtain the following dispersion equation for the phase velocity u':

$$\begin{aligned} u'^2 = U_s^2 - (u'^2 - v_s^2)\,\frac{k_F V}{\Omega}\,\frac{1}{4kR}\,\ln\,\{[\{[v_{\min} + \hbar k/2m)^2 - u'^2] \times \\ \times [u'^2 - (v_{\min} - \hbar k/2m)^2] - u'^4\gamma^2\}^2 + 4\gamma^2 v^2{}_{\min} u'^4 (\hbar k/2m)^2] \times \\ \times [\{u'^2 - (v_{\min} - \hbar k/2m)^2\}^2 + u'^4\gamma^2]^{-2}\}. \end{aligned} \qquad (5.16)$$

The contribution from γ^2 in this equation is significant near the boundary of the transparency window. At the same boundary when $u'^2 = (v_{\min} - \hbar k/2m)^2$ we may set $b \approx 1/2$ for γ in expression (5.15) and when $U_s \ll v_{\min}$, $(V/v_{\min})\ln(v_{\min}^3/Vu_0^2 \ll 1$ the logarithmic term in (5.16) is small compared to U_s^2, so we have:

$$u'^2 = U_s^2 - u_0^2\,\frac{k_F V^2}{2\Omega v_{\min}}\,\ln\left(\frac{16\Omega v_{\min}^2}{\pi k_F V^2 u_0^2}\right). \qquad (5.17)$$

Thus, in this case the dispersion curve originating from the damping region reaches the boundary $u' = \upsilon_{min} - \hbar k/2m$ when $u' \approx U_S$, $k \approx 2m(\upsilon_{min} - U_S)/\hbar$ at the same time that on the transparency window side the point of intersection is determined by the equalities $u = \upsilon_S$, $k = 2m(\upsilon_{min} - \upsilon_S)/\hbar$ and, consequently, there is a significant jump of the phase velocity on the boundary. The damping of the acoustic wave experiences a jump from zero to the value determined by the expression $\gamma = -\pi u_0^2 V^2 k_F/4\Omega\upsilon_{min}U_s^2$ in transiting the boundary.

The results provided for u' and γ at the point of boundary intersection relate to the magnetic field range in which the velocity υ_{min} is sufficiently high. With a reduction in velocity υ_{min} (approaching the peak of the state density oscillations corresponding to $\upsilon_{min} = 0$) the point of intersection of the dispersion curve of sound with the parabolic transparency window boundary shifts towards $kR \lesssim 1$. When $kR \sim 1$ on the boundary the value of γ is in the order of unity and the wave is significantly damped. At a certain field strength velocity υ_{min} coincides with υ_S and then the phase velocity of sound in the limit $k \to 0$ (see the examination of the longwave region above) also coincides with υ_S and the entire dispersion curve of sound lies within the damping region. In this case it turns out that when $kR \to 0$ the difference $u'^2 - \upsilon_S^2$ and the value of γ are proportional to kR. With a further reduction in the field strength we have $\upsilon_{min} < \upsilon_S$ and the dispersion curve intersects the boundary of the lobe transparency window, and experiences an anomaly here similar to that that exists on the boundary of a parabolic window when $\upsilon_{min} > \upsilon_S$.

At a fixed frequency ω the wave vector of the acoustic wave may correspond to both a transparency window and the damping region depending on the value of the magnetic field strength (velocity υ_{min}). Assume that when $\upsilon 2m_{in} \sim \sqrt{\hbar\Omega/m}$ the wave vector applies to a parabolic transparency window ($k < 2m\upsilon_{min}/\hbar$) and, consequently, there is no collisionless wave damping. With a drop of υ_{min} the boundary of the parabolic window shifts the direction of smaller k and at a certain value of υ_{min} the given wave vector k corresponding to a fixed frequency ω makes contact with the boundary. Wave damping arises as a jump at this point, and with a further reduction in υ_{min} it is nonzero to the υ_{min} value at which the wave vector makes contact with the boundary of the lobe transparency window. Even lower values of υ_{min} correspond to the wave vector in a lobe transparency window and to the absence of damping once again. Such a picture of an irregular change in acoustic damping with variation of the field strength repeats periodically as the successive critical velocities of the various quantization appear as υ_{min}. The phenomenon of giant quantum oscillations in acoustic absorption is manifest as this periodic repetition with changes in the magnetic field in regions with and without damping [5]. The phase velocity of the acoustic wave experiences anomalous dispersion at the boundaries of the damping regions consistent with the examination above. We emphasize that these phenomena may be attributed to a single acoustic wave with some degree of arbitrariness. We have demonstrated that in a certain range of variation in field strength the acoustic and quantum waves are not at all differentiable (they are strongly coupled). Hence, specifically, a wave in the damping region cannot in the general case be strictly correlated with either of the two waves in the transparency windows.

We will now return to a discussion of the dispersion curve of a wave in the damping region when $kR \gg 1$, $v_{\min} \gg U_s$ and will consider dispersion equation (5.16) far from the boundaries of the transparency windows. Taking $u'^4\gamma^2 \ll [u'^2 - (v_{\min} - \hbar k/2m)^2]^2$ we obtain from this equation

$$u'^2 = U_s^2 - (u'^2 - v_s^2)\frac{k_F V}{\Omega}\frac{1}{4kR}\ln\frac{(v_{\min} + \hbar k/2m)^2 - u'^2}{u'^2 - (v_{\min} - \hbar k/2m)^2} \approx$$
$$\approx U_s^2 - u_0^2\left(\frac{k_F V}{\Omega}\right)^2 \frac{2}{kR}\ln\frac{v_{\min} + \hbar k/2m}{|v_{\min} - \hbar k/2m|}. \tag{5.18}$$

Thus, the phase velocity of sound in this region is close to U_s and has an addition that is only weakly dependent on the wave vector k.
Damping in this case is determined by expression (5.15) in which we may set $u'^2 \approx U_s^2$, $b \approx 1$:

$$\gamma = -\frac{\pi}{kR}\left(\frac{k_F V}{\Omega}\right)^2 \frac{u_0^2}{U_s^2}. \tag{5.19}$$

We remember that with many occupied quantization levels $V \approx \Omega/2k_F$ and the velocity U_s coincides with the acoustic velocity in the absence of a magnetic field: $U_s \approx u_s$.

With an increase in the wave vector k the dispersion curve of sound reaches the boundary of a triangular transparency window $u' = \hbar k/2m - v_{\min}$. The dispersion of the phase velocity in the vicinity of this boundary is again described by equations (5.16) and (5.15). Here from the inequality $kR \gg 1$ the velocity u', as in the case examined above, when formula (5.17) is valid, remains close to U_s. After intersecting the boundary in the rectangular transparency window wave attenuation vanishes. It is easy to understand that with a further increase in k anomalies will appear along the dispersion curve at all boundaries of the triangular windows intersected by the curve, while damping will vanish in the windows. We may obtain a general formula for the dispersion relation far from the boundaries with large values of k directly from general dispersion equation (5.1). Indeed, in these conditions we may simply ignore the dependence on u in the sum over n, σ in (5.1), which yields:

$$u^2 = v_s^2 + u_0^2\left[\frac{1}{2kR}\sum_{\sigma}\sum_{n=0}^{N(\sigma)}\ln\frac{\hbar k + 2mv(n, \sigma)}{|\hbar k - 2mv(n, \sigma)|}\right]^{-1}. \tag{5.20}$$

The validity conditions of this formula when $\lambda k > 1$ are violated only in the narrow proximity of the boundaries of triangular transparency windows. We note that the singularities of acoustic velocity when $k \approx 2mv(n, \sigma)$ are analogous to the Kohn effect in the absence of a magnetic field, since the quantities $mv(n, \sigma)$ represent the Fermi momenta for electron groups on the n, σ levels. Such anomalies were examined in study [44], although here they were described by a different formula obtained from an incorrect dispersion equation that ignored the acoustic excitation of the electrical field, i.e., the Fröhlich model was assumed to be applicable in a quantizing magnetic field (see section 2 of this chapter). We have provided detailed results relating to the important case of sound interaction with the slowest quantum wave which corresponds to satisfaction of the condition $u_s < v_1$. If we have quantization in sufficiently weak fields where the velocity u_s is of the order $\sqrt{\hbar\Omega/m}$, the sound may be coupled to quantum waves having phase velocities that are also of the order $\sqrt{\hbar\Omega/m}$. We will now discuss this case which was the primary area of interest in studies [17, 24, 25].

The condition $u_s \sim \sqrt{\hbar\Omega/m} \sim v2F\sqrt{\hbar\Omega/}_F$ may be satisfied in the case of many occupied levels. As a result, as demonstrated in the preceding section, the phase velocities of the slow quantum waves are close to the velocities $v(n, \sigma)$ and with small values of k are found from dispersion equation (4.8). An analogous equation accounting for ion oscillations may be obtained from (5.2) and takes the form

$$\frac{v(n_0, +)}{v^2(n_0, +) - u^2} + \frac{v^2(n_0+1, -)}{v^2(n_0+1, -) - u^2} + \frac{2k_F}{\Omega}\left(1 - \frac{u_0^2}{u^2 - v_s^2}\right) = 0. \tag{5.21}$$

If the difference in the velocities $v(n_0, +)$ and $v(n_0, +1, -)$ is not small $(|v(n_0, +) - uj((n_0, + 1, -)| \gg \Omega/k_F)$, one of the first two components in equation (5.21) may be ignored. This equation describes two coupled waves. The significant difference from the case of sound coupling to the slowest quantum wave results from the smallness of the quantity $\Omega/2k_F$ compared to the velocities $v(n, \sigma)$ under consideration and may be traced to the fact that wave interaction occurs in the narrow proximity of velocities $v(n, \sigma)$ whose width is small compared to the phase velocities of the waves. We will consider, for example, two coupled waves with velocities near v_1. We obtain from equation (5.21)

$$u^2 = {}^1/_2\,\{u_1^2 + u_s^2 \pm \sqrt{(u_1^2 - u_s^2)^2 + 2u_0^2\Omega v_1/k_F}\}, \tag{5.22}$$

where $u_1 = v_1 + \Omega/4k_F$ is the phase velocity of the quantum wave in the absence of interaction (see (4.9)). It is clear from formula (5.22) that wave coupling is weak by virtue of the smallness of the ratio $\Omega/k_F u_1 \sim \Omega/k_F v_1$. In these conditions when the quantity u_s is close to v_1, two acoustic waves exist in the metal whose velocity differential is of the order $\sqrt{\Omega k_F}$.

In a more special case when the velocities $v(n_0, +)$ and $v(n_0, +1, -)$ (for example, v_1 and v_2) are very similar (i.e., $|v_1 - v_2| \lesssim \Omega/k_F$) and are close to u_s, three coupled waves arise in place of two. The formulae relating to this case and also accounting for exchange interaction whose role will be discussed below are given in study [17].

The dispersion curves of the coupled waves in the case $u_s \sim \sqrt{\hbar\Omega/m}$ are closely proximate across their entire range in a lobe transparency window. In other respects the nature of their behavior in the shortwave region is analogous to the case of $u_s \ll \sqrt{\hbar\Omega/m}$ (Fig. 8c, d). Specifically, one of the dispersion curves of the waves intersects the boundary of a lobe transparency window, and experiences an anomaly similar to that discussed above.

The theory of coupled acoustic and quantum waves (electron spin-acoustic waves) in an electron gas interacting with lattice ions is given above based on results from studies [17, 24-26]. In this regard we should note that an article was published in 1977 [45] that attempted to reconsider existing conclusions of the theory of sound interaction with quantum waves in metals obtained in studies [17, 24, 25]. We will briefly analyze the contents of article [45]. The quantum/acoustic wave coupling effect is the subject of section 4 of this article which provides: a dispersion equation for the coupled waves (equation (4.6) in study [45]); approximate formulae for the velocity and damping of sound (formulae (4.7), (4.8), and (4.11) in study [45]) and a graph analysis of solutions of the dispersion equation (Figs. 5-7 in study [45]). If we account for the condition $k < k_D$ which must be satisfied for these effects to occur, we may easily determine that the dispersion equation derived in study [45] corresponds to the idealized case $\Omega = 0$ and coincides with dispersion equation (5.1) given above; this equation was investigated long before article [45] in studies [17, 25]. Further, formula (4.7) in study [45] is an expression for acoustic velocity for the case of sound that does not interact with quantum waves and matches (although this is not noted in study [45]) the familiar expression obtained in study [34] (see section 2 above). Another formula (4.11) in study [45] relating to acoustic velocity describes its dispersion in the attenuation region and also has no relation to the wave coupling effect since it does not contain the influence of quantization of orbital motion and relates to the region in which there is no quantum wave. Thus, the discussion of the coupling of quantum and acoustic waves appears in article [45] only in the qualitative graphical analysis of the dispersion equation. Such a consideration is analogous to that carried out in study [25] and is characterized only by the fact that it accounts for the finite velocity while it was assumed that $v_s = 0$ in study [25]. We may determine by referring to Figs. 5-7 in article [45] and its accompanying text that they clearly illustrate the existence of roots corresponding to quantum waves and of an additional root caused by lattice vibrations. This fact was known long before article [45] and analytic expressions had already been considered in studies [17, 24] for the frequencies of the coupled waves. The only qualitative aspect discussed in study [45] not identified previously was the possibility for the simultaneous existence of two coupled waves with velocities greater than v_{min} from sound interaction with the slowest quantum wave. We emphasize that this fact is not related to the significant change in sound velocity from interaction with the slowest quantum wave predicted in study [26] using our analytic expressions for the velocities of the coupled waves. Thus, study [45] contains only a partial discussion of the familiar qualitative picture of the wave spectrum in nonrealizable conditions, when $\Omega_0 = 0$. Nonetheless the authors of article [45] simply did not cite studies [17, 19, 25] in which the primary results from the theory of interacting acoustic and quantum waves were developed. The contents of these studies that were carried out long before article [45] appeared were known to the authors of article [45], since the appropriate information was provided to the author in February of 1977 before the final version of article [45] was submitted for publication. The text of article [45] makes reference only to survey article [24] without noting that this article contained all the fundamental results of quantum wave theory and the theory of quantum wave interaction with sound. Moreover, the authors of study [45] state that in article [24] the problem of coupling of acoustic and quantum waves was examined phenomenologically based on concepts of the deformation interaction between electrons and the lattice. This statement is false and it distorts the contents of article [24]. First the initial studies [17, 19] devoted to these issues whose results are given in study [24] did not incorporate deformation interaction at all, and a consistent incorporation of deformation interaction in study [24] did not result in significant changes in the primary results. Second, the examination method called a "phenomenological" method by the authors of study [45] was improperly contrasted with their "microscopic" approach in study [45]. It is well known that if the wavelength is large compared to the dimensions of the elementary crystal cell (and quantum waves satisfy specifically this condition) both widely-used approaches discussed here are equivalent for analyzing the issue of coupled waves. We demonstrated this above in Section 2 and in Appendix 1. In both approaches the electrons are described by a kinetic equation or equivalent equations of motion, while the ion lattice is described by equations from elasticity theory or equivalent equations. Constants are introduced characterizing electron/ion interaction (the deformation interaction constant, see Section 2) together with Fermi-liquid electron interaction. In their article the authors of study [45] in fact demonstrated once again the equivalency of the approaches by using another method to obtain the already-familiar dispersion equation. This method in no way supports greater validity of the dispersion equation. We again note that a significant portion of the material presented in article [45] is devoted to a critique of the Fröhlich model. This critique whose primary concepts are well-known and have been presented in detail (for example, see study [46] has no relation to studies [17, 19, 24, 25] devoted to the interaction of sound with quantum waves). Moreover, the results from study [44] which employed the Fröhlich model to examine sound in a quantizing magnetic field were first discovered to be incorrect in study [24] long before the appearance of article [45].

In concluding our discussion of the interaction of sound with quantum waves in an electron gas we emphasize that in these conditions spin oscillations may be excited by sound. Such an effect was examined in studies [17, 19]. The relationship between the amplitudes of lattice oscillations and the spin electron density is provided by formula (2.27). Here if the sound-generated quantum wave corresponds to oscillations of electrons with only one spin direction, i.e., only one of the polarizabilities X^+, X^- has a resonance contribution, the resulting coupled waves will be accompanied by significant oscillations in the spin density. Specifically, in the example discussed above involving two occupied levels with spin $\sigma = -1$ we have $s^z(k, \omega) = ikQu(k, \omega)/2e$. In other words it turns out that the change in conductor volume is accompanied by a spin density corresponding to the polarized electrons in the variable volume.

6. Fermi-liquid effects in the quantum wave spectrum

Since quantum wave propagation is accompanied by significant oscillations in spin density, exchange interaction of electrons has a significant influence on their properties. The role of exchange and Fermi-liquid interaction in the quantum wave spectrum has been considered in studies [16, 17, 19, 21, 24]. In this section we will discuss the unique effects resulting from interelectron interaction that cannot be understood by means of the free electron model.

We will first consider waves with velocities that significantly exceed the velocity of sound when we may ignore the relation to lattice vibrations. The dispersion equation for such waves $\chi(\omega, k) = 0$ accounting for the explicit expression for polarizability (3.1) may be written as

$$X^+(\omega, k) + X^-(\omega, k) + 4\psi X^+(\omega, k) X^-(\omega, k) = 0. \tag{6.1}$$

In equation (6.1) obtained in study [16] the Fermi-liquid interaction is described by the constant ψ within the scope of simple approximation (2.21). With small k ($kR \ll 1$) this equation appears as follows:

$$\sum_{n=0}^{N(+)} \frac{v(n, +)}{v^2(n, +) - u^2} + \sum_{n=0}^{N(-)} \frac{v(n, -)}{v^2(n, -) - u^2} + $$

$$+ 2B_0 \frac{\Omega}{k_F} \sum_{n=0}^{N(+)} \frac{v(n, +)}{v^2(n, +) - u^2} \sum_{n'=0}^{N(-)} \frac{v(n', -)}{v^2(n', -) - u^2} = 0. \tag{6.2}$$

The primary qualitative effects resulting from exchange interaction of electrons (the components with the constants ψ and B_0 in (6.1), (6.2)) may be illustrated using simple particular cases for which explicit expressions were developed in preceding sections for the phase velocities of quantum waves. One such effect becomes manifest immediately in analyzing the case of two occupied quantization levels and results from the significant difference between the two possible situations (resulting from electron interaction) where the spin projections on the occupied levels are identical and different. If two levels with identical spin projections are occupied (for example, 0, –; 1, 0 when $\Omega_0 > 0$: the situation represented by formula (4.3) and (4.12)), exchange interaction will have no influence on the phase velocity of the wave, since equations (6.2) and (4.1) coincide since one of the polarizabilities $X^{\sigma}(\omega, k)(X^{+}(\omega, k) = 0$ (when the levels 0, - and 1,- are occupied) vanishes. If, on the other hand, the spin projections are different on the occupied levels (the levels 0, - and 0, + are occupied when $\Omega_0 < \Omega$) we obtain from (6.2) the expression for the squared phase velocity in the limit of small k:

$$u^2 = v(0,-)\, v(0,+)\left[1+\frac{\Omega}{k_F}\frac{2B_0}{v(0,-)+v(0,+)}\right]. \qquad (6.3)$$

In this case the exchange interaction changes the magnitude of the phase velocity of the quantum wave. Ignoring interaction ($B_0 = 0$) velocity u is equal to $\sqrt{v(0,+)\, v(0,-)}$ and lies in the gap between $v(0,+)$ and $v(0,-)$, and therefore the dispersion curve falls in a lobe transparency window. The interaction influence causes the dispersion curve to fall within the lobe window only at sufficiently small values of the constant B_0 for which the following inequalities must be satisfied:

$$\frac{v(0,+)}{v(0,-)} < 1+\frac{\Omega}{k_F}\frac{2B_0}{v(0,+)+v(0,-)} < \frac{v(0,-)}{v(0,+)}. \qquad (6.4)$$

Even when these inequalities are satisfied, the shape of the dispersion curve may be qualitatively different than when $B_0 = 0$. Specifically, with weak occupation of a level with a higher transverse energy when $v(0,+) \ll v(0,-)$ and when

$$1 < 2B_0\frac{\Omega}{k_F\, v(0,-)} < \frac{v(0,-)}{v(0,+)} \qquad (6.5)$$

it turns out that the dispersion curve does not terminate at the point of intersection of the upper and lower boundaries of the lobe window as in the case when $B_0 = 0$ (see section 4 and formulae (3.8), (3.9)), but rather at the upper boundary of the window $u = v(0,-) - \hbar k/2m$ at the point

$$k = k'_{\mathrm{rp}} = \frac{m}{\hbar v(0,-)}\left[v^2(0,-) - 2B_0\frac{\Omega}{k_F}v(0,+)\right], \qquad (6.6)$$

$$\omega = \omega'_{\mathrm{rp}} = \frac{m}{2\hbar v^2(0,-)}\left\{v^4(0,-) - \left[2B_0\frac{\Omega}{k_F}\, v(0,+)\right]^2\right\}. \qquad (6.7)$$

In the vicinity of this point the dispersion curve is described by the following approximate formula deriving from equation (6.1):

$$u = v(0,-) - \frac{\hbar k}{2m} + v(0,-)\frac{\hbar k'_{\mathrm{rp}}}{2m}\exp\left\{-\frac{2v(0,+)}{v(0,-)}\frac{k'_{\mathrm{rp}}}{k'_{\mathrm{rp}}-k}\right\}. \qquad (6.8)$$

If one of the inequalities in (6.4) is violated (which may occur since the component with B_0 in (6.3), (6.4) is in the order of unity when $B_0 \leqslant 1$), the dispersion curve of the quantum wave may fall outside the lobe window due to exchange interaction. The phase velocity of the wave will exceed $v(0,+)$ in accordance with (6.4) (1.3) for a metal with a positive constant B_0 when

$$B_0\frac{v(0,-)}{v_F} > \frac{m}{2\hbar\Omega}\left[v^2(0,+) - v^2(0,-)\right] = \frac{\Omega_0}{\Omega}. \qquad (6.9)$$

In this case the dispersion curve travels above the far boundary of the entire damping region $u = v(0, +) + \hbar k/2m$ and makes contact with this region at the point with $v(0, +) \ll v(0, -)$ as determined by the equalities:

$$k = k''_{\mathrm{rp}} = \frac{m}{\hbar v(0,-)}\left\{v(0,+)\left[v(0,+) + 2B_0\frac{\Omega}{k_F v(0,-)}\right] - v^2(0,-)\right\}, \qquad (6.10)$$

$$\omega = \omega''_{\mathrm{rp}} = \frac{m}{2\hbar v^2(0,-)}\left\{v^2(0,+)\left[v(0,+) + 2B_0\frac{\Omega}{k_F v(0,-)}\right]^2 - v^4(0,-)\right\}. \qquad (6.11)$$

The curve approaches this point exponentially analogous to (6.8):

$$u = v(0,-) + \frac{\hbar k}{2m} - 2v(0,-)\frac{\hbar k''_{\mathrm{rp}}}{m}\exp\left\{-\frac{2v(0,+)}{v(0,-)}\frac{k''_{\mathrm{rp}}}{k''_{\mathrm{rp}} - k}\right\}. \qquad (6.12)$$

The existence of a wave described by formulae (6.3), (6.12) in conditions of (6.9) is fundamental, since it demonstrates the possibility of quantum wave existence outside the transparency windows and at higher frequencies than the entire damping region. This reveals a significant reconfiguration of the spectrum of electron oscillations due to significant exchange interaction.

With a negative B_0 constant exchange interaction reduces the phase velocity of the quantum wave. In order for the quantity u_2 to remain positive the following inequality resulting from (6.3) must be satisfied:

$$1 - \frac{\Omega}{k_F}\frac{2|B_0|}{v(0,+) + v(0,-)} > 0. \qquad (6.13)$$

The violation of this inequality would cause instability of the ground state of the metal with respect to a transition to another state characterized by the spin ordering. However, even when inequality (6.13) is satisfied the first inequality in (6.4) may nonetheless be violated. Then if

$$0 < 1 - \frac{\Omega}{k_F}\frac{2|B_0|}{v(0,-) + (0,+)} < \frac{v(0,+)}{v(0,-)}, \qquad (6.14)$$

the dispersion curve of the quantum wave will lie in a parabolic transparency window rather than a lobe window. With growth of k the curve approaches the boundary of the parabolic window exponentially $u = v(0, +) - \hbar k/2m$ and makes contact; as described by the formula analogous to (6.8), (6.12).

Thus, when any condition noted above in (6.5), (6.9) or (6.14) is satisfied, Fermi-liquid electron interaction will not only produce a quantitative renormalization of the phase velocity of the quantum wave, but also will qualitatively change the entire nature of the dispersion curves. At present these results refer to the case where only two quantization levels are occupied. We will now proceed to an analysis of the case of more occupied levels.

We will consider the influence of exchange interaction on slow waves whose phase velocities are near the velocities $v(n_0, +)$, $v(n_0, +1, -)$. Equation (4.8) and formulae (4.9), (4.10) obtained ignoring exchange interaction with small k apply to these waves. Applying the same method of consideration to equation (6.2) as used in deriving (4.8), we obtain for the case $|v(n_0, +) - v(n_0, +1, -)| \gg \Omega/k_F$ a generalization of formulae (4.9):

$$u_1 = v(n_0, +) + \frac{1+2B_0}{1+B_0}\frac{\Omega}{4k_F}; \qquad u_2 = v(n_0+1, -) + \frac{1+2B_0}{1+B_0}\frac{\Omega}{4k_F}. \quad (6.15)$$

These formulae corresponding to the comparatively large velocity differential $v(n_0, +)$ and $v(n_0, +1, -)$ determine the change in the velocities of quantum waves due to exchange interaction. We note that if the constant B_0 is negative and its modulus value is greater than 1/2, the velocities u_1 and u_2 will be less (and not greater than as is the case when $B_0 = 0$) than the corresponding velocities $v(n, \sigma)$. In this case the dispersion curves of these waves will appear in different transparency windows than in the case where $B_0 = 0$. If $B_0 < 0$ and the modulus of B_0 exceeds 1/2, yet is not too close to unity, the dispersion curves will lie close to the upper boundaries of the lobe transparency windows rather than the lower boundaries, and with small k formulae (6.15) are valid. The dispersion curves of quantum waves may drift far from the window boundaries with strong exchange interaction when $1 - |B_0| \ll 1$.

In another limiting case of proximate velocities $v(n_0, +)$ and $v(n_0, +1, -)$ ($|v(n_0, +) - v(n_0, +1, -)| \ll \Omega/k_F$ ignoring exchange interaction there exists only a single branch of quantum waves with a velocity determined by formula (4.10). Exchange electron interaction produces one additional different branch. We will provide the corresponding formula for the phase velocity somewhat later, and will now show how such a branch appears by employing general dispersion equation (6.1). In these conditions the dependence of the functions $X^\sigma(\omega, k)$ on u is significant only in one of the components of the sum over n in (3.11) or (3.12). After isolating this component and labeling it $X_r^\sigma(u)$, we may write

$$X^\sigma(\omega, k) = \overline{X}^\sigma + X_r^\sigma(u), \quad (6.16)$$

where $\overline{X}^\sigma$ is independent of u and is the sum over n in (3.11) of the components whose dependence on u is insignificant. It turns out that $X_r^+(u) \approx X_r^-(u) \equiv X_r(u)$ by virtue of the approximate correlation of the critical velocities that are close to the phase velocity u of these waves ($v(n_0, +)$ and $v(n_0, +1, -)$). In this case dispersion equation (6.1) is transformed to

$$[\overline{X}^+ + \overline{X}^- + 2X_r(u)][1 + \psi(\overline{X}^+ + \overline{X}^-) + 2\psi X_r(u)] - \psi(\overline{X}^+ - \overline{X}^-)^2 = 0. \quad (6.17)$$

As a result with a large number of occupied levels $\overline{X}^+ \approx \overline{X}^- \approx \eta/2$ the left half of equation (6.17) is decomposed into two factors. The zero equivalency of one of these factors is reduced to the dispersion equation χ_0 $(\omega, k) = 0$ ignoring exchange interaction and yields a velocity u_1 coinciding with (4.10):

$$u_1 = v(n_0, +) + \frac{\Omega}{2k_F}. \tag{6.18}$$

The following equation is obtained from the other factor:

$$1 + \psi\chi_0(\omega, k) = 0, \tag{6.19}$$

and this also determines the new branch of quantum waves resulting from exchange interaction identified above. Its phase velocity is equal to

$$u_2 = v(n_0, +) + \frac{B_0}{1 + B_0}\frac{\Omega}{2k_F}. \tag{6.20}$$

We may easily understand the meaning of equation (6.19) since this is the form of the dispersion equation for the longitudinal spin waves in the absence of quantization. Waves described by equation (6.19) may be considered to be quantum longitudinal spin waves. Such waves are generated by spin oscillations in each of the one-dimensional subsystems of the electron system. They may occur when electrons with opposite spin directions have a single energy value in equilibrium (they belong to a single subsystem), so that pure spin oscillations are possible within the scope of a single subsystem without changing the orbital degrees of freedom. In order to demonstrate this specific nature of the waves we will estimate the corresponding amplitude of the spin electron density proceeding from expression (2.17):

$$\begin{aligned} s^z(k, \omega) &= -\frac{1}{2}e\Phi(k, \omega)\frac{X^+(\omega, k) - X^-(\omega, k)}{1 + \psi[X^+(\omega, k) + X^-(\omega, k)]} = \\ &= -\frac{1}{2\psi}e\Phi(k, \omega)\frac{X^+(\omega, k) + X^-(\omega, k)}{X^+(\omega, k) - X^-(\omega, k)}. \end{aligned} \tag{6.21}$$

The second equality was obtained here using dispersion equation (6.1). In order to estimate the amplitude $s^z(k, \omega)$ we must first determine the order of the difference $X^+(\omega, k) - X^-(\omega, k) \approx X - X^-$. Above we already took advantage of the fact that with many occupied levels this difference is small and we may determine by means of expression (3.12) that $X^+ - X^- \approx (\hbar\Omega/\varepsilon_F)\eta$. Hence for waves whose dispersion equation takes the form of (6.19) we obtain from the second expression in (6.21) $s^z(k, \omega) \sim (-\varepsilon_F/B_0^2\hbar\Omega)\eta e\Phi(k, \omega)$, while for waves with the dispersion equation $\chi_0(\omega, k) = 0$ we obtain from the first expression in (6.21): $s^z(k, \omega) \sim (-\hbar\Omega/\varepsilon_F)\eta e\Phi(k, \omega)$. Thus, for waves of the type in (6.19) the amplitude of the spin density is $(\varepsilon_F/\hbar\Omega B_0)^2$ times greater and we may determine by means of formula (2.18) that the amplitude of the charge density in these waves is smaller to an equal magnitude compared to waves for which the equation $\chi_0(\omega, k) = 0$ is valid. This is how we confirm the spin nature of waves described by dispersion equation (6.19). At this point we add that in the other limiting case examined above when the velocities $v(n_0, +)$ and $v(n_0, +1, -)$ are significantly different $(|v(n_0, +) - v(n_0, -1, -)| \gg \Omega/k_F)$ while the phase velocities of the waves are determined by formulae (6.15), the amplitude of the spin density and the charge density is of the same order: $s^z(k, \omega) \sim n(k, \omega) \sim -\eta e\Phi(k, \omega)$, and exchange interaction has no significant influence on this estimation. It follows that in the case of a significant energy differential between electrons with opposite spin projections, the quantum waves correspond to strongly coupled oscillations in charge and spin densities.

The existence of quantum longitudinal spin waves in metals in which the energies of states with different spin projections approximately coincide is a rather common feature. Indeed, in order for the left half of dispersion equation (6.1) to decompose into the product of the two factors $\chi_0(\omega, k)$ and $1 + \psi\chi_0(\omega, k)$, it is necessary for the difference to be small for these waves. This is specifically the case when the energies of states for different σ are close, i.e., the frequencies Ω and Ω_0 approximately coincide or are multiples of one another. Specifically, with equality of the frequencies Ω_0 and Ω all components of the sum over n in the functions $X^-(\omega, k)$ except the first component (with $n = 0$) are equal to the corresponding components in $X^+(\omega, k)$ (all energy levels aside from the lowest energy level are spin-degenerate), and hence the approximate equality $X^-(\omega, k) = X^+(\omega, k)$ will be valid for many branches of quantum waves. Thus, both slow and fast longitudinal quantum spin waves may exist. The component with $n = 0$ in $X^-(\omega, k)$ that is not compensated in the difference $X^+(\omega, k) - X^-(\omega, k)$ is significant only for waves with velocities close to the critical velocity $\upsilon(0, -)$ and the fastest such waves are always coupled. The wave examined above in the case of two occupied levels is analogous to precisely these perpetually coupled waves. Our examination of this case has referred to the situation where the energies of the states 0, + and 1, - are different and the state 1, - in unoccupied. If $\Omega_0 \approx \Omega$ then such energies are proximate. Thus, if we take the levels 0, -; 0, +, and 1, - as occupied when $\upsilon(0, +) \approx \upsilon(1, -) \ll \upsilon(0, -)$, we may identify the longitudinal spin wave. This corresponds to the case of weak occupation of the level with the highest energy which will be discussed below.

We will now return to formula (6.20) for the phase velocity of a longitudinal spin wave with small k and we note that this velocity may be both greater than and less than $\upsilon(n_0, +)$ depending on the sign of B_0. This in turn means that the dispersion curve will be located either at the lower boundary of one window (when $B_0 > 0$) or at the upper boundary of another window (when $B_0 < 0$). It is important that this event depends on only the sign and not the absolute value of the coefficient B_0. The dispersion curve of a longitudinal spin wave across the entire transparency window is described by equation (6.19). It terminates, like the dispersion curve of the charge density wave, at the point $\omega = \omega_{\Gamma p}$, $k = k_{\Gamma p}$.

This discussion of slow quantum waves with velocities close to $\upsilon(n_0, +)$ and $\upsilon(n_0, +1, -)$ has referred to the limiting cases $|y(n_0, +) - \upsilon(n_0, +1, -)| \gg \Omega/k_F$ and $|y(n_0, +) - \upsilon(n_0, +1, -)| \ll \Omega/k_F$. We will not provide the more general formulae that also describe the intermediate case $|\mathbf{y}(n_0, +) - \upsilon(n_0, +1, -)| \sim \Omega/k_F$; these were derived for small k in study [16] accounting for exchange interaction. We will briefly note that there are two coupled waves with phase velocities u_1

and u_2 in the limiting cases described by formulae (6.15) and (6.18), (6.20). Exchange interaction has a significant influence on the values u_1, u_2. At the same time that in ignoring exchange interaction each transparency window may contain a single dispersion curve of the corresponding quantum wave branch, the influence of exchange interaction may cause this to not be the case and the dispersion curves of two branches may fall within a single lobe window.

We will now consider the slowest quantum waves in the case of weak occupation of the level with the highest transverse energy. Analogous to the particular cases outlined above we may identify two characteristic situations differentiated by whether the electrons on the last occupied quantization level have one or two opposite spin projections. Let the energy of a weakly-occupied quantization level first correspond to only a single spin projection. Then, proceeding in the same manner as in the derivation of formula (4.5) we obtain from equation (6.2) its generalization:

$$u_{\min}^2 = v_{\min}\left[\frac{V_+(V_- + 2B_0\Omega/k_F)}{V_+ + V_- + 2B_0\Omega/k_F} + v_{\min}\right], \tag{6.22}$$

where the velocities V_σ are determined by the formulae

$$\frac{1}{V_\sigma} = \sum_{n=0}^{N(\sigma)} \frac{1}{v(n,\sigma)}, \tag{6.23}$$

we exclude the term containing $v_{\min}$ from the summation here. With many occupied levels $V_\sigma \approx \Omega/k_F$, formula (6.22) also takes the form

$$u_{\min}^2 = v_{\min}\left(\frac{\Omega}{2k_F}\,\frac{1+2B_0}{1+B_0} + v_{\min}\right). \tag{6.24}$$

If the constant B_0 is positive or negative yet has a sufficiently small modulus value, exchange interaction will result only in a quantitative renormalization of the phase velocity $u_{\min}$. On the other hand with a negative and sufficiently large absolute value of the constant B_0 exchange interaction may cause an instability of the ground state of the electron system in the range of variation of $v_{\min}$ in which the value of $u_{\min}^2$ becomes negative in accordance with formulae (6.22), (6.24). This is the case when $|B_0| > k_F V_-/2\Omega$, which takes the form $|B_0| > 1/2$ with many occupied levels. A metal with a constant B_0 satisfying this condition is unstable with magnetic field strengths for which $v_{\min} < (2|B_0| - 1)\Omega/k_F(1 - |B_0|)$, i.e., in a certain vicinity of the peak of the density of states ($v_{\min} = 0$). We note that study [48] also discovered violation of the stability of ground state of the electrons in a quantizing magnetic field with respect to magnetic perturbations when $|B_0| > 1/2$, $B_0 < 0$.

In another characteristic situation electrons with opposite spin projections have minimal energy corresponding to a weakly-occupied level. Here proceeding analogously to the case examined above we may isolate the branch of the quantum wave spectrum corresponding to longitudinal spin oscillations. The slowest longitudinal spin wave in the limit of small k has a phase velocity determined by the formula

$$u^2 = v_{\min}\left(\frac{B_0}{1+B_0}\frac{\Omega}{k_F} + v_{\min}\right). \tag{6.25}$$

Again the most important qualitative effect is the appearance of instability with negative B_0. A large absolute value of B_0 is not required for this type of instability to arise; only the range of values v_{min} ($v_{min} < |B_0|\Omega/k_F(1 - |B_0|)$ in which the system is unstable is dependent on the absolute value of B_0.

These results relating to the slowest wave describe the simplest limiting cases of a large differential between Ω and Ω_0 as well as their coincidence and are valid for small k. In the case of similar values of Ω and Ω_0 a level with an opposite spin projection may have an energy level similar to the weakly-occupied quantization level under consideration. In this case the slowest wave is coupled to the other slow quantum wave and Fermi-liquid interaction has a significant influence on the frequencies of these coupled waves.

In order to describe the dispersion curve of the slowest quantum wave with large k in a lobe transparency window it is necessary to account for components containing the velocities v_{min} and v_1 in the polarizabilities. A simple result may be obtained in the case where the frequencies Ω_0 and Ω are significantly separated and, moreover, the velocity v_1 following v_{min} corresponds to level n with the same spin projection as v_{min}. Then exchange interaction may be accounted for by substituting the quantity $4kR$ with $4kR(1 + B_0)/1 + 2B_0)$ in equation (1.13) and formulae (4.14) and (4.15).

A unique situation exists when the frequencies Ω_0 and Ω coincide with negative values of B_0 when for small k formula (6.25) is valid and the stability condition $v_{min} > |B_0|(\Omega/k_F)/(1 - |B_0|)$ is satisfied. In this case the dispersion curve of the slowest spin wave lies in the parabolic transparency window, and its shape may be described by the approximate dispersion equation:

$$1 - \frac{|B_0|}{1-|B_0|}\frac{\Omega}{k_F}\frac{m}{2\hbar k}\ln\frac{(v_{\min} + \hbar k/2m)^2 - u^2}{(v_{\min} - \hbar k/2m)^2 - u^2} = 0, \tag{6.26}$$

whose solution may be written in explicit form.

Concluding our examination of the influence of exchange interaction on the slowest quantum waves we will briefly discuss the issue of the nature of coupling between acoustic and quantum waves in accounting for exchange interaction. The complete dispersion equation of quantum waves incorporating the contribution of lattice oscillations consistent with (2.28), (3.1) may be written as

$$-\frac{u_0^2\eta}{u^2 - v_s^2}[1 + \psi(X^+(\omega,k) + X^-(\omega,k))] + X^+(\omega,k) + X^-(\omega,k) + 4\psi X^+(\omega,k)X^-(\omega,k) = 0. \tag{6.27}$$

We will provide two characteristic results to illustrate the influence of Fermi-liquid interaction on the spectrum of coupled acoustic and quantum waves. We will first consider the coupling of sound to the slowest quantum wave with a significant differential between Ω_0 and Ω and with small k, when formula (6.24) is valid for the phase velocity of the quantum wave u_{min}. In this case a generalization of formula (5.5) derives from equation (6.27) for multiple occupied levels for the phase velocities of coupled waves:

$$u_{1,2}^2 = \frac{1}{2}\left\{ U_s^2 + u_{min}^2 \pm \left[(U_s^2 - u_{min}^2)^2 + 8\frac{(1+2B_0)(1-B_0)}{1+B_0} u_0^2 V^2 \frac{k_F v_{min}}{\Omega} \right]^{1/2} \right\},$$

where $U_s^2 = v_s^2 + u_0^2[1 - 2B_0/(1 - B_0)]$, $V = (V_+^{-1} + V^{-1})^{-1}$. The influence of exchange interaction appears as renormalization of the phase velocities u_1 and u_2. When $B_0 < 0$ and $|B_0| > 1/2$ formula (6.28) as well as (6.24) produce instability in the ground state with sufficiently small v_{min}.

Another result applies to the case of an approximate correlation of the frequencies Ω and Ω_0 when sound interacts with a charge density wave and a longitudinal spin wave. Then the generalization of equation (6.17) accounting for formula (6.16) used in these conditions is valid:

$$\left[-\frac{u_0^2\eta}{u^2 - v_s^2} + \overline{X}^+ + \overline{X}^- + 2X_r(u) \right] [1 + \psi(\overline{X}^+ + \overline{X}^-) + 2\psi X_r(u)] - \\ - \psi(\overline{X}^+ - \overline{X}^-)^2 = 0. \qquad (6.29)$$

In the general case this expression describes three waves. It is used particularly to describe the interaction of sound with the slowest quantum waves with any number of occupied levels. However, when $N(\sigma) \gg 1$, $\overline{X}^+ \approx \overline{X}^- \approx \eta/2$ the interaction of sound with the longitudinal spin wave becomes negligible, and interaction with the density wave is described in the same manner as in the preceding section.

So far all results given in this section have referred to quantum waves in lobe transparency windows. However it was already noted above that one of the effects produced by Fermi-liquid interaction is that quantum waves may be generated in triangular transparency windows. Such waves may also exist with a negative constant ψ as we may determine with equation (6.1). In order to analyze the corresponding dispersion curves we will first consider the extensively discussed particular case of a correlation of energy levels with opposite spin projections. Then the triangular transparency windows contain longitudinal quantum spin waves described by dispersion equation (6.19) that may be written as

$$1 + \frac{B_0}{2kR} \sum_{\sigma} \sum_{n=0}^{N(\sigma)} \ln \left| \frac{[v(n,\sigma) + \hbar k/2m]^2 - u^2}{[v(n,\sigma) - \hbar k/2m]^2 - u^2} \right| = 0. \qquad (6.30)$$

With many occupied levels the value of kR in this range of k is large compared to unity, and therefore when $|B_0|/kR \ll 1$ the dispersion curves are close to the window boundaries. Only one term of the sum in (6.30) is significant near the lower corner of any window on the ω, k plane ($u \approx 0$, $\hbar k/m \approx v(n_0, +)$), and we obtain the following formula for the phase velocity:

$$u^2 = \left[v(n_0, +) - \frac{\hbar k}{2m}\right]^2 - 4v^2(n_0, +)\exp\left\{\frac{2kR}{B_0}(1 + B_0)\right\}, \tag{6.31}$$

valid when the exponent is negative and greater than unity, while the modulus of the difference $v(n_0, +) - \hbar k/2m$ is small compared to the value of $\hbar k/2m$ corresponding to the second edge of the window when $\omega = 0$. Formula (6.31) describes the two parts of the dispersion curve originating near the lower corners of a triangular transparency window. Their continuations merge near the upper angle, so the dispersion curve appears as shown schematically in Fig. 3b.

Similar results were obtained in analyzing the case of a significant differential between Ω_0 and Ω, although in this case for quantum waves to appear in the triangular windows it is necessary for the parameter B_0 to not only be negative but to exceed a modulus value of 1/2, thereby placing a negative sign on the combination $(1 + B_0)/(1 + 2B_0)$. If this constraint is satisfied, the form of the dispersion curves is analogous to that shown in Fig. 3b, while the exponent is substituted by $4kR(1 + B_0)/(1 + 2B_0)$ in formula (6.31).

We note that the sound may interact with quantum waves in triangular transparency windows if the sound velocity u_0 is less than $\sqrt{\hbar\Omega/m}$. Such interaction is significantly different in nature than in the case of lobe windows, since the different dispersion dependence of the wave frequencies causes interaction to occur in a narrow range of wave vectors.

7. The conditions for the existence and possibilities for observing quantum spin-acoustic waves

Quantum waves may exist in a metal if the electron collision frequency $1/\tau$ and the temperature T are sufficiently small. In discussing the wave spectrum above it was assumed for simplicity that $1/\tau$ and T could be assumed to be zero. We now will consider the constraints imposed by the influence of electron collisions and thermal motion and the manifestation of predicted effects within this scope. Moreover, we will briefly discuss the role of Fermi surface anisotropy in the formation of the quantum wave spectrum and will also identify possible methods of observing quantum spin-acoustic waves.

In order to obtain smallness of the influence of collisions on quantum waves we must rely on expression (3.2) for the polarizability $X^{\sigma}(\omega, k)$. It follows from this expression that for slightly damped waves to exist the quantity $1/\tau k$ must be small compared to the characteristic differences $u - w(n, \sigma, \pm k)$. In this regard the necessary condition for the existence of quantum waves is satisfaction of the inequality $1/\tau k \ll u$ or $\omega\tau \gg 1$. This condition is rather obvious, since the frequency dispersion of the polarizability will be manifest only when it is satisfied. It produces a lower frequency limit on the wave while an upper limit is placed on possible values of ω by the maximum value of $\omega_{\Gamma p}$ in the transparency window. Thus, the necessary condition for the existence of a quantum wave is written as

$$1/\tau \ll \omega < \omega_{\Gamma p}. \tag{7.1}$$

This condition is also sufficient if the characteristic values $|u - w(n, \sigma, \pm k)|$ are of the order or greater than u, which may occur when the dispersion curve is not close to the boundaries of the transparency window, and the difference of the velocities $v(n, \sigma)$ corresponding to the upper and lower boundaries of the lobe window is in the order of the velocities themselves. This situation exists for precisely the slowest quantum wave with weak occupation of the level with the highest transverse energy (see sections 4, 5). If the lowest velocity v_{min} from $v(n, \sigma)$ is much less than the next successive velocity v_1 (which is of the order $\sqrt{\hbar\Omega/m}$), the phase velocity of the quantum wave u_{min} is significantly greater than v_{min} and is less than v_1. Hence $u_{min} - v_{min} - \hbar k/2m \sim u_{min}$ while $v_1 + \hbar k/2m - u_{min} \gg u_{min}$ in the majority of the transparency window. Taking the critical frequency $\omega_{\Gamma p}$ to be in the order of Ω, we obtain from (7.1) the inequality

$$1/\tau \ll k u_{min} < \Omega, \tag{7.2}$$

determining the range of existence of the slowest quantum wave. In the case of interaction between this wave and sound (see section 5) the velocities of coupled waves in the order of the velocity of sound must be substituted into inequality (7.2) in place of u_{min}. Then if we take $1/\tau \sim 10^9\ s^{-1}$, the existence condition $1/\tau \ll \omega < \Omega$ is satisfied for real acoustic frequencies in the order of a few gigahertz and magnetic field strengths in the order of a few tens of kilogausses. It is only necessary to remember that when $v_s < v_{min}$ one of the dispersion curves of the coupled waves is in a parabolic transparency window (see section 5) and by virtue of the smallness of v_{min} for this branch the frequency $\omega_{\Gamma p}$ is comparatively small, while at frequencies exceeding $\omega_{\Gamma p}$ the dispersion curve of sound falls within the Landau damping region.

When a few quantization levels are occupied by electrons, the sufficient condition for the existence of quantum waves is also reduced to (7.1). Indeed, in this case the dispersion curves are not close to the boundaries of the transparency windows, while the differences between the various velocities $v(n, \sigma)$ are in the order of $\sqrt{\hbar\Omega/m} \sim v_F$. We note that condition (7.1) allows the existence of fast waves ($u \sim v_F$) for comparatively small k exceeding only the inverse free path length.

If many levels are occupied, for all branches of quantum waves aside from the slowest wave the quantities $|u - w(n, \sigma, \pm k)|$ are of the order Ω/k_F and are small compared to the phase velocities u. As a result the condition for the existence of such waves $\Omega\tau > k_F/k$ is more rigid than (7.1). Specifically, for relatively slow waves ($u \sim \sqrt{\hbar\Omega/m}$) inequalities (7.1) are replaced by the following:

$$\sqrt{\frac{\varepsilon_F}{\hbar\Omega}}\frac{1}{\tau}<\omega<\Omega, \qquad (7.3)$$

and $\Omega\tau$ must be greater than $\sqrt{\varepsilon_F/\hbar\Omega}$. For fast waves we must substitute $(1/\tau)\sqrt{\varepsilon_F/\hbar\Omega}$ with $(1/\tau)(\varepsilon_F/\hbar\Omega$ in the left half of (7.3). Condition (7.3) was used as the existence condition of quantum waves in studies [8, 19]. We again emphasize that less rigid condition (7.1) must be satisfied for the existence of the slowest wave and waves in the case of few occupied quantization levels.

The possibilities for the existence of quantum waves are also limited by the influence of thermal electron motion. For a qualitative evaluation of the role of this factor we note that thermal motion produces a spread of the electron energies near the Fermi energy ζ over an interval of $\varkappa T$ ($\varkappa$ is Boltzmann's constant), and we may therefore assume that the velocities $\upsilon(n, \sigma)$ have a spread of the order $\varkappa T/m\upsilon(n, \sigma)/\partial\zeta = \varkappa T/m\upsilon(n, \sigma)$. This estimate of the spread of velocities is valid if $\varkappa T/m\upsilon(n, \sigma) \ll \upsilon(n, \sigma)$ otherwise there will be no corresponding transparency windows; specifically, when $\varkappa T/m\upsilon_{min} > \upsilon_{min}$ the parabolic transparency window is "blurred" due to thermal motion, while the spread of electron velocities on the weakly occupied level is of the order $\sqrt{\varkappa T/m}$. In the general case the thermal spread of critical velocities produces an effective shift of the boundaries on the Landau damping region, i.e., the boundaries of the transparency windows. A quantum wave is only slightly damped if this shift does not exceed the difference between its phase velocity and the phase velocity corresponding to the edge of the damping region when $T = 0$. The condition for weak damping of the slowest quantum wave when $\varkappa T < m\upsilon^2_{min}$ is reduced to the inequality $\varkappa T/m\upsilon_{min} < u_{min}$. With strong coupling of an acoustic and the slowest quantum wave their phase shifts, as in the case of υ_{min}, will be on the order of the sound velocity and hence the condition for the existence of such coupled waves will take the form

$$\varkappa T < mu_s^2. \qquad (7.4)$$

This condition may be satisfied at temperatures $T \lesssim 1$ K. Significantly fewer constraints are imposed on the existence of fast quantum waves in the case of several occupied quantization levels. In this case the quantity u_s^2 is substituted with $\hbar\Omega/m$ in inequality (7.4) for waves whose velocity (and those corresponding to $\upsilon(n, \sigma)$) is of the order $\sqrt{\hbar\Omega/m}$; the condition $\varkappa T < \hbar\Omega$ is then obtained and this condition must be satisfied for all quantum effects to be manifest. On the other hand, the more rigid condition must be satisfied in order to assure the existence of quantum waves with velocities on the order of $\sqrt{\hbar\Omega/m}$ in the case of multiple occupied levels. The corresponding inequality takes the form $\varkappa T/m\upsilon(n, \sigma) < \Omega/k_F$ and for slow waves ($u \sim \upsilon(n, \sigma) \sim \sqrt{\hbar\Omega/m}$) it is reduced to the following:

$$\varkappa T/\hbar\Omega < \sqrt{\hbar\Omega/\varepsilon_F}. \qquad (7.5)$$

Such a condition for the existence of quantum waves was obtained in studies [8, 19].

Thus, an analysis of the influence of electron collisions and thermal motion shows that they limit the existence of fast quantum waves to the least extent in the case of several occupied quantization levels. Such conditions for the existence of the slowest quantum waves interacting with sound **corresponds** to existing experimental capabilities. Regarding the many branches of quantum waves with many occupied quantization levels ($\hbar\Omega \ll \varepsilon_F$) comparatively extensive efforts would seem to be required to observe these waves.

The quantum wave theory presented above applies to a metal with a spherical Fermi surface. If the Fermi surface is anisotropic, yet the quasi-classical electron trajectories are closed, the electron energy spectrum in a quantizing magnetic field is described by the function $E(n, p_z, \sigma)$ that is in the general case significantly more complex than for free electrons [2]. However, analogous to the process used in our simple model based on formula (1.1) we may incorporate the critical longitudinal velocities $\upsilon(n, \sigma)$ and represent the electron system as a set of one-dimensional subsystems. Hence the physical causes behind the quantum waves remain in force and, consequently, the quantum waves may exist in a metal with an arbitrary closed Fermi surface. This may be confirmed by the formulation of the corresponding dispersion equation analogous to that used in section 2. There are, however, two facts that will cause the picture of quantum wave propagation in an anisotropic metal to differ from the picture described above for the isotropic case.

If the direction of the magnetic field is not parallel to the axis of symmetry of the Fermi surface, the longitudinal oscillations in electron density will not be isolated from oscillations of the transverse electromagnetic field. Hence in the general case quantum waves will be coupled to the transverse waves, specifically helicons.

Another circumstance is the dependence of the transparency windows on the form of the function $E(n, p_z, \sigma)$ with large k; this dependence may be different from that obtained in the case of an isotropic dispersion law. An example illustrating this may be found in survey [14]. We may conclude on this basis that the anisotropy of the Fermi surface will influence the shape of the dispersion curves of quantum waves in the shortwave range. Thus the anisotropy of the energy spectrum of electrons in the metal will produce two qualitative effects: a change in the dispersion of the quantum waves with large k and their coupling to the transverse waves.

We will discuss possible methods of observing quantum spin-acoustic waves. The results presented here show that quantum waves in appropriate conditions are coupled to lattice oscillations. Hence a quantum wave may be observed by mechanical excitation of such oscillations. Specifically theory predicts the generation of two waves with phase velocities other than the acoustic velocity in a metal at the given frequency of the external force. This effect may be observed when the conditions for the existence of a slow quantum wave interacting with sound are satisfied.

A direct method of exciting quantum waves is based on, for example, placing a wafer made from the test material in a plane-parallel capacitor. Such a method was examined in study [49] with respect to the excitation of plasma acoustic waves. The excitation of longitudinal waves in the wafer is manifest in the oscillating dependence of the impedance of the capacitor on the magnetic ,field frequency or field strength. The longitudinal quantum waves become coupled to oscillations in the transverse electromagnetic field when the constant magnetic field is inclined with respect to the wafer surface; this also results from the limited sample size or possible anisotropy of the Fermi surface. Hence it is fundamentally possible to excite quantum waves by a high-frequency electromagnetic field in a cavity (see study [4]).

The thermal excitation of quantum waves in a system of conduction electrons creates a new light scattering mechanism. The method of detecting quantum waves based on this method is proposed in study [50] for application to a degenerate semiconductor. The method is based on observing light scattering by spin density oscillations. The corresponding scattering cross-section is expressed through the longitudinal spin electron susceptibility that is dependent on the frequency and the wave vector. Study [50] provides an estimation of the effective spin-acoustic wave scattering cross-section for the case of two electron-occupied levels in InSb.

In concluding our discussion of the possibilities of observing quantum spin-acoustic waves we may state that the primary difficulties involve maintaining the existence conditions of such waves as established in this section. If such conditions are satisfied, existing observation methods (acoustic or electromagnetic) allow experimental investigation of the properties of quantum waves.

APPENDIX 1

COMPARISON OF VARIOUS APPROACHES

In this section we will discuss the correlation between our approach to the theory of electron and lattice oscillations and the formulations employed in several other studies. We will first write relations that generalize formulae (2.26)-(2.28) to conditions in which satisfaction of the inequality $k \ll k_D$ is note required. Then we have the equation system:

$$\left[1+\left(\frac{4\pi e^2}{k^2}+\varphi\right)\chi(\omega, k)\right] n(k,\omega)=\left(\frac{4\pi eQ}{k^2}-\Lambda\right)\chi(\omega, k)\, iku(k,\omega), \tag{P1.1}$$

$$\left(\omega^2-\Omega_p^2-\frac{\lambda^0}{\rho_m}k^2\right)u(k,\omega)=\frac{1}{\rho_m}\left(\frac{4\pi eQ}{k^2}-\Lambda\right) ikn(k,\omega). \tag{P1.2}$$

The solvability condition of this system corresponds to the following dispersion equation that generalizes (2.28):

$$1-\frac{(4\pi eQ/k^2-\Lambda)^2(k^2/\rho_m)}{(4\pi e^2/k^2+\varphi)(\omega^2-\Omega_p^2-\lambda^0k^2/\rho_m)+(4\pi eQ/k^2-\Lambda)^2(k^2/\rho_m)}+$$
$$+\left(\frac{4\pi e^2}{k^2}+\varphi\right)\chi(\omega,k)=0. \tag{P1.3}$$

It follows from equations (P1.1) and (P1.2) that the constant Λ plays the role of "contact" part of the electron and ion interaction potential. Indeed this becomes obvious if we note that the quantity

$$V_{ei}(k)\equiv\left(\frac{4\pi eQ}{k^2}-\Lambda\right)\frac{1}{n_i} \tag{P1.4}$$

infers the Fourier transform of electron/ion interaction energy. Analogously in accordance with equation (P1.1) the quantity

$$V_{ee}(k)=\frac{4\pi e^2}{k^2}+\varphi \tag{P1.5}$$

infers the Fourier transform of electron/electron interaction energy. Using the conventions of (P1.4), (P1.5) dispersion equation (P1.3) takes the following form:

$$\omega^2=\frac{\lambda^0}{\rho_m}k^2+\Omega_p^2-\frac{k^2n_i}{M}\frac{V_{ei}(k)}{V_{ee}(k)}\left[1-\frac{1}{\varepsilon_e(\omega,k)}\right], \tag{P1.6}$$

where

$$\varepsilon_e(\omega,k)=1+V_{ee}(k)\chi(\omega,k) \tag{P1.7}$$

is the longitudinal dielectric constant of the electrons. Equation (P1.6) corresponds to the picture in the case of longwave sound propagation in models that examine the metal as a system of interacting electrons and hard ions (see, for example, [31, Chapter 5; 46, Chapter 3; and 47]). Such models normally incorporate the dispersion of the plasma frequency of the ions, taking $\Omega_p^2(k) = \Omega_p^2 - ak^2$. In our approach such a relation corresponds to one of the possible contributions to (λ^0/ρ_m) that is not related to the conduction electrons. Then equation (P1.6) may be rewritten in the following manner (compare formula (3.55) from study [46]):

$$\omega^2=\frac{\lambda'}{\rho_m}k^2+\Omega_p^2(k)\left\{1-\frac{g^2(k)}{V_{ee}(k)}\left[1-\frac{1}{\varepsilon_e(\omega,k)}\right]\right\}, \tag{P1.8}$$

where $\lambda' = \lambda^0 + a\rho_m$ while the quantity

$$g^2(k)=\frac{k^2n_iV_{ei}(k)}{M\Omega_p^2(k)} \tag{P1.9}$$

is the squared electron-phonon interaction constant. Finally we note that the left half of dispersion equation (P1.3) is the effective longitudinal dielectric constant of the electron-ion system in a quantizing magnetic field:

$$\varepsilon_{eff}(\omega,k)=\varepsilon_e(\omega,k)-\frac{g^2(k)}{V_{ee}(k)}\frac{\Omega_p^2(k)}{\omega^2-(\lambda' k^2/\rho_m)-\Omega_p^2(k)\{1-[g^2(k)/V_{ee}(k)]\}}. \quad \text{(P1.10)}$$

Formulae (P1.6), (P1.8), and (P1.10) demonstrate, on the one hand, the interrelationship between the electron-phonon interaction theory formulated in our studies [17, 19, 24, 42] and another approach presented in studies [46, 47], and on the other hand these formulae make it possible to evaluate the capabilities of the Fröhlich model. We note that the complexity of using the Fröhlich model in quantum wave theory is attributable to the same difficulty that makes this model inaccessible in analyzing lattice instability. A detailed discussion of the latter problem may be found in study [46, Chapter 3].

In order to make the conversion to the Fröhlich model it is sufficient to carry out the following transformation:

$$\Omega_p^2(k)\left\{1-\frac{g^2(k)}{V_{ee}(k)}\left[1-\frac{1}{\varepsilon_e(\omega,k)}\right]\right\}\equiv$$
$$\equiv\frac{\Omega_p^2(k)}{\varepsilon_e(\omega,k)}\{1+[V_{ee}(k)-g^2(k)]\chi(\omega,k)\}\equiv\omega_0^2(k)\{1+$$
$$+[V_{ee}(k)-g^2(k)]\chi(\omega,k)\}. \quad \text{(P1.11)}$$

In this regard (P1.8) is transformed into the dispersion equation of the Fröhlich model [43, 46]:

$$\omega^2=\frac{\lambda'}{\rho_m}k^2+\omega_0^2(k)\{1+[V_{ee}(k)-g^2(k)]\chi(\omega,k)\}, \quad \text{(P1.12)}$$

if we take $\omega^2{}_0(k)$ in this expression to be a given (bare) quantity. In this formulation the entire electron renormalization of the squared acoustic velocity is related only to the electron polarizability $\psi(\omega, k)$ explicitly contained in the right half of (P1.12). Formula (P1.11) clearly reveals the unsuitability of the Fröhlich model in all cases where the determinant role is played by the dispersion of the electron dielectric constant $\varepsilon_e(\omega, k)$. This is due to the fact that part of the electron renormalization of the acoustic velocity is included in the bare frequency $\omega_0(k)$ and is exacerbated in investigating quantum waves, since the representation of $\omega_0^2(k)$ eliminates the most significant quantum effects contained in $\omega_0^2(k) = \Omega_p^2(k)/\varepsilon_e(\omega, k)$ in accordance with (P1.11) from the examination due to the dispersion of the electron dielectric constant. We discovered the resulting mistakes made in article [44] as discussed in study [24].

APPENDIX 2

THE INTERACTION OF QUANTUM SPIN-ACOUSTIC WAVES WITH HELICONS

Quantum spin-acoustic waves may propagate not only in the direction of the magnetic field B but also at a certain angle θ (not too close to $\pi/2$) to the field. In this case a qualitatively new effect may arise due to the fact that with inclined propagation the longitudinal oscillations of the electrons are coupled to the transverse oscillations, and this infers coupling between the quantum waves and electromagnetic waves of a different type. Slow quantum waves are coupled to helicons having "spiral" polarization. Studies [11, 20, 51] are devoted to the investigation of such coupled waves. The most comprehensive theory that accounts for spin oscillations and electron exchange interaction may be found in study [20]. In order to discuss this theory within the scope of this chapter we must generalize the fundamental equations formulated in section 2. Without dwelling on the corresponding cumbersome derivation we will provide only the complete system of equations describing the interaction of helicons and quantum spin-acoustic waves in this Appendix using simple qualitative concepts to explain its form. We will then briefly discuss the primary manifestations of the wave coupling effect.

The interaction of quantum waves with helicons is largely manifest in the longwave region at low frequencies ($\omega \ll \Omega$). In the case of transverse propagation the Maxwell equation for the amplitude of the transverse electrical field $E_x(k, \omega)$ takes the form

$$\left(1-\frac{u^2}{u_\Gamma^2}\right) E_x(k,\omega)=0, \tag{P2.1}$$

where $u_\Gamma = ckB/4\pi|e|n_e$ is the phase velocity of the helicon. The equation for the longitudinal field examined above is the zero value condition of the nonequilibrium charge density. Now let the wave vector have y- and z-components: $k = \{0, k_y, k_z\} = \{0, k \sin\theta, k\cos\theta\}$; the angle θ in this case will be significantly different from $\pi/2$, as required by the existence conditions of both helicons and quantum waves. Then as before one of the equations is reduced to the electroneutrality condition, although the electron charge density now contains an additional contribution proportional to the amplitude of the z-component of magnetic induction $b_z(k, \omega) = -ck_y(\omega)E_x(k, \omega)$. The amplitude of the density of electrons with a single spin direction is described in the following manner:

$$n^\sigma(k,\omega) = -X^\sigma(k_z,\omega)\left[-\frac{e}{ik_z}E_z(k,\omega)+2\psi s^z(k,\omega)\right]+$$
$$+\chi_b^\sigma(\omega,k_z)\left(-\frac{ck_y}{\omega}\right)E_x(k,\omega). \tag{P2.2}$$

A portion of this formula including terms proportional to $E_z(k, \omega)$ and $s^z(k, \omega)$ is differentiated from the complete expression for $n^\sigma(k, \omega)$ when $k_y = 0$ (formulae (2.17) and (2.18)) simply by the substitution of k with k_0 (in longitudinal propagation $E_z(k, \omega) = -ik\Phi(k, \omega)$). A qualitative new component in expression (P2.2) is the term containing the function $\chi_b^\sigma(\omega, k_z)$ that may be represented as

$$\chi_b^\sigma(\omega,k_z)=\frac{n_\sigma}{B}+X_\mu^\sigma(\omega,k_z), \tag{P2.3}$$

$$X_\mu^\sigma(\omega,k_z)=\frac{\Omega}{4k_F}\eta\sum_{n=0}^{N(\sigma)}\mu(n,\sigma)\left\{\frac{2v(n,\sigma)}{v^2(n,\sigma)-u^2}+i\pi\frac{m}{\hbar k_z}\int\limits_{w(n,\sigma,-k)}^{w(n,\sigma,k)}dv\,\delta(v-u)\right\}. \tag{P2.4}$$

Here n_σ is the number of electrons with a given spin projection in equilibrium ($n_e = n_+ + n_-$); $\mu(n, \sigma) = -(|e|\hbar/mc\ (n + 1/2)) + \mu_0\sigma$ is the magnetic moment of the electron in state n, σ (μ_0 is the spin magnetic moment of a free electron); $u = \omega/k_z$. We note that the function $X^\sigma_\mu(\omega,\ k_z)$ is differentiated from $X^\sigma(\omega,\ k_z)$ (see formulae (3.12), (3.4)) only in the factor $\mu(n,\ \sigma)$ under the sum sign. Expression (6.4) was written assuming $\omega\tau \gg 1$.

It is easy to understand the physical meaning of the function $X^\sigma_b(\omega,\ k_z)$ if we consider a formula analogous to (P2.2) in the static limit $\omega/k_z \to 0$ (in expression (6.4) we must set $u = 0$ and drop the imaginary part). We may determine that the quantity $\chi^\sigma_b(0,\ 0)$ is the derivative $\partial n_\sigma/\partial B$ evaluated ignoring the influence of exchange interaction on the electron magnetic moment (the contribution of exchange interaction is accounted for in formula (P2.2) by the term proportional to ψ). Analogously when $\omega/k_z \to 0$ we have $X^\sigma(0,\ 0) = \partial n_\sigma/\partial\zeta$. Thus, the function $\chi^\sigma_b(\omega,\ k_z)$ describes the response of electron density to the magnetic induction $b_z(\omega,\ k_z)$; its incorporation is substantiated by the equalities (P2.2) and (P2.4) for $\omega \ll \Omega$, $kR \ll 1$.

The electroneutrality condition is the zero value of the sum of the amplitudes of the densities $n^+(k,\ \omega)$ and $n^-(k,\ \omega)$ determined by formula (P2.2) (lattice oscillations here are not accounted for for simplicity). At the same time the half-difference of the quantities $n^+(k,\ \omega)$ and $n^-(k,\ \omega)$ represent the amplitude of the spin density $s^z(k,\ \omega)$. Thus, two equations are written using equality (P2.2) for $E_z(k,\ \omega)$, $W_x(k,\ \omega)$, $s^z(k,\ \omega)$:

$$n^+(k,\omega) + n^-(k,\omega) = 0,\quad s^z(k,\omega) = {}^1/_2\,[n^+(k,\omega) - n^-(k,\omega)]. \qquad \text{(P2.5)}$$

The third equation is described by a generalization of equation (P2.1) for a transverse field to the case of inclined propagation:

$$\left(1 - \frac{u^2}{u_\Gamma^2}\right) E_x = (k,\omega) = \frac{4\pi i\omega}{ck^2}\, j^m_x = \frac{4\pi i\omega}{c^2k^2}\, ick_y M_z(k,\omega). \qquad \text{(P2.6)}$$

Here the quantity $j^m_x(k,\ \omega)$ infers the amplitude of the x-component of the magnetization current, while $M_z(k,\ \omega)$ infers the z-component of the nonequilibrium magnetization of the electrons. For $M_z(k,\ \omega)$ we obtain an expression analogous to (P2.2):

$$M_z(k,\omega) = \varkappa(\omega, k_z)\left(-\frac{ck_y}{\omega}\right) E_x(k,\omega) - \\ - \sum_\sigma \chi^\sigma_b(\omega, k_z)\left(-\frac{e}{ik_z} E_z(k,\omega) + 2\psi\sigma s^z(k,\omega)\right), \qquad \text{(P2.7)}$$

where $\varkappa(\omega,\ k_z)$ is the generalized magnetic susceptibility of the electrons:

$$\varkappa(\omega, k_z) = \frac{2M}{B} + \sum_\sigma X^\sigma_{\mu\mu}(\omega, k_z). \qquad \text{(P2.8)}$$

Here M is the equilibrium magnetization, and the expression for the function $X^{\sigma}_{\mu\mu}(\omega, k_z)$ when $\omega\tau \gg 1$ differs from (P2.4) only by the additional multiplier $\mu(n, \sigma)$ under the sign of the sum over n. We obtain the static magnetic susceptibility assuming $u = 0$ and we drop the imaginary part of $\varkappa(\omega, k_z)$. Here we arrive at the familiar thermodynamic expression for electron susceptibility in a quantizing magnetic field ignoring the contributions of changes in the chemical potential and the exchange interaction (these contributions are described by the second term in (P2.7)).

The second component for magnetization in (P2.7) contains the same function $\chi^{\sigma}_b(\omega, k_z)$ that enters into (P2.2). This may be explained by again referring to the static limit. Analogous to (P2.2) we may easily conclude that the contribution to magnetization attributable to the longitudinal field must be proportional to the derivative $\partial M/\partial\zeta$. However the thermodynamic relation $\partial M/\partial\zeta = \partial n/\partial B$ exists and we find from here that the quantity $\partial M/\partial\zeta$ coincides with $\chi^{+}_b(0, 0) + \chi^{-}_b(0, 0)$. An analogous relation is valid in the range of ω and k considered here.

Equations (P2.5), (P2.6) completely describe low-frequency waves for the case of inclined propagation. The field components that do not enter into these equations are simply expressed through $E_x(k, \omega)$, $E_z(k, \omega)$ from Maxwell's equations:

$$E_y(k, \omega)\cos\theta = \frac{iu}{u_r} E_x(k, \omega) + E_z(k, \omega)\sin\theta,$$

$$\mathbf{b}(k, \omega) = \frac{c}{\omega}[\mathbf{k} \times \mathbf{E}(k, \omega)], \qquad \mathbf{kb}(k, \omega) = 0. \tag{P2.9}$$

In the quasi-classical limiting case the contribution of the z-component of the electrical field and the spin density in expression (P2.7) is not significant and we may also ignore the real part of the susceptibility $\varkappa(\omega, k_z)$. The imaginary part of the susceptibility:

$$\operatorname{Im}\varkappa_{\text{кл}}(\omega, k_z) = \frac{3}{32}\frac{u}{u_r} kR \tag{P2.10}$$

describes collisionless helicon damping (so-called magnetic Landau damping [52, 53]). Quantization of electron orbital motion results in transparency windows appearing in the solid region of magnetic damping that are identical to those for longitudinal waves (lobe windows with small k; see Fig. 2), therefore producing giant quantum oscillations in helicon absorption [54]. This effect is described by the imaginary part of the susceptibility $\varkappa(\omega, k_z)$ in the quantum range (the imaginary part of expression (P2.4) with the substitution of $\mu(n, \sigma)$ with $m^2(n, \sigma)$). Another quantum effect is the interaction of helicons with quantum waves in the transparency windows. This is due to the coupling of the transverse helicon field to the z-components of the field and the spin density. Here the frequency dispersion of the real parts of the functions $\varkappa(\omega, k_z)$ and $\chi^{\sigma}_b(\omega, k_z)$ become significant.

The general nature of the dependencies of the coupled wave frequencies on k_z is determined by the fact that the dispersion curve of the helicon $\omega = k_z u_\Gamma$ decomposes near the lines $\omega = k_z u_{KB}(n, \sigma)$ describing the quantum waves. Near the intersection points $(u_\lambda = u_{KB}(n, \sigma))$ the dispersion relations have a form characteristic of the picture of interacting waves. Simple formulae for these regions may be obtained using the same method as used above for slow quantum waves, i.e., accounting for the dependence on y in only one or two terms of the sum over n in the functions $X^\sigma(\omega, k_z)$, $\chi^\sigma_b(\omega, k_z)$ and $\varkappa(\omega, k_z)$. For example, in the case of helicon/wave interaction (6.15) we obtain the following two roots of the dispersion equation for $u = \omega/k_z$:

$$u_{1,2} = v(n_0, +) + \frac{\Omega}{8k_F}\{1 + \beta + \Delta_1 \pm \sqrt{(1 + \beta - \Delta_1)^2 + 4A}\}, \qquad \text{(P2.11)}$$

where $\beta = B_0/(1 + B_0)$; $\Delta_1 = (4k_F/\Omega)(u_\Gamma - v(n_0, +))$; $A = 8\pi\varkappa_0 (u_\Gamma/v_F) \times$ $\times (\varepsilon_F/\hbar\Omega)^3 \sin^2\theta$; $\varkappa_0\eta$ — is the paramagnetic susceptibility of free electrons. formula (P2.11) is valid when the quantities Δ and A are not too great, so that both roots are close to $v(n_0, +)$. In the limit of large Δ_1 we obtain from this formula the phase velocities of the noninteracting waves. In the central portions of the transparency windows interaction with quantum waves has a weak influence on the frequency dispersion of the helicon.

The propagation of coupled helicon-quantum waves is accompanied by oscillations in the longitudinal component of the spin density of electrons [20]. Since such waves may be excited by a transverse field, it is interesting to compare the amplitude of the spin magnetization $\mu_0 s^z(k, \omega)$ to the amplitude of the transverse magnetic induction $b_x(k, \omega)$. The relation $\mu_0 S^z(k, \omega)/b_x(k, \omega)$ achieves comparatively high values in the case where the difference in the energies of the spin half-levels $(|\Omega_0 - \Omega| \sim \Omega)$ is high. According to the estimate obtained in study [20] in this case the modulus $|\mu_0 s^z(k, \omega)/b_x(k, \omega)|$ in the range of interaction is of the order $(\varkappa/(1 + B_0))(\varepsilon/\hbar\Omega) \sin\theta$ and increases significantly at points where $\Delta_1 \approx \sqrt{(1 + \beta - \Delta_1)^2 + 4A}$ (see formula (P2.11)).

BIBLIOGRAPHY

1. Landau, L.D., Lifshits, Ye.M. "Kvantovaya mekhanika: Nerelyativistskaya teoriya" [Quantum mechanics: nonrelativistic theory] Moscow: Nauka, 1974, 112, p. 522-528.

2. Lifshits, Ye.M., Pitaevskiy, L.P. "Statisticheskaya fizika" [Statistical physics] Chapter 2, Moscow: Nauka, 1978, Ch. VI, p. 265-335.

3. Stephen, M.J. "Oscillations of a plasma in a magnetic field" PHYS. REV., 1963, V. 129, p. 997-1004.

4. McWhorter, A.L., May, W.G. "Acoustic plasma waves in semimetals." IBM J. RES. AND DEVELOP., 1964, V. 8, p. 285-290.

5. Gurevich, V.L., Skobov, V.G., Firsov, Yu.A. "Giant quantum oscillations in acoustic absorption by metals in a magnetic field" ZhETF, 1961, V. 40, p. 787-791.

6. Pines, D., Schrieffer, J. "Collective behavior in solid state plasmas" PHYS. REV., 1961, V. 124, p. 1387-1400.

7. Ginzburg, S.L., Konstantinov, O.V., Perel', V.I. "The acoustic branches of plasma electron oscillations in metals in a quantizing magnetic field" FTT, 1967, V. 9, p. 2139-2142.

8. Konstantinov, O.V., Perel', V.I. "Acoustic waves in the electron plasma of metals in a quantizing magnetic field" ZhETF, 1967, V. 53, p. 2034-2040.

9. Anokhin, S.B., Kondrat'ev, A.S. "On zero sound in charged Fermi-systems" In: "Voprosy elektroniki tverdogo tela" [Issues in solid state electronics] Leningrad, 1968, No. 345, p. 24-30.

10. Benford, G., Book, D. "Acoustic plasma modes in high magnetic fields" PHYS. REV. LETT., 1968, V. 21, p. 898-900.

11. Kaner, E.A., Skobov, V.G. "On the possibility of existence of quantum electromagnetic waves in metals" PHYS. REV., 1968, V. 169, p. 530-540.

12. Glick, A.J., Callen, E. "Quantum structure in the dielectric function of metals and anomalous propagation modes" PHYS. REV., 1968, V. 169, p. 530-540.

13. Kaner, E.A., Skobov, V.T. "Quantum waves in semimetals" FTP, 1967, V. 1, p. 1367-1374.

14. Demikhovskiy, V.Ya., Protogenov, A.P. "Electromagnetic oscillations in metals and semimetals in a strong magnetic field" UFN, 1976, V. 118, p. 101-138.

15. Zyryanov, P.S., Okulov, V.I., Silin, V.P. "Quantum spin waves" ZhETF, 1968, V. 8, p. 489-492.

16. Zyryanov, P.S., Okulov, V.I., Silin, V.P. "Quantum electron spin-acoustic waves2" PIS'MA I ZhETF, 1969, V. 9, p. 371-374.

17. Zyryanov, P.S., Okulov, V.I., Silin, V.P. "Low velocity quantum electron spin-acoustic waves" FMM, 1969, V. 28, p. 558-560.

18. Zyryanov, P.S., Okulov, V.I., Silin, V.P. "Parametric absorption of ultrasound by conduction electrons in a quantizing magnetic field" PIS'MA I ZhETF, 1969, V. 10, p. 98-100.

19. Zyryanov, P.S., Okulov, V.I., Silin, V.P., "Quantum waves in the

degenerate electron liquid of a metal" ZhETF, 1970, V. 58, p. 1295-1309.

20. Zyryanov, P.S., Okulov, V.I., Silin, V.P., "Coupled spiral and quantum spin-acoustic waves in the electron liquid of a metal" FMM, 1970, V. 30, p. 1093-1095.

21. Okulov, V.I., Pamyatnykh. Ye.A. "The spectrum of quantum waves in the electron liquid of a metal" FMM, 1974, V. 38, p. 279-288.

22. Okulov, V.I., Pamyatnykh, Ye.A. "The influence of quantum effects on the spectrum of spin waves in a nonferromagnetic metal" FTT, 1974, V. 16, p. 1611-1619.

23. Pamyatnykh, Ye.A. "Oscillations in the electron liquid of a metal in a quantizing magnetic field;" Dissertation for candidate for Doctor of Physics and Mathematics 1974.

24. Zyryanova, •N.P., Okulov, V.I., Silin, V.P. "Waves in the quantum plasma of a metal" In: Problemy fiziki tverdogo tela [Issues in solid state physics] Sverdlovsk: UNTs AN SSSR, 1975, p. 38-86.

25. Chock, D.P., Lee, J.G. "Collective oscillations in solids along a quantizing magnetic field" PHYSICA, 1970, V. 50, p. 317-330.

26. Okulov, V.I., Silin, V.P. "Sound interaction with quantum waves" KVAT. SOOBSHCH. PO FIZIKE, 1981, No. 2, p. 43-47.

27. Demikhovskiy, V.Ya., Protozenov, A.P. "The interaction of optical phonons with electrons in a quantizing magnetic field" ZhETF, 1970, V. 58, p. 651-565.

28. Kondrat'ev, A.S., Kuchma, A.Ye. "Elektronnaya zhidkost' normal'nykh metallov" [The electron liquid of normal metals] Leningrad: Izd-vo LGU, 199 p.

29. Akhiezer, A.I., Bar'yakhtar, V.G., Peletminskiy, S.V. "Spinovye volny" [Spin waves] Moscow: Nauka, 1967, 368 p, see also: Silin, V.P. "Spinovye volny v neferromatnitnykh metallakh" [Spin waves in nonferromagnetic metals].

30. Zyryanov, P.S. "Quantum theory of acoustic oscillations of an electron-ion plasma in a magnetic field" ZhETF, 1961, V. 40, p. 1353-1359.

31. Pains, V. "Elementarnye vozbuzhdeniya v tverdykh telakh" [Elementary excitations in solids] Moscow: Mir, 1965, 382 p.

32. Bohm, D., Staver, T. "Application of collective treatment of electron and ion vibrations to theories of conductivity and superconductivity" PHYS. REV., 1951, V. 84, p. 836-837.

33. Silin, V.P. "The spectrum of excitations of an electron and ion system" ZhETF, 1952, V. 23, p. 649-659.

34. Rodriguez, S. "Oscillations of the velocity of sound in metals in a magnetic field" PHYS. REV., 1963, V. 132, p. 535-541.

35. Quinn, J.J., Rodriguez, S. "Electromagnetic properties of a quantum plasma in a uniform magnetic field" PHYS. REV., 1962, V. 128, p. 2487-2493.

36. Filippov, B.N., Zyryanov, P.S. "Theory of ultrasonic absorption by free electrons in a metal in a quantizing magnetic field. Part 2. Acoustic propagation along the magnetic field" FMM, 1967, V. 24, p. 18-26.

37. Silin, V.P. "Theory of degenerate electron fluid and electromagnetic waves in metals" FMM, 1970, V. 29, p. 681-734.

38. Akhiezer, A.I., Kaganov, M.I., Lyubarskiy, G.Ya. "Ultrasound absorption in metals" ZhETF, 1957, V. 32, p. 837-841.

39. Silin, V.P. "Theory of ultrasound absorption in metals" ZhETF, 1960, V. 38, p. 977-983.

40. Kontorovich, V.M. "Equations of elasticity theory and acoustic dispersion in metals" ZhETF, 1963, V. 45, p. 1638-1653; 1970, V. 59, p. 2116-2129.

41. Vasov, K.B., Filippov, V.N. "Rotation of the polarization plane of ultrasound in metals in a strong magnetic field" ZhETF, 1964, V. 46, p. 223-231.

42. Zimbovskaya, N.A., Okulov, V.I. "The Fermi-liquid quantum theory of ultrasound propagation in metals" Manuscript deposited at The All Union Institute of Scientific and Technical Information, No. 2750-77 Dep.

43. Fröhlich, H. "Theory of the superconducting state. The ground state at the absolute zero of temperature" PHYS. REV., 1950, V. 79, p. 845-856.

44. Blank, A.Ya., Kaner, E.A. "The phonon spectrum of metals in a magnetic field" ZhETF, 1966, V. 50, p. 1013-1023.

45. Kaner, E.A. Chebotarev, L.B. "Nonadiabatic effects in the phonon spectrum of metals in a magnetic field" ZhETF, 1977, V. 73, p. 1811-1830.

46. "Problema vysokotemperaturnoy sverkhprovodimosti" [The problem of high-temperature superconductivity] Ed. by L.N. Bulaevskiy, V.L. Ginzburg, G.F. Zharkov, et al., Moscow: Nauka, 1977, 400 p.

47. Brovman, Ye.G., Kagan, Yu. "Phonons in nontransition metals" UFN, 1974, V. 112, p. 369-426.

48. Bagaev, V.N., Okulov, V.I., Pamyatnykh, Ye.A. "Low-temperature quantum oscillations in the magnetic susceptibility of the electron liquid of a metal" FNT, 1978, V. 4, p. 742-752.

49. Konstantinov, O.V., Perel', V.I. "The theory of acoustic plasma waves in bismuth" FTT, 1961, V. 9, p. 3051-3058.

50. Wolf, P.A. "Theory of light scattering from magnetoacoustic waves in solid-state plasmas" PHYS. REV., 1970, V. 1, p. 164-168.

51. Kaner, E.A., Lyubimov, O.I., Skobov, V.G. "Theory of quantum electromagnetic waves in metals in a magnetic field" ZhETF, 1970, V. 58, p. 730-738.

52. Kaner, E.A., Skobov, V.G. "Theory of resonance excitation of weak-damped electromagnetic waves in metals" ZhETF, 1963, V. 45, p. 610-627.

53. Buchsbaum, S.J., Platzman, P.M. "Nonlocal damping of helicon waves" PHYS. REV., 1967, V. 154, p. 395-398.

54. Skobov, V.G., Kaner, E.A. "Quantum theory of electromagnetic wave propagation in metals in a magnetic field" ZhETF, 1964, V. 46, p. 1809-1819.

Theory of Surface Quantum Resonance Properties of Electron Liquid in a Magnetic Field

O.M. Tolkachev

Abstract: This study formulates for the first time kinetic equations suitable for describing surface electrons accounting for their Coulomb interaction. A new propagation effect of surface spin waves along a magnetic field parallel to the sample surface for metals with an arbitrary Fermi surface is predicted. The existence conditions for such excitations are noted. The conditions for collisionless damping of such waves are investigated.

A method within the scope of Silin–Landau electron liquid theory is developed that makes it possible for the first time to determine the parameters of interelectron interaction based on an analysis of experimental data on oscillations in the surface impedance in a magnetic field. Processing of experimental data made it possible to find the values of the electron interaction constants in bismuth.

INTRODUCTION

The problem of accounting for the influence of Coulomb electron interaction which always occurs in actual conductors on the high-frequency resonance properties of metals has long been the focus of attention of theoreticians and experimenters. A comprehensive theory of the electron Fermi-liquid of metals such as that discussed in study [1] allows description of the response of a system of interacting electrons to an external perturbation with frequency ω and wave vector $\mathbf{k}$. In discussing the importance of accounting for interelectron interaction, it is always necessary to identify the possibility of spin wave propagation in normal metals resulting solely from the correlation of conduction electrons. The results from the theory of such a phenomenon developed by V.P. Silin [2-5] based on electron liquid concepts were confirmed in experiments by Dunifer, Platzman, and Schultz [6-8] who investigated the selective transparency of alkali metals near the electron spin resonance frequency and discovered a series of transparency lines corresponding to the excitation of spin waves in the metal. If a spin wave propagates in the direction of a magnetic field with wave vector **k**, the frequency spectrum of such a wave, consistent with study [1], may be written as

$$\omega = \omega_s \left[1 - \frac{k^2 v_F^2}{3\omega_s^2} \frac{(1+B_0)^2 (1+B_1)}{B_0 - B_1} \right], \tag{1}$$

where the Bloch spin resonance frequency of the electrons is equal to $\omega_S = 2\mu_0 B/\hbar$; μ_0 is the magnetic moment of a noninteracting electron; B is the constant magnetic field; k is the modulus of the wave vector of the spin wave; v_F is the Fermi-velocity ($kv_F \ll \omega_S$); B_0 and B_1 are the zeroeth and first coefficients of the expansion of the spin-dependent part of the Landau interaction function in terms of Legendre polynomials (see also [9, 10]). The experimental discovery of spin waves in sodium and potassium in a magnetic field [6-8] with precisely such a dispersion law (1) demonstrated a qualitative difference between the conduction electrons and the noninteracting particle gas.

Another effect showing the difference in the properties of the electron liquid of metals from gas is the propagation of cyclotron waves in metals located in a constant magnetic field. In the longwave limit the eigenfrequencies of such cyclotron waves, as indicated by the theory of such a phenomenon [11, 12], unlike the case of the electron gas are equal to

$$\omega_{lm} = -m\Omega\,(1 + A_l) + O\,(k^2), \tag{2}$$

where k is the modulus of the wave vector of the cyclotron wave, A_l are the coefficients of the expansion of the spin-independent part of the correlation function in terms of Legendre polynomials, $|m| \leqslant l$, Ω is the Larmor frequency of electron gyroscopic rotation. The dependence of frequency (2) on the coefficient A_l is caused by the Fermi-liquid interaction of electrons. The shift of critical frequencies (2) by $m\Omega A_l$ from the Larmor frequency Ω makes possible, as demonstrated in study [1], the propagation of cyclotron waves in an electron liquid in the absence of collisionless Landau damping. The possibility for collisionless propagation of undamped cyclotron waves in metals is one significant characteristic that differentiates electron liquid from electron gas.

The effects associated with accounting for the difference in the momentum of particles on different quantum levels and those having an energy equal to the Fermi energy become significant in strong magnetic fields and at low temperatures. Here it turns out [13=19] that the Fermi-liquid interaction of electrons plays a determinant role in determining the spectrum of electromagnetic and spin waves in such conditions.

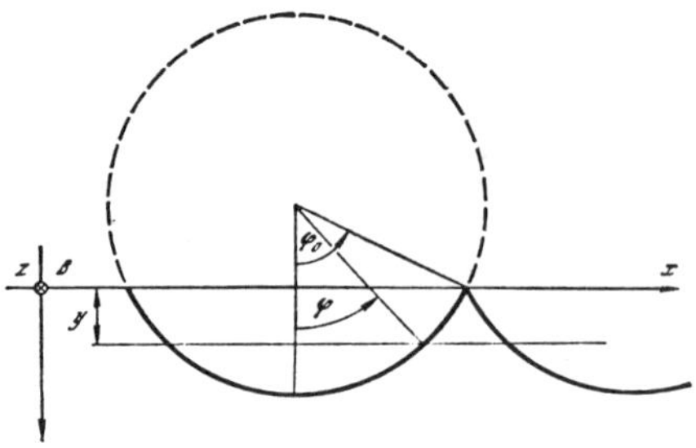

Fig. Schematic representation of the skipping electron trajectory.

Until recently the problem of formulating a Fermi-liquid theory suitable for describing the resonance properties of a metal associated with the generation of a system of magnetic surface levels near the surface of a metal in a magnetic field was ignored. Magnetic surface electron levels were discovered by M.S. Khaykin in 1960 in investigating surface impedance oscillations in tin [20] in weak magnetic fields (~10 ergs) at frequencies of $\omega/2\pi \sim 10$ GHz. The theoretical substantiation of this phenomenon was provided in 1967 in the study by Nee and Prange [21]. In this case the surface impedance oscillations of the metal were attributed to resonance absorption of the electromagnetic wave incident on the sample due to transitions between the quantum levels of electrons colliding with the sample surface, while electron surface reflection was assumed to be near-specular. The oscillations in surface impedance in weak magnetic fields in the RF range was investigated experimentally for cadmium [20], indium [20, 22], aluminum [22], copper and gallium [23, 24], bismuth [20, 25-29], and tungsten [30]. This effect was examined theoretically within the scope of gas theory in studies [31-35]. The most detailed development of the theory of cyclotron resonance at the magnetic surface levels ignoring electron interaction may be traced to a series of studies by Kaner and Markov [36-38] as well as studies by Kaner et al. [39-42].

The appearance of a system of surface electron levels may be explained based on a simple metal sample with a cylindrical Fermi surface located in a magnetic field parallel to the z axis and lying along the sample boundary (Fig. 1). The axis y is perpendicular to the surface and extends into the metal horizontally, and electron reflection off the interface is assumed to be specular. The momentum dependence of the energy is determined by the formula

$$\varepsilon\,(p_x, p_y) = (2m)^{-1}(p_x^2 + p_y^2),$$

where m is the effective electron mass corresponding to the cylindrical Fermi surface. In addition to the electrons in the metal bulk for which the center of the Larmor orbit is much greater than the radius of this orbit, there exist electrons that collide with the sample boundary. The most significant role in this case is played by the electrons in which the projection of the center of the Larmor orbit onto the y axis is negative and is approximately twice the orbital radius in magnitude. The motion of such electrons in the magnetic field on the x axis in Fig. 1 is infinite, and is finite on the y axis, and hence is quantized. Expanding the Hamiltonian $\hat{H}$ corresponding to the Schrodinger equation

$$\hat{H}\psi_n = E_n\psi_n, \tag{3}$$

in which we take the ratio of the coordinate y to the radius of the Larmor electron orbit $|cp_x/eB|$ to be small, where B is a constant magnetic field, we obtain

$$\hat{H}\,(p_x + eBy/c,\; p_y) = \varepsilon\,(p_x, 0) + eByv_x/c, \tag{4}$$

while $\upsilon_x = \partial\varepsilon(p_x, 0)/\partial p_x$. Schrodinger equation (3) with Hamiltonian (4) correspond to eigenfunctions that with constant factor accuracy are Airy functions:

$$\psi_n \sim \mathrm{Ai}\,(-\xi - a_n)$$

and the eigenvalues

$$E_n = \varepsilon\,(p_x, 0) + eBv_x y_n/c. \tag{5}$$

Here $y_n = a_n \mid 2v_x eBm/c\hbar^2 \mid^{-1/3}$, while

$$\xi + a_n = [-y + (E_n - \varepsilon\,(p_x, 0))c/v_x eB][2v_x eBm/c\hbar^2]^{1/3}.$$

We obtain the quantization condition by requiring the vanishing of the wave function at the metal boundary when $y = 0$ ($\xi = 0$):

$$\mathrm{Ai}\,(-a_n) = 0,$$

inferring that a_n is the nth root of the Airy function. The values of these roots are given in the tables of study [43] and may be approximated with good accuracy by the asymptotic formula:

$$a_n = [(3\pi/2)(n - {}^1/_4)]^{2/3}.$$

Substituting the derived values into (5) we find the following expression for the energy eigenvalues:

$$E_n = \frac{p_x^2}{2m} + \left[\frac{3\pi}{2}\,\frac{eB}{mc}\left(n - \frac{1}{4}\right)\right]^{2/3}\left(\frac{p_x^2}{2m}\right)^{1/3}. \tag{6}$$

If follows that the system of energy levels in (6) arises near the surface of a metal in a magnetic field, while the following frequency corresponds to the transitions between these levels

$$\omega_{n'n} = \frac{E_{n'} - E_n}{\hbar} = \left(\frac{3\pi}{2}\,\frac{eB}{mc}\right)^{2/3}\left(\frac{\varepsilon_F}{\hbar}\right)^{1/3}\left[\left(n' - \frac{1}{4}\right)^{2/3} - \left(n - \frac{1}{4}\right)^{2/3}\right]. \tag{7}$$

Due to the fact that, as a rule, for normal metals in achievable magnetic fields ($\hbar eB/m_c$) $\ll \varepsilon_F$, where ε_F is the Fermi electron energy, the fact that

$$\hbar\,\frac{|e|B}{mc} \ll \hbar\omega_{n'n} \ll \varepsilon_F,$$

was incorporated in deriving (7) which the momentum p_x was assumed to be near-Fermi. Resonance occurs when the frequency of the emission impacting the sample is close to one of the frequencies in (7) which makes possible absorption of the energy of the incident emission. Examination of formula (7) shows that, first, the surface levels are equidistant throughout the magnetic field, and has a dependence on the field as $B^{2/3}$ and, second, even for magnetic fields of ~10 Ergs frequency (7) falls in the microwave range of ~10 GHz for normal metals.

It is precisely in this range of magnetic fields and for the frequencies discussed above that the initial experimental measurements of the effect were performed.

We assumed above that electron reflection off the metallic surface was specular. Actual existing surface irregularities in the sample and electron collisions cause the surface levels to "wash out." Here, as demonstrated in studies [44, 45] for a metal with a surface having irregularities of less than or in the order of 10^{-6} cm the "wash out" of the levels due to electron collisions with the irregularities is negligible compared to the distance between the surface levels (7). Physically this conclusion is related to the fact that the greater the effective electron wavelength in direction y, the closer electron reflection is to specular reflection. These studies state that with surface irregularities less than 10^{-6} cm the damping decrement associated with surface collisions will be less than 10^{9} cm^{-1}. This means that the sample surface in this case may be considered a specular surface and wave damping in the vicinity of the transition frequencies between the magnetic surface levels will be determined primarily by volumetric collisions rather than the distance in the surface irregularities.

The gas theory of cyclotron resonance at skipping orbits discussed above satisfied the inquiries of initial experiments, since within existing experimental accuracy this theory explained the positions of the resonance peaks in the surface impedance of normal metals measured in the experiments. The situation changed in the 1970's when, on the one hand, experimental accuracy improved significantly and, on the other, rather pure samples with a longer electron free path length appeared. The experimental research [25, 26] contains data that cannot be explained by the gas theory, while experimental results [27] directly indicate the need for further development of the theory to account for electron interaction and quantum corrections of $\sim\hbar\omega/\varepsilon_F$ to the frequency spectra of surface waves for adequate description of experiments. This additional development is the purpose of the present study.

The primary purpose of this study is to formulate an investigatory theory of the surface resonance properties of normal conductors located in a magnetic field accounting for electron interaction near the transition frequencies between the magnetic surface levels. Such a theory would be suitable, first, to explain the new effects in the spectra of surface spin and cyclotron waves and, second, the theoretical formulae must be modified for comparison to existing experimental data. In order to solve this problem in the first chapter we formulate a quantum kinetic equation for the spin density matrix of skipping electrons [46, 47]. This equation accounts for the interaction of the skipping electrons both with surface electrons and volumetric electrons. Based on an investigation of the resonance properties of the density matrix obtained from a solution of the quantum kinetic equation, a new surface quantum wave propagation effect in a magnetic field in metals with an arbitrary Fermi surface with frequencies near the transition frequencies between the levels of the skipping electrons is predicted theoretically. This chapter considers the contribution of volumetric, nonresonance surface and resonance electrons (for the latter case the value of the transition frequency is near ω) to the kernel of the integral equation for the Fourier-component of the electron spin density. The conditions in which the primary contribution to the kernel of the integral equation comes from resonance electrons in the case of weak collisions ($\omega\tau \gg 1$, τ is the momentum relaxation time) are identified. Unlike metals with a cylindrical Fermi surface for which there is no collisionless damping of surface spin waves, such collisionless damping of spin surface waves will always exist in metals with a noncylindrical Fermi surface. In the first chapter a metal with a spherical Fermi surface is used as an example to indicate the conditions in which collisionless Landau damping of surface waves is not significant. The influence of nonsphericity of the Fermi surface and the influence of accounting for the finite wavelength of the spin oscillations on the surface wave spectra are considered.

The second chapter is devoted to formulating a quasi-classical theory of cyclotron resonance at the skipping orbits of surface electrons in the electron liquid of metals with an arbitrary Fermi surface [48-50]. The quasi-classical kinetic equation formulated in this case for the distribution function of skipping electrons makes it possible, first, to easily incorporate parameters characterizing interaction and, second to consistently incorporate boundary conditions on the metallic surface. The dispersion equation of surface cyclotron waves is obtained by solving the kinetic equation accounting for the boundary condition and then by quantizing the derived solution. In the second chapter we investigate the onset of collisionless Landau damping of surface cyclotron waves in detail and identify the conditions in which such damping is small. Due to the fact that cyclotron resonance at skipping orbits has been investigated extensively in experiments on bismuth, the theoretical formulae are modified for the case of an ellipsoidal Fermi surface. Expressions are derived for the frequency spectra of surface cyclotron waves that are suitable for comparison of theory to experiment and for determining the electron interaction parameters in bismuth based on such a comparison [48]. This chapter also shows that when the Fermi-liquid electron interaction constant is not small compared to unity, accounting for the electrodynamic effects in order to determine the spectra of surface cyclotron waves is not significant. We also carry out a detailed substantiation of the impossibility of surface cyclotron wave propagation near the transition frequencies between the skipping electron levels in the noninteracting particle model for metals with convex Fermi surfaces.

In the third chapter we compare the developed theory to experimental data on cyclotron resonance at the skipping orbits in bismuth obtained for the microwave range [51, 52]. In this case in order to provide the most accurate quantitative analysis of experimental data a method is developed for accounting for minor deviations of the Fermi surface of the bismuth from an ellipsoidal form. The Fermi-liquid electron interaction constants corresponding to processing of data on the transitions between levels labeled n' and n where $n' - n \ll n$ were calculated based on an analysis of the experimental values of the resonance magnetic fields obtained by Doezema et al. (1975) [27] in bismuth at $\omega/2\pi$ = 36.26 GHz. Graphs plotting the arbitrary real part of the impedance against the magnetic field at finite values of $\omega\tau$ are drafted and these graphs are used to compare the experimentally-measured extrema of the impedance derivative to values predicted by electron liquid and electron gas theories. In this case it is established that for bismuth it is possible to match the experimentally-measured positions of the resonance peaks of the surface impedance to the theoretically calculated values with a Fermi-liquid electron interaction constant that is small compared to unity. This chapter also proposes a new method of determining the electron interaction constant by measuring absolute values of the resonance peaks of the impedance or its derivative.

The fourth chapter provides results from quantum theory of cyclotron resonance [53, 54] suitable for interpreting experimental data referring to the infrared (IR) range and obtained for the transitions between levels with arbitrary labels n' and n and not only satisfying the quasi-classicality condition $n' - n \ll n$. Such a theory accounts for the corrections to the cyclotron wave spectra of $\sim\hbar\omega/\varepsilon_F$ that arise due to differences in the critical values of electron momentum on the levels before and after the transition. The values of the interelectron interaction constant in bismuth obtained by processing our experimental data using our formulae are given in the tables in the Appendix. These values based on the experimental data for the IR range are consistent with the values for the electron interaction constant in bismuth obtained in the third chapter from processing microwave data. By accounting for the various values of $\omega\tau$ for four IR-frequencies [54] graphs are plotted for the magnetic field derivatives of the real part of the impedance. An examination of these graphs reveals that ignoring electron interaction makes it impossible to match the positions of the extrema of the curve calculated theoretically to the experimentally measured values. Incorporating electron interaction results, first, in increasing the amplitude of the resonance peaks of the impedance derivative and, secondly, causes a shift in the system of extrema towards lower fields corresponding to experiment. In drafting these graphs in the IR range formulae were used that account for the quantum corrections $\sim\hbar\omega/\varepsilon_F$.

In addition to the tables of interaction constants identified above, the appendices also include quasi-classical formulae for calculating the matrix elements that are integrals of the Airy functions together with results from the numerical calculation of such integrals necessary for interpreting experimental data. The possibility for determining the parameters of interelectron interaction in copper using experimental data on cyclotron resonance at the skipping orbits is discussed. We may estimate only the upper boundary for the interaction constant in copper based on existing available experimental data for the microwave range. A method is identified for more precisely determining the interaction constants, related primarily to increasing the frequency of emission used in the experiments.

Chapter 1

SURFACE QUANTUM SPIN WAVES IN THE DEGENERATE ELECTRON LIQUID OF METALS IN A MAGNETIC FIELD NEAR THE TRANSITION FREQUENCIES BETWEEN SKIPPING ELECTRON LEVELS

1. Quantum kinetic equation for the spin density matrix of skipping electrons

In order to describe the weakly-excited states of the electron liquid it is productive to use a quantum kinetic equation for the small deviation of the density matrix from an equilibrium Fermi distribution [1]. In solving the problem of interest to us we must first carry out a generalization of the quantum kinetic equation from electron liquid theory (see, for example, [1, 13]) to the case of the surface levels of the electrons. Bearing in mind that the electron state is characterized by a set of quantum numbers ν and taking the density matrix of the ground state of the electrons to the diagonal, with a time dependence of $\sim\exp(-i\omega t)$, we may write the nonequilibrium density matrix in the energy representation (dependent on the quantum numbers ν' and ν) as [46, 47]

$$\rho_{\nu'\nu} = \rho^0(\nu)\,\delta_{\nu'\nu} + \delta\rho_{\nu'\nu}\exp(-i\omega t). \tag{1.1}$$

Here $\delta_{\nu'\nu}$ is the Kronecker symbol. Sinçe we are interested in linear problems for the nonequilibrium addition to the density matrix, we may write the following approximate equation [47]:

$$[\hbar\omega - E(\nu') + E(\nu)]\,\delta\rho_{\nu'\nu} + \\ + [\rho^0(\nu') - \rho^0(\nu)]\left\{-\sum_{\mathbf{q}} \frac{1}{c}\,\mathbf{j}^0_{\nu'\nu}(-\mathbf{q})\,\delta\mathbf{A}_{\mathbf{q}\omega} + \sum_{\nu_1'\nu_1} F^{\nu_1\nu_1'}_{\nu'\nu}\,\delta\rho_{\nu_1'\nu_1}\right\} = J^{st}_{\nu'\nu}. \tag{1.2}$$

This equation is a quantum analog of the quasi-classical equation of the theory of degenerate electron liquid [1]. Here $E(\nu)$ are the eigenvalues of the electron energy on the surface level with a set of quantum numbers ν; $\delta\mathbf{A}$ is the amplitude of the vector potential of a nonequilibrium electromagnetic field; $J^{st}_{\nu'\nu}$ is the collision integral. The Fourier component of the electron current density operator for an electron that does not interact with other electrons (compared to study [14, 15]) is equal to

$$\mathbf{j}^0_{\nu'\nu}(-\mathbf{q}) = \delta_{\sigma'\sigma}e\mathbf{v}^0_{\alpha'\alpha}(-\mathbf{q}) - \mu_0 ci\,[\mathbf{q}\sigma]_{\sigma'\sigma}I_{\alpha'\alpha}(-\mathbf{q}),$$

where

$$\mathbf{v}^0_{\alpha'\alpha}(-\mathbf{q}) = \frac{1}{2m}\left\langle\alpha'\left|e^{i\mathbf{q}\mathbf{r}}\left(\hat{\mathbf{p}} - \frac{e}{c}\mathbf{A}\right) + \left(\hat{\mathbf{p}} - \frac{e}{c}\mathbf{A}_0\right)e^{i\mathbf{q}\mathbf{r}}\right|\alpha\right\rangle,$$

while the matrix element of the exponent in the wave functions of surface states is equal to

$$I_{\alpha'\alpha}(-\mathbf{q}) = \langle \alpha' \mid e^{i\mathbf{qr}} \mid \alpha \rangle.$$

Here we account for the fact that the complete set of quantum numbers includes the orbital numbers α and the spin indices $\sigma(\nu = \alpha, \sigma)$, $e = -|e|$ is electron charge; μ_0 is the magnetic moment of the electron; σ is the vector of the Pauli matrices; $\mathbf{A}_0$ is the vector potential of the static magnetic field; m is the effective electron mass corresponding to the squared dependence of electron energy on momentum $\mathbf{p}$. We note that the generalization to the more complex electron dispersion law, as demonstrated below, does not encounter any difficulties.

The last component of the left half of (1.2) describes the Fermi-liquid interaction of electrons. Ignoring spin-orbital interaction

$$F^{\nu_1'\nu_1}_{\nu'\nu} = \varphi^{\alpha_1\alpha_1'}_{\alpha'\alpha}\delta_{\sigma'\sigma}\delta_{\sigma_1'\sigma_1} + \Psi^{\alpha_1\alpha_1'}_{\alpha'\alpha}(\sigma_{\sigma'\sigma}\sigma_{\sigma_1\sigma_1'}). \tag{1.3}$$

In attempting to identify effects related to spin waves as done in study [47], in this chapter we will limit our examination to the simplest model of Fermi-liquid electron interaction (compare to study [13]), when

$$\varphi^{\alpha_1\alpha_1'}_{\alpha'\alpha} = 0, \quad \psi^{\alpha_1\alpha_1'}_{\alpha'\alpha} = \psi\sum_{\mathbf{q}} I_{\alpha'\alpha}(-\mathbf{q})\, I_{\alpha_1\alpha_1'}(\mathbf{q}).$$

For subsequent use we may then specify the set of orbital quantum numbers α. These will infer: p_z is projection of the electron quasi momentum in the direction of the constant magnetic field B, p_x is the projection of the quasi momentum onto the x axis (or $y_0 = -cp_x/eB$) is the projection onto the y axis of the coordinate origin of the Larmor electron orbit and n is the energy quantum number. Here the magnetic field on the z axis lies on the plane of the metallic surface occupying the half space $y > 0$. We note that identifying the surface states involves satisfying the boundary condition of vanishing of the y-coordinate-dependent part of the wave function of the electron $|\alpha\rangle = (2\pi\hbar)^{-1}\cdot\exp\{ip_x x/\hbar + ip_z z/\hbar\}\psi_n(y)$ for $y = 0$:

$$\psi_n(y) = 0. \tag{1.4}$$

Such a boundary condition corresponding to specular electron reflection off the metal boundary indicates that the surface electrons complete infinite motion on the x axis in the magnetic field. On the other hand motion on the y axis is finite and hence is quantized. Using the fact that for skipping electrons $|y| \ll |cp_x/eB| = |y_0|$, while $|p_y| \ll |p_x|$, and representing the Hamiltonian of such electrons as:

$$\hat{H}\left(p_x + \frac{eB}{c}y, p_y, p_z\right) \simeq \hat{H}(p_x, 0, p_z) + \frac{eB}{c}v_x y + \frac{1}{2}\frac{\partial v_y}{\partial p_y}p_y^2, \tag{1.5}$$

where the derivatives $v_x = \partial H(p_x, 0, p_z)$ and $\partial v_y/\partial p_y = \partial^2 H(p_x, p_y, p_z)/\partial^2 p_y$ are taken for $p_y = 0$, we find that the dependence of the wave function on the coordinate is described by the Airy function $\mathrm{Ai}(-\xi - a_n)'$, where

$$\xi = -y\,|2v_x eB/c\hbar^2\,(\partial v_y/\partial p_y)\,|^{1/3},$$

while a_n is the nth root of the Airy function coinciding with good accuracy with expression $[(3\pi/2)(n - 1/4)]^{2/3}$. Here condition (1.4) is satisfied. The eigenvalues of the Schrodinger equation corresponding to Hamiltonian (1.5),

$$E_n\,(p_z, p_x) = \hat{H}\,(p_x, 0, p_z) + |\,e\,|Bv_x y_n/c,$$

where

$$y_n = a_n\,|\,2v_x eB/c\hbar^2\,(\partial v_y/\partial p_y)\,|^{-1/3},$$

in the case of interest to us of a quadratic isotropic dependence of the energy on the momentum are equal to

$$E_n\,(p_z, p_x) = \frac{p_z^2}{2m} + \frac{p_x^2}{2m} + \left(\frac{3\pi}{2}\right)^{2/3}\left(\frac{p_x^2}{2m}\right)^{1/3}(\hbar\Omega)^{2/3}\left(n - \frac{1}{4}\right)^{2/3}. \qquad (1.6)$$

From here we see that the spectrum of surface levels (1.6) is not equidistant in the magnetic field unlike the familiar spectrum of the levels of the volumetric electron states in a magnetic field that do not collide with the metallic surface

$$E_n\,(p_z, p_x) = \frac{p_z^2}{2m} + \left(n - \frac{1}{2}\right)\hbar\,\frac{|\,e\,|B}{mc}.$$

In deriving the kinetic equation we will consider the following conditions to be satisfied

$$n' - n \ll n, \qquad |\,p_z' - p_z\,| \equiv \hbar\,|\,k_z\,| \ll |\,p_z\,|.$$

We will also be interested in perturbations for which $\hbar k_x = p_x' - p_x = 0$, corresponding to the equality $y_0' = y_0$. In this regard we may write the nonequilibrium addition to the density matrix as

$$\delta\rho_{\nu'\nu} = \delta\rho\,(n', n, p_z, p_x, k_z, \sigma', \sigma).$$

Henceforth it will be convenient to use the vector spin distribution function which is the convolution of the density matrix with the vector of the Pauli matrices:

$$\delta\sigma\,(n', n, p_z, p_x, k_z) = \sum_{\sigma'\sigma} \sigma_{\sigma'\sigma}\delta\rho\,(n', n, p_z, p_x, k_z, \sigma', \sigma).$$

The quantum kinetic equation for the spin distribution function $\delta\sigma^{\pm}(n'\ n,\ p_z,\ p_x,\ k_z)$ corresponding to the surface states of the electrons ($\delta\sigma^{\pm} = \delta\sigma^{+} \pm i\delta\sigma^{y}$) will be written based on equation (1.2) and accounting for (1.3) as

$$[\omega - k_z v_z - \omega_{n'n}(p_z) \pm \Omega_0 + i/\tau + i/T]\, \delta\sigma^{\pm}(n', n, p_z, p_x, k_z) +$$
$$+ (\partial f_0/\partial\varepsilon)\,[k_z v_z + \omega_{n'n}(p_z) \mp \Omega_0 - i/T] \sum_{q,\,0} I_{n'n}(-q, p_z) \times$$
$$\times \{-\mu_0 \delta B^{\pm}(0, q, k_z) + \psi \sum_{p_{1z} p_{1x} r' r} I_{r'r}(q, p_{1z})\, \delta\sigma^{\pm}(r', r, p_{1z}, p_{1x}, k_z)\} =$$
$$= (i/\tau)(\partial f_0/\partial\varepsilon)\,[\sum_{\mathbf{p}} \partial f_0/\partial\varepsilon]^{-1} \sum_{q>0} I_{n'n}(-q, p_z) \times$$
$$\times \sum_{p_{1z} p_{1x} r' r} I_{r'r}(q, p_{1z})\, \delta\sigma^{\pm}(r', r, p_{1z}, p_{1x}, k_z). \qquad (1.7)$$

Here $f_0(\varepsilon)$ is the Fermi distribution function; ε_F is the Fermi value of the electron energy; $\delta B^{\pm}(\mathbf{k})$ is the Fourier component of the non-equilibrium magnetic field, τ is the electron momentum relaxation time, T is the spin relaxation time. In deriving the collision integral we utilized expression (6.7) from study [1]. For the electron spin resonance $T \gg \tau \gg \omega^{-1}$. In quasi-classical theory it is precisely such time T that determines the linewidth of electron paramagnetic resonance. The effective magnetic moment of the electron γ determines the frequency $\Omega_0 = 2\gamma B/\hbar$. Summation over r' *and* r in (1.7) is carried out over both the surface and the volumetric states and we label the electron velocity in the direction of the constant magnetic field as v_z, while the transition frequency between the levels of the skipping electrons with indices n' and n takes the form

$$\omega_{n'n}(p_z) = [E_{n'}(p_z, p_x) - E_n(p_z, p_x)]/\hbar = (3\pi/2)^{2/3}\hbar^{-1/3}\Omega^{2/3}[\varepsilon_F - p_z^2/2m]^{1/3} \times$$
$$\times [(n' - 1/4)^{2/3} - (n - 1/4)^{2/3}], \quad I_{n'n}(q, p_z) = \langle n', p_z \mid \exp(-iqy) \mid n, p_z \rangle. \qquad (1.8)$$

Here the cyclotron frequency is equal to $\Omega = |e|B/mc$. In obtaining (1.7) and (1.8) we have accounted for the fact that in the case $\hbar\Omega \ll \varepsilon_F$ of energy-variable-dependent δ-functions in the right and left halves of (1.7) indicates that for momentum $\mathbf{p}$ lying on a spherical Fermi surface the following relation is satisfied:

$$p_x^2 + p_z^2 = 2m\varepsilon_F \equiv p_F^2. \qquad (1.9)$$

Formula (1.9) is the definition of p_F: the magnitude of the Fermi momentum of the electron in a metal with a spherical Fermi surface.

2. Eigenfrequencies of spin density oscillations in a metal with an isotropic electron dispersion law

In this section we will focus on the consequences of kinetic equation (1.7) for the spin distribution function when we ignore the influence of the alternating magnetic field. Then for the quantity

$$\delta S^{\pm}(0, k_y, k_z) = \sum_{p_z p_x n' n} I_{n'n}(k_y, p_z)\, \delta\sigma^{\pm}(n', n, p_z, p_x, k_z),$$

which is the Fourier-component of the spatial electron spin density, we may write the following equation:

$$\delta S^{\pm}(0, k_y, k_z) + \sum_{k'_y>0} Q(k_z, k_y, k'_y)\,\delta S^{\pm}(0, k'_y, k_z) = 0, \tag{1.10}$$

where

$$Q(k_z, k_y, k'_y) = \sum_{p_z p_x n' n} (\partial f_0/\partial\varepsilon)\, I_{n'n}(k_y, p_z)\, I_{n'n}(-k'_y, p_z)\,[\omega - k_z v_z - \tilde{\omega}_{n'n}(p_z) \mp$$
$$\pm \Omega_0 + i/\tau + i/T]^{-1}\{\psi[k_z v_z + \tilde{\omega}_{n'n}(p_z) \mp \Omega_0 - i/T] - i\,[\sum_{\mathbf{p}} \partial f_0/\partial\varepsilon]^{-1}\tau^{-1}\}.$$

Here $\bar{\omega}_{n'n}(p_z)$ is equal to $(n' - n)$ for the volumetric states and to expression (1.8) for the surface electrons. It follows from an examination of the electron gas model [55] that in metals with protrusions including spherical Fermi surfaces, surface waves cannot propagate in the limit of weak collisions. In order to demonstrate fundamentally new possibilities which result from accounting for electron interaction, we will focus on an examination of a metal with a spherical Fermi surface. For such a model we will obtain fundamental results that may be generalized to the case of an arbitrary Fermi surface.

We will represent the kernel Q in the form of three components:

$$Q = Q_{\text{об}} + Q_{\text{нр}} + Q_{\text{р}},$$

generated by the volumetric electrons, the nonresonance surface electrons and the resonance electrons, respectively, with a maximal value of the transition frequency $\omega_{n'n}$ close to the frequency ω. We are interested in the case of sufficiently weak fields when $\omega \gg \Omega$, Ω_0 and we may completely ignore the influence of the magnetic field on the volumetric electrons. Then

$$Q_{\text{об}}(k_z, k_y, k'_y) = \sum_{p_z p_x n' n} \frac{\partial f_0}{\partial\varepsilon}\,\frac{\pi^2\hbar^3 v}{p_F^2}\,B_0 \times$$
$$\times \frac{k_z v_z + (n' - n)\Omega + i\tau^{-1}B_0^{-1}}{\omega - (n' - n)\Omega - k_z v_z + i\tau^{-1} + iT^{-1}}\, J_{|n'-n|}\Big(k'_y\sqrt{2c\hbar\big(n + \tfrac{1}{2}\big)/|e|B}\Big) \times$$
$$\times J_{|n'-n|}\Big(k_y\sqrt{2c\hbar\big(n + \tfrac{1}{2}\big)/|e|B}\Big) \simeq \Big\{B_0 - \Big[\frac{B_0\omega}{2kv} + \frac{i}{\omega\tau}\Big]\ln\frac{\omega + kv}{\omega - kv}\Big\}\,\delta_{k_y k'_y}. \tag{1.11}$$

Here $J_m(x)$ is the Bessel function; $B_0 = \psi m^2 v/\pi^2\hbar^3$; v is the modulus of electron velocity on the spherical Fermi surface, $k = (k_y^2 + k_z^2)^{1/2}$. The contribution of the resonance electrons to the kernel Q is given by the expression

$$Q_{\text{р}}(k_z, k_y, k'_y) = \frac{\hbar}{4mp_F}\int dp_z dp_x \frac{\partial f_0}{\partial\varepsilon}\, I_{n'n}(-k'_y, p_z) \times$$
$$\times I_{n'n}(k_y, p_z)\,\frac{B_0\,[k_z v_z + \omega_{n'n}(p_z) \mp \Omega_0] + i/\tau}{\omega \pm \Omega_0 - \omega_{n'n}(p_z) - k_z v_z + i\tau^{-1} + iT^{-1}}. \tag{1.12}$$

The possibility of isolating the resonance component in the kernel Q is related to the fact that the solution of equation (1.10) in the form

$$|\omega - \omega_{n'n}| \ll \omega, \tag{1.13}$$

is of interest to us, where $\omega_{n'n} = \omega_{n'n}(0)$ is the maximal value of the frequency $\omega_{n'n}(p_z)$ which is achieved in the central cross-section of the Fermi surface. In view of the fact that $\Omega_0 \ll \omega_{n'n}$ both in this condition (1.13) and below we will not account for the dependence on Ω_0, and we will drop the signs $\pm$ on $\delta S^{\pm}$. Condition (1.13) determines the indices n and n' in formula (1.12).

After using relation (1.8) we may go to integration in terms of the angle θ in formula (1.12) by representing for the spherical Fermi surface $p_z = p_F \cos\theta$

$$Q_{\text{p}}(k_z, k_y, k_y') = -\frac{\hbar}{4p_F}\int_0^{\pi} d\theta I_{n'n}(-k_y', p_F\cos\theta)\, I_{n'n}(k_y, p_F\cos\theta) \times$$
$$\times \frac{B_0[k_z v\cos\theta + \omega_{n'n}\sin^{2/3}\theta] + i\tau^{-1}}{\omega - \omega_{n'n}\sin^{2/3}\theta - k_z v\cos\theta + i\tau^{-1} + iT^{-1}}. \tag{1.14}$$

By virtue of condition (1.13) the primary contribution to integral (1.14) comes from the range of angles near $\theta = \pi/2$ and hence assuming everywhere aside from the resonance denominator $\theta \approx \pi/2$ we have

$$Q_{\text{p}}(k_z, k_y, k_y') = -I_{n'n}(-k_y', p_F)\, I_{n'n}(k_y, p_F)(\hbar\sqrt{3}/2p_F) \times$$
$$\times [B_0 + i/\omega_{n'n}\tau](\Delta + i\gamma)^{-1/2}\operatorname{arctg}[\pi/(\Delta + i\gamma)^{1/2}\, 2\sqrt{3}]. \tag{1.15}$$

Here

$$\Delta = \frac{\omega - \omega_{n'n}}{\omega_{n'n}} - \frac{3}{4}\left(\frac{k_z v}{\omega_{n'n}}\right)^2, \qquad \gamma = \frac{1}{\omega_{n'n}\tau} + \frac{1}{\omega_{n'n}T}.$$

It follows from (1.15) that in the case of small k_z, when the argument of the arc tangent is large compared to unity, the resonance multiplier appears in formula (1.15) $|\Delta + i\gamma|^{-13} \gg 1$ when $|\Delta| \ll 1$. Below we will be interested in the case where the conditions $k_z v \ll \omega_{n'n}$ and $|\Delta| \ll 1$ are satisfied. We will consider the contribution of non-resonance electrons to the kernel of equation (1.10)

$$Q_{\text{HP}}(k_z, k_y, k_y') = -\frac{\hbar}{4mp_F}\sum_{r'r}{}' \int dp_z\, dp_x I_{r'r}(-k_y', p_z) \times$$
$$\times I_{r'r}(k_y, p_z)\,\delta(\varepsilon - \varepsilon_F)\frac{B_0[k_z v_z + \omega_{r'r}(p_z)] + i\tau^{-1}}{\omega - \omega_{r'r}(p_z) - k_z v_z + i\tau^{-1} + iT^{-1}} =$$
$$= -\frac{\hbar}{4p_F}\sum_{r'r}\int_0^{\pi} d\theta I_{r'r}(-k_y', p_F\cos\theta)\, I_{r'r}(k_y, \cos\theta) \times$$
$$\times \frac{B_0(\omega_{r'r}\sin^{2/3}\theta + k_z v\cos\theta) + i\tau^{-1}}{\omega - \omega_{r'r}\sin^{2/3}\theta - k_z v\cos\theta + i\tau^{-1} + iT^{-1}}. \tag{1.16}$$

In formula (1.16) the prime on the sum designates that $r' \neq n'$, $r \neq n$. The fact that in certain conditions the contribution from the sum of nonresonance components may be small compared to the contribution of the resonance component is most easily demonstrated in the limit $k_z = 0$ and $\omega\tau \gg 1$. Calculating in these conditions the imaginary and real parts of $Q_{Hp}(0, k_y, k_y')$ separately we have, accounting for condition (1.13), the following expression for the imaginary part:

$$\operatorname{Im} Q_{\text{Hp}}(0, k_y, k_y') = \sum_{r'r}{}' (3\pi\hbar/4p_F) B_0 \int d(\sin^{2/3}\theta) \operatorname{tg}\theta \times$$
$$\times I_{r'r}(-k_y', p_F\cos\theta) I_{r'r}(k_y, p_F\cos\theta)\,\delta(\omega/\omega_{r'r} - \sin^{2/3}\theta) =$$
$$= (3\pi\hbar/4p_F) \sum_{r'r}{}' [\omega_{n'n}^3/(\omega_{r'r}^3 - \omega_{n'n}^3)]^{1/2} B_0 \times$$
$$\times I_{r'r}(-k_y', p_F[1 - (\omega_{n'n}/\omega_{r'r})^3]^{1/2}) I_{r'r}(k_y, p_F[1 - (\omega_{n'n}/\omega_{r'r})^3]^{1/2}). \tag{1.17}$$

The appearance of the imaginary part of the kernel Q in equation (1.10) corresponds to the collisionless Landau damping effect which always arises in a metal with a spherical Fermi surface due to the fact that the dependence of the momentum projection $p_z = p_F\cos\theta$ on the angle θ makes possible satisfaction of the equality

$$\omega_{n'n} = \omega_{r'r}(p_F\cos\theta)$$

for sine angle values θ equal to

$$\sin\theta = (\omega_{n'n}/\omega_{r'r})^{3/2},$$

if $\omega_{n'n} < \omega_{r'r}$. Such a situation is fundamentally different from the case of a metal with a cylindrical Fermi surface for which the transition frequency $\omega_{n'n}$ is independent of the momentum p_z and as a result the equality $\omega_{r'r} = \omega_{n'n}$ is not satisfied for various $r' \neq n'$ and $r = n$. This means that summation in (1.17) will be carried out over such r' and r for which the condition

$$(r' - 1/4)^{2/3} - (r - 1/4)^{2/3} > (n' - 1/4)^{2/3} - (n - 1/4)^{2/3}.$$

is satisfied. The real part of $Q_{Hp}(0, k_y, k_y')$ is given by the following expression:

$$\operatorname{Re} Q_{\text{Hp}}(0, k_y, k_y') = -\sum_{r'r}{}' \frac{\pi\hbar\sqrt{3}}{4p_F} B_0 I_{r'r}(-k_y', p_F) I_{r'r}(k_y, p_F)\left(\frac{\omega_{r'r}}{\omega_{r'r} - \omega_{n'n}}\right)^{1/2}, \tag{1.18}$$

where summation is carried out over

$$(r' - 1/4)^{2/3} - (r - 1/4)^{2/3} < (n' - 1/4)^{2/3} - (n - 1/4)^{2/3}.$$

Accounting for (1.11), (1.15), (1.17), and (1.18) we will represent integral equation (1.10) as

$$\delta S(0, k_y, 0)\left[1 + B_0 - B_0\frac{\omega}{k_y v}\ln\frac{\omega + kv}{\omega - kv}\right] -$$
$$- I_{n'n}(k_y, p_F)\int_0^\infty dk_y' I_{n'n}(-k_y', p_F)\,\delta S(0, k_y', 0)\, B_0\frac{\hbar\sqrt{3}}{4p_F}\left(\frac{\omega_{n'n}}{\omega - \omega_{n'n}}\right)^{1/2} =$$

$$= \sum_{\omega_{r'r}<\omega_{n'n}} \frac{\hbar\sqrt{3}B_0}{4p_F} I_{r'r}(k_y, p_F) \int_0^\infty dk'_y I_{r'r}(-k'_y, p_F) \times$$
$$\times \delta S(0, k'_y, 0)\left[\frac{\omega_{r'r}}{\omega_{n'n}-\omega_{r'r}}\right]^{1/2} - i\frac{3\hbar}{4p_F} \sum_{\omega_{r'r}>\omega_{n'n}} I_{r'r}\left(k_y, p_F\left[1-\left(\frac{\omega_{n'n}}{\omega_{r'r}}\right)^3\right]^{1/2}\right)\times$$
$$\times B_0 \int_0^\infty dk'_y \delta S(0, k'_y, 0) I_{r'r}\left(-k'_y, p_F\left[1-\left(\frac{\omega_{n'n}}{\omega_{r'r}}\right)^3\right]^{1/2}\right)\left[\frac{\omega^3_{n'n}}{\omega^3_{r'r}-\omega^3_{n'n}}\right]^{1/2}. \tag{1.19}$$

In equation (1.19) integration with respect to k'_y is carried out within the limits from 0 to δ, which corresponds to the even continuation of the nonequilibrium spin density to the region $y < 0$. As demonstrated in Appendix 1, the use of the quasi-classical asymptotics of the wave functions of the surface states (the Airy functions) makes it possible to represent the matrix element $I_{n'n}(k_y, p_F)$ in the following form

$$I_{n'n}(k_y, p_F) = \int_{-1}^{1} dx\left[1+\frac{2(n'-n)}{3(n-1/4)}x^2\right]^{-1/4} \cos[\pi(n'-n)x]\cos[k_y y_n(1-x^2)], \tag{1.20}$$

where the depth of the segment of the nth orbit for a metal with a spherical Fermi surface is equal to:

$$y_n = \frac{(3\pi)^{2/3}}{2}\left(n-\frac{1}{4}\right)^{2/3}\left(\frac{\hbar}{p_x}\right)^{2/3}\left(\frac{v}{\Omega}\right)^{1/3}.$$

From here for characteristic values of $k_y \sim y_n^{-1}$ consistent with (1.20) subject to (1.13) we find that the ratio

$$k_y v/\omega = v/\omega_{n'n}y_n \sim (n-1/4)^{2/3}[(n'-1/4)^{2/3}-(n-1/4)^{2/3}](\hbar\Omega/\varepsilon_F)^{1/3}$$

is small compared to unity if the quantum numbers n and n' are not large (within the first ten). Below we will be interested in specifically these values of n' and n ignoring in the left half of equation (1.19) the last term in brackets. Multiplying both halves of equation (1.19) by $I_{n'n}(-k_y, p_F)$ and integrating with respect to k_y, we obtain

$$\left[1+B_0-N_{n'n}B_0\frac{\hbar\sqrt{3}}{4p_F}\frac{1}{\sqrt{\Delta}}\right]\int_0^\infty dk'_y I_{n'n}(-k'_y, p_F)\,\delta S(0, k'_y, 0) =$$
$$= \sum_{\omega_{r'r}<\omega_{n'n}} \frac{B_0\pi\hbar\sqrt{3}}{4p_F} N^{r'r}_{n'n}\left[\frac{\omega_{r'r}}{\omega_{n'n}-\omega_{r'r}}\right]^{1/2}\times$$
$$\times \int_0^\infty dk'_y I_{r'r}(-k'_y, p_F)\,\delta S(0, k'_y, 0) - i\frac{3\hbar}{4p_F}\sum_{\omega_{r'r}>\omega_{n'n}} B_0 M^{r'r}_{n'n}\times$$
$$\times\left[\frac{\omega^3_{n'n}}{\omega^3_{r'r}-\omega^3_{n'n}}\right]^{1/2}\int_0^\infty dk'_y I_{r'r}\left(-k'_y, p_F\left[1-\left(\frac{\omega_{n'n}}{\omega_{r'r}}\right)^3\right]^{1/2}\right)\delta S(0, k'_y, 0). \tag{1.21}$$

Here we use the conventions:

$$N_{n'n}=\int_0^\infty dk_y I_{n'n}(-k_y, p_F)\, I_{n'n}(k_y, p_F),$$

$$N_{n'n}^{r'r}=\int_0^\infty dk_y I_{n'n}(-k_y, p_F)\, I_{r'r}(k_y, p_F),$$

$$M_{n'n}^{r'r}=\int_0^\infty dk_y I_{n'n}(-k_y, p_F)\, I_{r'r}\left(k_y, p_F\left[1-\left(\frac{\omega_{n'n}}{\omega_{r'r}}\right)^3\right]^{1/2}\right).$$

The second component in the right half of (1.21) describes the collisionless Landau damping effect in a metal with a spherical Fermi surface. Using the explicit form of matrix element (1.20) $I_{n'n}(k, p_F)$, we will write the following expressions (see Appendix 1) for the integrals of the products of the matrix elements:

$$N_{n'n}=\frac{\pi}{y_n}\int_0^1 dx \cos^2[\pi(n'-n)x]\left[x^2+\frac{2(n'-n)}{3(n-1/4)}\right]^{-1/2}=\frac{\pi}{2y_n}L_{n,\, n'-n}, \tag{1.22}$$

where

$$L_{n, n'-n}=\ln[6(n-{}^1/_4)/(n'-n)],\quad n'-n\leqslant n,$$

$$N_{n'n}^{r'r}=[\pi/2y_n]\, L_{n,\, n'-n}^{r,\, r'-r},\quad L_{n,\, n'-n}^{r,\, r'-r}=\sqrt{y_n/y_r}\ln\{4/[y_n/y_r-1]\},$$

$$M_{n'n}^{r'r}=[\pi/2y_n][(n-{}^1/_4)/(r-{}^1/_4)]\ln[4/|(n-{}^1/_4)/(r-{}^1/_4)-1|].$$

Introducing the conventions $s = n' - n$ and $m = r' - r$ and ignoring the dependence of $M_{n'n}^{r'r}$ on the numbers r', r and n', n, we may estimate the magnitude of collisionless damping in formula (1.21)

$$\sum_{r,\, m}\left[\left(\frac{m}{r}\right)^3\frac{n-1/4}{r-1/4}-1\right]^{-1/2}\simeq\sum_{r=0}^{n-1/4}\sum_{x=1}^{\infty}[x^3-1]^{-1/2}\simeq$$

$$\simeq\left(n-\frac{1}{4}\right)3^{-1/4}F(\pi, \sin 15°)\simeq 2{,}4\left(n-\frac{1}{4}\right).$$

Here F is a first order elliptical integral. With small values of the numbers n the smallness of this contribution in (1.21) compared to the contribution of the resonance component $[\omega/\omega\pi_{n'n} - 1]^{-1/2} \gg 1$ makes it possible to ignore collisionless damping. In such a case of small values of n' and n the contribution of the first sum of the right half of (1.21) is also small compared to the contribution of the resonance component. We may obtain this derivation by estimating the first sum ignoring the dependence on r' and r:

$$\sum_{\omega_{r'r}<\omega_{n'n}}\left[\frac{\omega_{n'n}}{\omega_{r'r}}-1\right]^{-1}=\sum_{rm}\left[\frac{s(r-1/4)^{1/3}}{m(n-1/4)^{1/3}}-1\right]^{-1/2}\simeq s\sum_{x=0}^{1}\frac{\sqrt{x}}{[1-x]^{1/2}}\simeq 2s.$$

Accounting for the arguments above and ignoring collisionless damping and using equation (1.21) we may write the dispersion relation for the surface spin oscillations near the transition frequency with numbers n' and n less than 10 in the form

$$1+B_0=B_0 N_{n'n}\hbar\sqrt{3}/4p_F\sqrt{\Delta}. \tag{1.23}$$

From here subject to (1.22) we obtain the critical value of the frequency of surface oscillations when $k_z = 0$ and $(\omega\tau)^{-1} = 0$:

$$\omega = \omega_{n'n}\left[1 + \frac{\pi^{2/3}}{3^{7/3}2^{14/3}}\left(\frac{\hbar\Omega}{\varepsilon_F}\right)^{2/3}\frac{L^2_{n,\,n'-n}}{(n-1/4)^{4/3}}\left(\frac{B_0}{1+B_0}\right)^2\right]. \tag{1.24}$$

The deviation of the frequency of surface spin oscillations from the values $\omega_{n'n}$ is due to the Fermi-liquid electron interaction. It follows from (1.23) that in the limit $k_z = 0$ and $(\omega\tau)^{-1} = 0$ there exists a solution of the dispersion equation when the following condition is satisfied:

$$B_0 N_{n'n}\hbar\sqrt{3}/(1+B_0)4p_F > 0. \tag{1.25}$$

In reality for normal metals with a spherical Fermi surface condition (1.25) represents the requirement that $B > 0$.

3. The influence of dissipative effects, the finite wavelength, and form of the Fermi surface on the spectrum and collisionless damping of surface spin waves

In this section we will consider the consequences deriving from integral equation (1.10) in which we drop the sum of the nonresonance components which, as demonstrated above, may be done for values of n' and n that do not exceed 10 or 20,

$$\begin{aligned}(1+B_0)\,\delta S(0, k_y, k_z) = \pi^{-1}\int_0^\infty dk'_y\,\delta S(0, k'_y, k_z)\, I_{n'n}(-k'_y, p_F)\times\\ \times I_{n'n}(k_y, p_F)(\hbar\sqrt{3}/2p_F)[B_0 + i/\omega_{n'n}\tau]\times\\ \times(\Delta + i\gamma)^{-1/2}\operatorname{arctg}[\pi(\Delta+i\gamma)^{-1/2}/2\sqrt{3}].\end{aligned} \tag{1.26}$$

The conventions in (1.26) coincide with those used in writing formula (1.15). Taking as before $|\Delta| \ll 1$ and $\omega_{n'n}\tau \gg 1$ we obtain the following dispersion equation as the solvability condition of (1.26):

$$1 + B_0 = [B_0 + i/\omega_{n'n}\tau]N_{n'n}\hbar\sqrt{3}/4p_F\,(\Delta + i\gamma)^{1/2}. \tag{1.27}$$

Introducing the conventions $\mathrm{Re}\,\Delta = \tilde{\Delta}$ and $\mathrm{Im}\,\Delta + \gamma = \tilde{\gamma}$ we obtain the following system of equations by means of (1.27):

$$\begin{aligned}&[\tilde{\Delta} + \sqrt{\tilde{\Delta}^2 + \tilde{\gamma}^2}]^{1/2} = \sqrt{6}B_0 N_{n'n}\hbar/(1+B_0)4p_F,\\ &[-\tilde{\Delta} + \sqrt{\tilde{\Delta}^2 + \tilde{\gamma}^2}]^{1/2} = \sqrt{6}N_{n'n}\hbar/\omega_{n'n}\tau(1+B_0)4p_F.\end{aligned} \tag{1.28}$$

From here we will see, specifically, that in accounting for $(\omega\tau)^{-1} \neq 0$ the solution of the system of dispersion equations (1.28) written for a metal with a spherical Fermi surface exists when condition (1.25) noted above is satisfied. Solving equation system (1.28) and substituting the explicit value of $N_{n'n}$, we find:

$$\mathrm{Re}\,\Delta=\left[B_0^2-\frac{1}{(\omega_{n'n}\tau)^2}\right]\frac{\pi^{2/3}}{3^{1/3}2^{14/3}}\left(\frac{\hbar\Omega}{\varepsilon_F}\right)^{2/3}\frac{L^2_{n,\,n'-n}}{(n-1/4)^{4/3}}\frac{1}{(1+B_0)^2}, \tag{1.29}$$

$$\mathrm{Im}\,\Delta=-\frac{1}{\omega_{n'n}\tau}-\frac{1}{\omega_{n'n}T}+\frac{2B_0}{\omega_{n'n}\tau}\frac{1}{(1+B_0)^2}\frac{\pi^{2/3}}{3^{1/3}2^{14/3}}\left(\frac{\hbar\Omega}{\varepsilon_F}\right)^{2/3}\frac{L^2_{n,\,n'-n}}{(n-1/4)^{4/3}}. \tag{1.30}$$

It follows from formula (1.29) that in sufficiently pure samples such that $\omega\tau > B_0^{-1}$, the eigenfrequencies of the spin waves lie above the surface transition frequency.

Now the frequency spectrum and the damping increment of the surface spin oscillations may be represented as

$$\omega=\omega_{n'n}\left\{1+\frac{\pi^{2/3}}{3^{1/3}2^{14/3}}\left(\frac{\hbar\Omega}{\varepsilon_F}\right)^{2/3}\frac{L^2_{n,\,n'-n}}{(n-1/4)^{4/3}(1+B_0)^2}\cdot\left[B_0^2-\frac{1}{(\omega_{n'n}\tau)^2}\right]-\frac{i}{\omega_{n'n}T}-\right.$$
$$\left.-\frac{i}{\omega_{n'n}\tau}\left[1-\frac{2B_0}{(1+B_0)^2}\frac{\pi^{2/3}}{3^{1/3}2^{14/3}}\left(\frac{\hbar\Omega}{\varepsilon_F}\right)^{2/3}\frac{L^2_{n,\,n'-n}}{(n-1/4)^{4/3}}\right]+\frac{3}{4}\left(\frac{k_zv}{\omega_{n'n}}\right)^2\right\}. \tag{1.31}$$

It follows from here that the linewidth of the surface oscillations is determined by the momentum relaxation time unlike the electron spin resonance line whose linewidth is determined by the spin relaxation time [1]. Taking $\omega\tau \gg B_0^{-1}$ and ignoring the small corrections, we may write expression (1.31) by expressing the quantities through the ratio $\hbar\omega/\varepsilon_F$ subject to (1.8) and (1.13) in the form

$$\omega=\omega_{n'n}\left\{1+\frac{1}{2^5}\frac{L^2_{n,\,n'-n}}{(n'-n)(n-1/4)}\frac{\hbar\omega}{\varepsilon_F}\left(\frac{B_0}{1+B_0}\right)^2+\frac{3}{4}\left(\frac{k_zv}{\omega_{n'n}}\right)^2\right\}-\frac{i}{\tau}. \tag{1.32}$$

Expression (1.32) for the case $n'-n \ll n$ applies critical formula (1.24) to the case of finite wavelength and finite values of $\omega\tau$. As in the preceding section we see that the spectrum of spin waves in a metal with a spherical Fermi surface lies above the transition frequency between the electron surface levels.

In order to better understand the dependence of the spectrum of surface oscillations on the form of the Fermi surface, we will examine equation (1.10) for a metal with a Fermi surface in the form of an ellipsoid of revolution. We will ignore the contribution of the nonresonance components, and will also take $k_0 = 0$. In order to solve equation (1.10) we will first find the value of the transition frequency between these surface levels. We will place the origin of the coordinate system in momentum space in the center of the ellipsoid so that the axis of revolution makes an angle of ξ with the p_z axis. Using expansion (1.5) in the case of an inverted ellipsoid we find the following expression for the surface transition frequency dependent on the angle ξ:

$$\omega_{n'n}(p_z,\xi)=(n'-n)\left\{\frac{2\pi^2e^2B^2}{3\hbar c^2m_1^2m_3(n-1/4)}\left[\varepsilon_F(m_3\cos^2\xi+m_1\sin^2\xi)-\frac{p_z^2}{2}\right]\right\}^{1/3}. \tag{1.33}$$

here m_1 is the transverse effective mass of the electron on an ellipsoidal Fermi surface, while m_3 is the longitudinal mass. If $m_1 = m_3$, expression (1.33) becomes (1.8) ($n' - n \ll n$) as would be expected. It follows from (1.33) that the surface transition frequency as a function of p_z reaches a maximum at the central cross-section of the Fermi surface when $p_z = 0$. In the case $m_3 > m_1$ (an "extended" ellipsoid) the maximorum transition frequency is achieved when $\xi = 0$ and is equal to

$$\omega_{n'n}^{\text{maximorum}} = (n' - n)[2\pi^2 e^2 B^2 \varepsilon_F / 3\hbar c^2 m_1^2 (n - 1/4)]^{1/3}. \tag{1.34}$$

If the opposite case $m_1 > m_3$ is realized (a "flattened" ellipsoid of revolution), the maximoral transition frequency corresponds to the angle $\xi = \pi/2$:

$$\omega_{n'n}^{\text{maximorum}} = (n' - n)[2\pi^2 e^2 B^2 \varepsilon_F / 3\hbar c^2 m_1 m_3 (n - 1/4)]^{1/3}. \tag{1.35}$$

A comparison of formulae (1.34) and (1.35) shows that the highest value of the surface transition frequency in the case of a metal with an ellipsoidal Fermi surface is always achieved in the central cross-section, and in both cases such a cross-section is minimal as a function of the angle ξ.

Here we will provide the dispersion equation of surface oscillations deriving from (1.10) for a metal with an arbitrary Fermi surface whose surface transition frequency is extremal on the central cross-section. Specifically, such an equation is valid for the case of a Fermi-surface in the form of an ellipsoid of revolution inclined with respect to the magnetic field direction as examined above. The dispersion equation takes the form

$$1 = \frac{B_0}{1 + B_0} N_{n'n} \left[\int \frac{d\sigma}{v} \right]^{-1} \int \frac{dp_z}{v_x(p_z, \xi)} \frac{\omega_{n'n}(p_z, \xi)}{\omega' - \omega_{n'n}(p_z, \xi)}, \tag{1.36}$$

where $B_0 = 2(2\pi\hbar)^{-3} \int d\sigma v^{-1}$; v is the modulus of electron velocity on the Fermi surface; $d\sigma$ is the element of this surface; $\omega' = \omega + i/\tau$; $v_x = \partial\varepsilon/\partial p_x$. Expanding $\omega_{n'n}(p_z, \xi)$ near the extremum on the central cross-section $\omega_{n'n}(p_z, \xi) = \omega_{n'n}(0, \xi) + \omega''_{n'n}(0, \xi) p_z^2/2$, where $\omega''_{n'n}(0, \xi) = d^2\omega_{n'n}(p_z, \xi)/dp_z^2|_{p_z=0}$, we represent (1.36) as

$$1 = \frac{B_0}{1 + B_0} N_{n'n} \left[\int \frac{d\sigma}{\hbar v} \right]^{-1} \frac{1}{v_x(0, \xi)} \frac{2^{3/2} \omega_{n'n}(0, \xi)}{\sqrt{\omega''_{n'n}(0, \xi)[\omega_{n'n}(0, \xi) - \omega']}} \times$$
$$\times \operatorname{arctg} \left[\frac{p_z^*(\xi) \sqrt{\omega''_{n'n}(0, \xi)}}{\sqrt{2[\omega_{n'n}(0, \xi) - \omega']}} \right]. \tag{1.37}$$

Here $p_z^*(\xi)$ is the coordinate of the point of intersection of the p_z axis with the Fermi surface. In the particular case corresponding to the inverted ellipsoid of revolution when $\omega_{n'n}(p_z, \xi)$ is given by formula (1.33) and the quantities entering into (1.37) are equal to

$$\begin{aligned} \omega''_{n'n}(0, \xi) &= -\omega_{n'n}(0, \xi)/3\varepsilon_F [m_3 \cos^2\xi + m_1 \sin^2\xi], \\ v_x(0, \xi) &= (2\varepsilon_F)^{1/2} (m_1^{-1} \cos^2\xi + m_3^{-1} \sin^2\xi)^{1/2}, \end{aligned} \tag{1.38}$$

$$p_z^*(\xi) = (2\varepsilon_F)^{1/2}(m_1^{-1}\sin^2\xi + m_3^{-1}\cos^2\xi)^{-1/2}, \tag{1.39}$$

we write the dispersion equation of surface oscillations of spin density as

$$1 = \frac{B_0}{1+B_0} N_{n'n} \frac{\sqrt{3}\hbar}{4\sqrt{2\varepsilon_F m_1}} \sqrt{\frac{\omega_{n'n}(0,\xi)}{\omega' - \omega_{n'n}(0,\xi)}} \frac{m_3\cos^2\xi + m_1\sin^2\xi}{m_1},$$

coinciding with equation (1.33) for a spherical Fermi surface in the case where $m_1 = m_3$. The following conditions must be satisfied for the existence of real solutions of equation (1.37) and for the absence of collisionless damping in the case $\omega\tau \gg 1$ when condition (1.13) is satisfied for several extended Fermi surfaces when the argument is large compared to unity:

$$\omega''_{n'n}(0,\xi)[\omega_{n'n}(0,\xi) - \omega] > 0, \tag{1.40}$$

$$\left[\int \frac{d\sigma}{v}\right]^{-1} N_{n'n} \frac{1}{v_x(0,\xi)} \frac{B_0}{1+B_0} > 0. \tag{1.41}$$

In this case the real part of the frequency spectrum of the surface spin oscillations is given by the expression in which the frequency shift from $\omega_{n'n}(0, \xi)$ has a quadratic dependence on $B_0/(1 + B_0)$,

$$\mathrm{Re}\,\omega = \omega_{n'n}(0,\xi)\left\{1 - \left(\frac{B_0}{1+B_0}\right)^2 N_{n'n}^2\pi^2\left(\int\frac{d\sigma}{\hbar v}\right)^{-2} v_x^{-2}(0,\xi)\frac{2\omega_{n'n}(0,\xi)}{\omega''_{n'n}(0,\xi)}\right\}, \tag{1.42}$$

while the imaginary part is

$$\mathrm{Im}\,\omega = 1/\tau.$$

We see from condition (1.40) that for metals with convex Fermi surfaces when $\omega''_{n'n}(0, \xi) < 0$ the spectrum of (1.42) lies above the surface transition frequency $\omega_{n'n}(0, \xi)$, while for metals with concave Fermi surfaces $\omega''_{n'n}(0, \xi)$ $\mathrm{Re}\,\omega < \omega_{n'n}(0, \xi)$. Condition (1.41) determines the sign of the constant $B_0/(1 + B_0)$ in which an undamped wave with frequency spectrum (1.42) may propagate. Specifically, in the model of a metal with an ellipsoidal Fermi surface discovery of spin waves with spectrum (1.42) in condition $\omega\tau \gg B_0^{-1}$ would indicate that $B_0/(1 + B_0) > 0$ for such a metal. For a metal with a concave Fermi surface in the case when $[v_x(0, \xi) \int d\sigma/v] < 0$ detection of spin waves with spectrum (4.24) would, on the other hand, indicate that $B_0/(1 + B_0) < 0$ for the case of weak collisions.

In the other limiting case of near-cylindrical Fermi surfaces when the transition frequency is independent of p_z, based on equation (1.37) we may write the spectrum of surface oscillations as

$$\omega = \omega_{n'n}(0,\xi)\left\{1 + \frac{2B_0}{1+B_0} N_{n'n}\left[\int\frac{d\sigma}{\hbar v}\right]^{-1}\frac{p_z^*(0,\xi)}{v_x(0,\xi)}\right\} + \frac{i}{\tau}, \tag{1.43}$$

where we will take $m_3 \gg m_1$ in determining the inclusive quantities based on formulae (1.38) and (1.39). The situation here is different from the case of (1.42), specifically: depending on the sign of the constant $B_0/(1 + B_0)$ spectrum (1.43) may lie both above and below the surface transition frequency. Formulae (1.42) and (1.43) solve the problem of the critical spectrum of surface spin oscillations in a model of a metal with an arbitrary and, specifically, an ellipsoidal Fermi surface inclined at an angle ξ to the magnetic field.

We note that to date spin waves with spectrum (1.32) have not been observed in experiment in alkali metals. The observation difficulties are related, first, to the smallness of the ratio $\hbar\omega/\varepsilon_F$ for the range of frequencies used by the experimenters and, second, due to the fact that the electron paramagnetism in alkali metals is comparatively weak, $B_0/(1 + B_0)$ is much less than unity.

Chapter 2

QUASI-CLASSICAL THEORY OF CYCLOTRON RESONANCE AT THE SKIPPING ORBITS IN THE ELECTRON LIQUID OF METALS AND SEMIMETALS WITH AN ARBITRARY FERMI SURFACE

1. Derivation of the integral equation for the skipping electron distribution function. Boundary conditions

In the first chapter we investigated the possibility of the propagation of surface quantum spin waves in the electron liquid of metals. The eigenfrequencies of such spin waves are close to the observed resonance absorption frequencies of the electrons skipping along the surface of a metal in a magnetic field [20, 21]. At the same time due to the comparative weakness of electron paramagnetism the observed impedance resonances in the weak magnetic field are not related to the spin waves, but rather to the cyclotron waves. The theory of surface cyclotron waves and resonances in an electron gas was developed in studies [21, 55, 56] in which, specifically, it was demonstrated that cylindrical sections of the Fermi surface must exist (see also [34]) for the existence of such waves near the transition frequencies of the skipping electrons. This statement, as demonstrated in study [48] is related to the use of a noninteracting particle model for the conduction electrons. In addition we note that according to study [47] the surface quantum spin waves may exist in metals whose Fermi-surface has no cylindrical sections when accounting for interelectron interaction.

Our study [48] developed the theory of cyclotron resonance at the skipping orbits of surface electrons in the electron liquid of metals with an arbitrary Fermi surface. We note that in the longwave limit the frequency of the surface waves investigated in studies [21, 55, 56] in an electron gas are less than the transition frequency between the surface levels. Unlike this case the critical frequency of the surface waves predicted by our theory [40-50] in metals with convex Fermi surfaces lies above the surface transition frequency. There are no such waves in the gas model. The new possibility for the propagation of surface waves near the transition frequency maxima is due to the interaction between conduction electrons. A comparison of the formulae from the theory developed in study [48] to experimental data on impedance oscillations in the metal makes it possible to establish the magnitude of interelectron interaction.

The effects discussed below are significantly quantum effects. However in striving for as simple a presentation as possible, we will not formulate the theory based directly on a quantum equation for the density matrix in this chapter (compare to study [1]), but rather will follow the quasi-classical approach used in the theory of cyclotron skipping electron resonance [55]. In this case using the quasi-classical equation it is easy to, first, incorporate parameters characterizing Fermi-liquid interaction of conduction electrons and, second, to incorporate the boundary conditions for electrons on the metallic surface in a comprehensive manner. The subsequent quantization of the classical solution, largely following the procedure in study [55] makes it possible to directly trace the new effects manifest in the cyclotron resonance at skipping orbits by accounting for interelectron interaction.

In electron liquid theory [1] ignoring the paramagnetic effects the weakly-excited states are characterized by a deviation δf in the distribution function from equilibrium. Assuming a time dependence of $\sim\exp(-i\omega t)$ and setting the constant magnetic field on the z axis and assuming the metal is in the half-space $y > 0$, we may write the following kinetic equation [48] (compare to [57]) for the case of a surface wave propagating in the direction of the constant magnetic field:

$$[i(\omega - k_z v_z) - v_y \partial/\partial y - 1/\tau + \Omega(p_z)\,\partial/\partial\varphi] \times$$
$$\times [\delta f(y, \varepsilon, p_z, \varphi) - (\partial f_0/\partial\varepsilon)\,\delta\varepsilon(y, \varepsilon, p_z, \varphi)] =$$
$$= (\partial f_0(\varepsilon)/\partial\varepsilon)\,[e\mathbf{v}\mathbf{E}(y) - i\omega\delta\varepsilon(y, \varepsilon, p_z, \varphi)]. \tag{2.1}$$

Here e is electron charge; $\Omega(p_z)$ is the cyclotron frequency of gyroscopic electron rotation; τ is the momentum relaxation time; φ is the angular variable characterizing the position of the electron in its orbit in the momentum space; ε is the electron energy; p_z is the projection of electron momentum in the direction of the magnetic field, while

$$\delta\varepsilon(y, \varepsilon, p_z, \varphi) = \int_0^\infty d\varepsilon' \int dp_z' d\varphi' \Phi(\varepsilon, p_z, \varphi; \varepsilon', p_z', \varphi')\,\delta f(y, \varepsilon', p_z', \varphi'). \tag{2.2}$$

Here $\Phi(\varepsilon, p_z, \varphi; \varepsilon', p_z', \varphi')$ is the function characterizing electron interaction. After designating $\delta\bar{f} = \delta f - \delta\varepsilon\,(\partial f_0/\partial\varepsilon)$, we may search the solution of equation (2.1) as $\delta f(y, p_z, \varphi) = g_0(y, p_z, \varphi)\,\partial f_0/\partial\varepsilon$, $\delta\bar{f}(y, p_z, \varphi) = g(y, p_z, \varphi) \times \partial f_0/\partial\varepsilon$, where $\partial f_0/\partial\varepsilon = -2(2\pi\hbar)^{-3}\delta(\varepsilon - \varepsilon_F)$, ε_F is the Fermi energy. The function Φ determines the resolvent operator of the equation

$$g(y, p_z, \varphi) = g_0(y, p_z, \varphi) + \frac{2}{(2\pi\hbar)^3}\int dp_z' d\varphi' \Phi(\varepsilon_F, p_z, \varphi; \varepsilon_F, p_z', \varphi') g_0(y, p_z', \varphi'), \tag{2.3}$$

and using this equation we have

$$g_0(y, p_z, \varphi) = \int dp_z' d\varphi' R(p_z, \varphi; p_z', \varphi') g(y, p_z', \varphi'). \tag{2.4}$$

Consistent with formula (2.4) we obtain from (2.2) the expression $\delta\varepsilon(y, \varepsilon_F, p_z, \varphi) = -\int dp_z' d\varphi' \alpha(p_z, \varphi; p_z', \varphi') g(y, p_z', \varphi')$, where we use the convention

$$\alpha(p_z, \varphi; p_z', \varphi') = \frac{2}{(2\pi\hbar)^3}\int dp_z'' d\varphi'' \Phi(\varepsilon_F, p_z, \varphi; \varepsilon_F, p_z'', \varphi'') R(p_z'', \varphi''; p_z', \varphi'). \tag{2.5}$$

We will not provide the argument ε equal to ε_F. The function α characterizes the interelectron interaction effects.

We will now address the problem of searching solutions of equation (2.1) describing the resonance electrons skipping along the surface of the metal and colliding with the surface at small angles. This means that in the characteristic equation

$$\Omega(p_z) dy = -d\varphi v_y(p_z, \varphi) \tag{2.6}$$

v_y will be considered small. Then recording the angle φ from its zero value in which $v_y(p_z, 0) = 0$, we may write $v_y(p_z, 0) = v_y'(p_z)\varphi$ for the skipping electrons. Consequently consistent with characteristic equation (2.6) we obtain for the orbits of the skipping electrons $y = R(p_z)[\varphi_0^2 - i^2]/2$, where φ_0 is the angle at which the electron impacts the surface, while $R(p_z) = v_y'(p_z)/\Omega(p_z)$ is the electron orbital radius.

In searching solutions of equation (2.1) we will use the specular reflection condition [58] of the skipping electrons $g(0, p_z, \varphi) = g(0, p_z, -\varphi)$. We then have from equation (2.1):

$$g(y, p_z, \varphi) = [2i\Omega(p_z)]^{-1}\Big\{\mathrm{ctg}[\beta(p_z)\varphi_0(|y|, p_z, \varphi)]\int_{-\varphi_0(|y|, p_z, \varphi)}^{\varphi_0(|y|, p_z, \varphi)} d\varphi' -$$
$$- i\Big[\int_{\varphi}^{\varphi_0(|y|, p_z, \varphi)} d\varphi' - \int_{-\varphi_0(|y|, p_z, \varphi)}^{\varphi} d\varphi'\Big]\Big\} \times$$
$$\times \exp[-i\beta(p_z)(\varphi - \varphi')]\Big\{e\mathbf{v}(p_z, 0)\mathbf{E}\Big(\frac{1}{2}R(p_z)[\varphi_0^2(|y|, p_z, \varphi) - \varphi'^2]\Big) -$$
$$- i\omega\delta\varepsilon\Big(\frac{1}{2}R(p_z)[\varphi_0^2(|y|, p_z, \varphi) - \varphi'^2], p_z, 0\Big)\Big\}. \tag{2.7}$$

Here we account for the smallness of φ compared to unity and use the conventions: $\varphi_0(y, p_z, \varphi) = [\varphi^2 + 2y/R(p_z)]^{1/2}$, $\beta(p_z) = [\omega - k_z v_z(p_z, 0) + i/\tau]/\Omega(p_z)$. Formula (2.7) is written in a form that allows us to see the possibility for its even continuation to the region of negative values of y.

Then we may use the Fourier expansion:

$$g(y, p_z, \varphi) = \pi^{-1}\int_0^\infty dk G(k, p_z, \varphi)\cos ky, \quad \mathbf{E}(y) = \pi^{-1}\int_0^\infty dk \mathbf{E}(k)\cos ky,$$

$$\delta\varepsilon(y, p_z, 0) = \pi^{-1}\int_0^\infty dk\,\delta\varepsilon(k, p_z)\cos ky.$$

By using equation (2.7) we may write the following expression for the Fourier transformant:

$$G(k, p_z, \varphi) = [R(p_z)/i\pi\Omega(p_z)]\int_0^\pi d\varphi_0\varphi_0\cos[kR(p_z)(\varphi_0^2 - \varphi^2)/2]\times$$

$$\times\left\{\operatorname{ctg}[\beta(p_z)\varphi_0]\int_{-\varphi_0}^{\varphi_0} d\varphi' - i\left[\int_{\varphi}^{\varphi_0} d\varphi' - \int_{-\varphi_0}^{\varphi} d\varphi'\right]\right\}\exp[i\beta(p_z)(\varphi - \varphi)]\times$$

$$\times\int_0^\infty dk'\cos\left[\frac{k'}{2}R(p_z)(\varphi_0^2 - \varphi'^2)\right]\{e\mathbf{v}(p_z, 0)\mathbf{E}(k') - i\omega\delta\varepsilon(k', p_z)\}. \tag{2.8}$$

In deriving this equation the contribution of large values of y that cannot correspond to skipping electrons is ignored and hence will not produce resonances at the surface transition frequencies. Moreover, we carry out substitution of the variable $y \to \varphi_0$. Now in large measure similar to the process used in study [15] we make the transformation from classical expression (2.8) to the corresponding quantum expression that allows identification of effects associated with transitions between the quantum levels of the surface electrons. We first account for

$$\exp[i\beta(p_z)(\varphi' - \varphi)] = \sum\frac{(-1)^s\sin[\beta(p_z)\varphi_0]}{\beta(p_z)\varphi_0 - \pi s}\exp\left[\frac{i\pi s}{\varphi_0}(\varphi' - \varphi_0)\right].$$

Further in quantum theory we must account for quantization of the angle at which the electron impacts the surface of the metal:

$$\varphi_n(p_z) = \left\{3\pi\left(n - \frac{1}{4}\right)\left|\hbar\Omega(p_z)/p_y'(p_z, \varphi = 0)\,v_y'(p_z, \varphi = 0)\right|\right\}^{1/3}.$$

Substituting φ_0 with this expression in (2.8) and converting from integration with respect to φ_0 to summation, accounting for $\Delta\varphi_0 = \Delta\varphi_n = \Delta n\varphi_n/3(n - 1/4)$, we obtain

$$G(k, p_z, \varphi) = [\hbar/ip_y'(0)]\sum_{ns}\cos[kR(p_z)(\varphi_n^2(p_z) - \varphi^2)/2]\times$$

$$\times\exp[-i\pi s\varphi/\varphi_n(p_z)][\omega - s\omega_n(p_z) - k_z v_z(p_z, 0) + i/\tau]^{-1}\times$$

$$\times(-1)^s\varphi_n^{-2}(p_z)\left\{\cos\beta\varphi_n(p_z)\int_{-\varphi_n(p_z)}^{\varphi_n(p_z)} d\varphi' - i\sin\beta\varphi_n(p_z)\times\right.$$

$$\times \Big[\int\limits_{\varphi}^{\varphi_n(p_z)} - \int\limits_{-\varphi_n(p_z)}^{\varphi} \Big] d\varphi' \Big\} \exp [i\pi s\varphi'/\varphi_n(p_z)] \times$$

$$\times \int\limits_0^\infty dk' \cos [k' R(p_z)(\varphi_n^2(p_z) - \varphi'^2)] \{e\mathbf{v}(p_z, 0) E(k') - i\omega\delta\varepsilon(k', p_z)\}, \qquad (2.9)$$

where the transition frequency between the quasi-classical levels is determined by the formula (compare to study [44]): $|E_{n+s}(p_z) - E_n(p_z)|/\hbar = s\omega_n(p_z) = s\,|\pi^2\Omega^2(p_z)p_y'(p_z)v_y'(p_z)/3(n - 1/4)\hbar|^{1/2}$. For further transformation of equation (2.9) we must take advantage of the fact that the resonances at the surface transitions exist only when one of the denominators of the sum of the right half is small, so that $|\omega - s\omega_n(p_z, 0) - k_z v_z(p_z)| \ll \omega$. Here it is important that the denominator does not vanish, since otherwise strong collisionless damping will occur which produces significant broadening which nearly liquidates resonance. Resonance is possible only when the variable frequency ω is close to the extremal transition frequency $s\omega_n$ when $p_z = 0$. In other words this is possible when $|p_z - p_0| \ll p_0$, where when $k_z \neq 0$ the "extremal" momentum $p_0(k_z)$ is determined by the equation $(d/dp_z)[s\omega_n(p_z) + k_z v_z(p_z, 0)] = 0$. The numbers n and s corresponding to the denominator that is small when $p_z = p_0$ will be called resonance. If n and s are not very large compared to unity, the resonance component is unique. We emphasize that the denominators of the other components ($m \neq n$, $r \neq s$) may be small, although the singularities corresponding to them may be far from the extremal momentum, and hence their contribution is comparatively small, although this may result in an effect such as collisionless damping.

Bearing in mind this effect we may take $\beta\varphi_n = \pi s$ in all components of formula (2.9). Then we obtain

$$G(k, p_z, \varphi) = [\hbar/ip_y'(p_0)] \Big\{ \Psi_{ns}^*(k, p_0, \varphi) [\omega +$$
$$+ i/\tau - s\omega_n(p_0) - k_z v_z(p_0, 0) - (p_z - p_0)^2 (s\omega_n'' + k_z v_z'')/2]^{-1} \times$$
$$\times \int\limits_0^\infty dk' [e\mathbf{v}(p_0, 0)\mathbf{E}(k') - i\omega\delta\varepsilon(k', p_0)] \int\limits_{-\varphi_n(p_0)}^{\varphi_n(p_0)} d\varphi' \Psi_{ns}(k', p_0, \varphi') -$$
$$- i\pi \sum_{r\neq s,\, m\neq n} \delta(\omega - r\omega_m(p_z) - k_z v_z(p_z, 0)) \Psi_{mr}^*(k, p_z, 0) \times$$
$$\times \int\limits_0^\infty dk' [e\mathbf{v}(p_z, 0)\mathbf{E}(k') - i\omega\delta\varepsilon(k', p_z)] \int\limits_{-\varphi_n(p_z)}^{\varphi_n(p_z)} d\varphi' \Psi_{mr}(k', p_z, \varphi') \Big\}, \qquad (2.10)$$

where $(s\omega_n'' + k_z v_z'') = \{(d^2/dp_z^2)[s\omega_n(p_z) + k_z v_z(p_z, 0)]\}_{p_z = p_0}$,

$$\Psi_{mr}(k, p_z, \varphi) = \varphi_m^{-1}(p_z) \exp [i\pi r\varphi/\varphi_m(p_z)] \cos [k R(p_z)(\varphi_m^2(p_z) - \varphi^2)/2].$$

In accordance with Appendix 1 we have the following expression for the quasi-classical matrix element:

$$\langle n+r|\cos ky|n\rangle = \int_{-\varphi_n(p_z)}^{\varphi_n(p_z)} d\varphi \Psi_{nr}^{KB}(k, p_z, \varphi),$$

where $\Psi_{nr}^{KB}(k, p_z, \varphi) = \psi_{nr}(k, p_z, \varphi)[1 + 2r\varphi_n^2/3\,(n - {}^1/_4)\varphi^2]^{-1/4}$.

Bearing in mind the significant contribution of small angles φ in place of φ_{nr} we will use its quantum analog below. Moreover, accounting for the fact that $\Psi_{nr}^{KB} = 0$, when $|\varphi| > \varphi_n(P_z)$.

It will then be sufficient for us to find the function

$$G(k, p_z) = \int_{-\pi}^{\pi} d\varphi G(k, p_z, \varphi),$$

for which consistent with (2.10) and the properties of Ψ_{ns}^{KB} we may write the following integral equation:

$$\begin{aligned} G(k, p_z) = {} & [\hbar/ip_y'(p_0)]\Big\{I_{ns}(k, p_z)\int_0^\infty dk' I_{ns}(k', p_0) \times \\ & \times \Big[e\mathbf{v}(p_0, 0)\mathbf{E}(k') + i\omega\int dp_z'\alpha(p_0, p_z)G(k', p_z') \times \\ & \times [\omega + i/\tau - s\omega_n(p_0) - k_z v_z(p_0, 0) - (p_z - p_0)^2(s\omega_n'' + k_z v_z'')/2]^{-1} - \\ & - i\pi \sum_{r\neq s,\, m\neq n} I_{mr}(k, p_z)\int_0^\infty dk' I_{mr}(k', p_z)\Big[e\mathbf{v}\mathbf{E}(k') + \\ & + i\omega\int dp_z'\alpha(p_z, p_z')G(k', p_z')\Big]\delta(\omega - r\omega_m(p_z) - k_z v_z(p_z, 0))\Big\}. \end{aligned} \qquad (2.11)$$

Here we use the conventions

$$I_{ns}(k, p_z) = \int_{-1}^{1} dx\,[1 + 2s/3\,(n - 1/4)\,x^2]^{-1/4}\cos\pi sx\cos[kR(p_z)\varphi_n^2(p_z)(1 - x^2)/2]. \qquad (2.12)$$

The resonance properties of the solution of quasi-classical equation (2.11), as we will see below, allows us to understand the regularities of cyclotron resonance at the skipping electrons.

2. Solution of the integral equation for the skipping electron distribution function. The influence of collisionless Landau damping on the spectrum of surface cyclotron waves

In this section we will obtain a solution of equation (2.11) for $k_z = 0$ and when k_z is nonzero. Earlier studies [21, 55, 56] devoted to investigating the properties of skipping electrons examined a model of a metal with a cylindrical Fermi surface that is characterized by a transition frequency independent of p_z. This means that the integration with respect to p_z involved in solving the integral equation is reduced to multiplication by the size of the cylindrical Fermi-surface in the p_z direction. The independence of the transition frequency from p_z caused neither frequency $r\omega_m$ when $m \neq n$, $r \neq s$ to satisfy the resonance condition $|\omega - r\omega_n| \ll \omega$. Consequently in the model of the metal with a cylindrical Fermi surface the sum of nonresonance components does not contain terms whose denominator would satisfy the resonance condition. Consistent with (2.11) the lack of a possibility for satisfying the resonance condition $|\omega - r\omega_m| \ll \omega$ when $r \neq s$ and $m \neq n$ indicates absence of collisionless damping in the model of a metal with a cylindrical Fermi surface.

A different situation exists in the case of a metal with a convex noncylindrical Fermi surface for which a dependence of the transition frequency $r\omega_m(p_z)$ on p_z is characteristic. Such a dependence causes a change in the value of $r\omega_m(p_z)$ from zero at the reference point to the maximal value $r\omega_m$ at the point corresponding to the maximum of the Fermi surface cross-section. In the components of the nonresonance sum for which the condition $r(m - 1/4)^{-1/3} > s(n - 1/4)^{-1/3}$ is satisfied the dependence of $r\omega_m(p_z)$ on p_z causes the following resonance condition to be satisfied at the point p_1

$$s\omega_n(p_0) - r\omega_m(p_1) = 0, \tag{2.13}$$

corresponding to the generation of collisionless Landau damping. Formula (2.13) determines the quantity $p_1 = p_1(\omega, m, r)$. The point $p_0 = p_0(0)$ corresponds to the extremum of the resonance frequency $s\omega_n(p_z)$. The first component of the right half of equation (2.11) makes the most significant contribution to the integral in terms of p_z from the region near the extremum of the denominator, which allows searching of the solution as

$$G(k, p_z) = g_{ns}(k)\,\delta(p_z - p_0) + \sum_{rm} q_{mr}(k)\,\delta(p_z - p_1(m, r)).$$

Bearing in mind that collisionless damping is a minor effect, we may write the following approximate expression for the function q_{mr}:

$$\begin{aligned} q_{mr}(k) = &- [\pi\hbar/p_y'(p_1(m, r))]\, I_{mr}(k, p_1(m, r)) \times \\ &\times \int_0^\infty dk' I_{mr}(k', p_1(m, r))\,[e\mathbf{v}(p_1(m, r), 0)\,\mathbf{E}(k') + \\ &+ i\omega\alpha(p_1(m, r), p_0)\, g_{ns}(k')]\,|r\omega_m'(p_1(m, r))|^{-1}. \end{aligned} \tag{2.14}$$

Then we obtain the integral equation for the function $g_{ns}(k)$:

$$\begin{aligned} g_{ns}(k) = &A_{ns} I_{ns}(k, p_0) \Big\{ \int_0^\infty dk' I_{ns}(k', p_0)\, g_{ns}(k') + \\ &+ \sum_{rm} B(n, s; m, r) \int_0^\infty dk' I_{mr}(k', p_1(m, r))\, g_{mr}(k') \Big\} + W_{ns}(k). \end{aligned} \tag{2.15}$$

Here we use the conventions:

$$A_{ns} = -\frac{\pi\sqrt{2}\hbar\omega\alpha}{s\omega_n''(p_0)\, p_y'(p_0)} \sqrt{\frac{s\omega_n''(p_0)}{s\omega_n(p_0) - \omega - i/\tau}},$$

$$\begin{aligned} B(n, s; m, r) = \\ = - i\pi\omega\hbar N_{ns}^{mr} \alpha(p_0, p_1(m, r))\, \alpha(p_1(m, r), p_0)/\alpha p_y'(p_1(m, r))\,|r\omega_m'(p_1(m, r))|, \end{aligned}$$

$$W_{ns}(k)=(A_{ns}/i\omega\alpha)I_{ns}(k,p_0)\int_0^\infty dk'\Big\{I_{ns}(k',p_0)e\mathbf{v}(p_0,0)-$$

$$-i\pi\hbar\omega\sum_{rm}N_{ns}^{mr}\frac{\alpha(p_0,p_1(m,r)e\mathbf{v}(p_1(m,r),0)}{p_y'(p_1(m,r))\,|r\omega_m(p_1(m,r))|}I_{mr}(k',p_1(m,r))\Big\}\mathbf{E}(k'),$$

where the conventions coincide with those used in Chapter 1:

$$N_{ns}=\int_0^\infty dkI_{ns}^2(k,p_0),\qquad N_{ns}^{mr}=\int_0^\infty dkI_{ns}(k,p_0)I_{mr}(k,p_1(m,r)),$$

а $\alpha=\alpha(p_0,p_0)$. При этом $p_0\equiv p_0(k_z=0)$.

Bearing in mind the smallness of the coefficients B, we may write the solution of equation (2.15) as

$$g_{ns}(k)=W_{ns}(k)+A_{ns}I_{ns}(k,p_0)\Big\{1+A_{ns}\sum_{mr}B(n,s;m,r)N_{ns}^{mr}\Big\}\times$$

$$\times\Big[1-A_{ns}N_{ns}-A_{ns}\sum_{m'r'}B(n,s;m',r')N_{ns}^{m'r'}\Big]^{-1}\int_0^\infty dk'W(k')\Big\{I_{ns}(k',p_0)+$$

$$+A_{ns}N_{ns}\sum_{m''r''}B(n,s;m''r'')I_{m''r''}(k',p_1(m'',r''))\Big\}. \tag{2.16}$$

It follows that the electron distribution has a resonance dependence on frequency, and the value of the frequency of such resonance itself is determined by the equation:

$$1=A_{ns}\{N_{ns}+\sum_{r\neq s,\ m\neq n}B(n,s;m,r)N_{ns}^{mr}\}. \tag{2.17}$$

Here we have

$$\omega=s\omega_n(p_0)\Big\{1-\frac{2\omega_n(p_0)}{\omega_n''(p_0)}\Big[\frac{\pi\alpha\hbar N_{ns}}{p_y'(p_0)}\Big]^2\Big\}-i\Big(\frac{1}{\tau}+\nu_L(n,s)\Big). \tag{2.18}$$

Here the contribution of collisionless damping is characterized by

$$\nu_L(n,s)=-\frac{4\pi^3\hbar^3s^2\omega_n^3(p_0)\alpha N_{ns}}{\omega_n''(p_0)\,[p_y'(p_0)]^2}\sum_{\substack{r\neq s\\ m\neq n}}\frac{\alpha(p_0,p_1(m,r))\,\alpha(p_1(m,r),p_0)\,[N_{ns}^{mr}]^2}{p_y'(p_1(m,r))\,|r\omega_m'(m,r)|}. \tag{2.19}$$

If the resonance frequency is close to the maximum transition frequency when $\omega_n''(p_0)<0$, then according to solution (2.18) $\omega > s\omega_n(p_0)$. If resonance exists near the minimum of the transition frequency, then $\omega < s\omega_n(p_0)$. In both cases in order for solution (2.18) to exist the shift in the resonance frequency from $s\omega_n(p_0)$ that arises due to electron interaction must be greater than the sum of collisionless damping related to the collisions

$$\alpha^2>\Big|[1/\tau+\nu_L(n,s)]\,\frac{\omega_n''(p_0)}{2\omega_n(p_0)}\Big[\frac{p_y'(p_0)}{\pi\hbar N_{ns}}\Big]^2\Big|.$$

Here the sign of the function α characterizing interelectron interaction is determined by the condition

$$\omega_n''(p_0)\, p_y'(p_0)\alpha < 0. \tag{2.20}$$

According to our estimates in the experiments on cyclotron skipping electron resonance $\nu_L\tau \ll 1$. Hence below we will focus on a consideration of the consequences in the case where collisionless damping is ignored. First we use (2.12) to write

$$N_{ns} = \int_0^\infty dk I_{ns}^2(k, p_0) = [2\pi/R(p_0)\varphi_n^2(p_0)] \times$$
$$\times \int_0^1 dx\, [x^2 + 2s/3(n - 1/4)]^{-1/2} \cos^2 \pi s x \simeq [\pi/R(p_0)\varphi_n^2(p_0)]\, L_{ns}, \tag{2.21}$$

where $L_{ns} = \ln [6(n - {}^1/_4)/s]$, $(n \gg s)$.

Consistent with this equation and ignoring collisionless damping formula (2.18) takes the following form:

$$\omega = s\omega_n(p_0)\left\{1 - \frac{2\pi^{8/3}\omega_n(p_0)\,\alpha^2}{3^{4/3}\omega_n''(p_0)}\left[\frac{\hbar\Omega(p_0)}{p_y'(p_0)\,v_y'(p_0)}\right]^{2/3} \frac{L_{ns}^2}{(n - 1/4)^{4/3}}\right\} - \frac{i}{\tau}. \tag{2.22}$$

This expression will be used below to analyze the experimental data.

We will provide the solutions of equation (2.11) for $k_z \neq 0$ without derivation. Here we ignore collisionless damping. As a result

$$G(k, p_z) = [\hbar/ip_y'(0)]\,[\omega - s\omega_n(p_0) - k_z v_z(p_0, 0) -$$
$$- (p_z - p_0)^2 (s\omega_n''(p_0) + k_z v_z''(p_0, 0))/2 +$$
$$+ i/\tau]^{-1} I_{ns}(k, p_0) \int_0^\infty dk' I_{ns}(k', p_0)\, e\mathbf{v}(p_0, 0)\,\mathbf{E}(k')\,\Lambda^{-1}. \tag{2.23}$$

Here $p_0 = p_0(k_z)$, while

$$\Lambda = 1 + \frac{2\pi\alpha\hbar\,[s\omega_n(p_0) + k_z v_z(p_0, 0)]\,N_{ns}}{[s\omega_n''(p_0) + k_z v_z''(p_0, 0)]\sqrt{\Delta(k_z)}\,p_y'(p_0)}, \tag{2.24}$$

using the convention

$$\Delta(k_z) = 2\,\frac{s\omega_n(p_0(k_z)) + k_z v_z(p_0(k_z), 0) - \omega - i/\tau}{s\omega_n(p_0(k_z)) + k_z v_z''(p_0(k_z), 0)}. \tag{2.25}$$

Consistent with (2.23) the resonance frequency is determined by the equation

$$\sqrt{\Delta(k_z)} = A, \tag{2.26}$$

where

$$A = -\frac{2\pi^{4/3}\alpha L_{ns}}{3^{2/3}(n - 1/4)^{2/3}}\left[\frac{\hbar\Omega(p_0)}{p_y'(p_0)\,v_y'(p_0)}\right]^{1/3} \frac{s\omega_n(p_0) + k_z v_z(p_0, 0)}{s\omega_n''(p_0) + k_z v_z''(p_0, 0)}. \tag{2.27}$$

When the shift of the resonance frequency determined by equation (2.26) exceeds damping, i.e., $A^2 \gg |s\omega_n''(p_0(k_z)) + k_z v_z''(p_0(k_z), 0)]\tau|^{-1}$, the solution of equation (2.26) corresponds to the resonance frequency when

$$\alpha[s\omega_n(p_0) + k_z v_z(p_0, 0)]/p_y'(0)[s\omega_n''(p_0) + k_z v_z''(p_0, 0)] < 0.$$

With a finite value of k_z resonance will be identified with the waves propagating along the metal surface with frequency

$$\omega = s\omega_n(p_0(k_z)) + k_z v_z(p_0(k_z), 0) - i/\tau - A^2[s\omega_n''(p_0(k_z)) + k_z v_z''(p_0(k_z), 0)]/2. \tag{2.28}$$

The permissible values of k_z are determined by the reality condition $\surd(\Delta k_z)$ when $\tau \to \infty$. Specifically, when $s\omega_n'' + k_z v_z'' < 0$ this condition takes the form

$$k_z v_z(p_0(k_z)\ 0) < \omega - s\omega_n(p_0(k_z)), \tag{2.29}$$

while in the opposite case

$$-k_z v_z(p_0(k_z), 0) < s\omega_n(p_0(k_z)) - \omega. \tag{2.30}$$

With small values of k_z when a power expansion is possible these inequalities are simplified. If resonance is related to the electrons in the noncentral cross-section of the Fermi surface, when $v_z(p_0(0)) \neq 0$, when $p_0(0)$ is determined, as usual, by the extremum surface transition frequency $\omega_n'(p_0(0)) = 0$, then consistent with (2.29) and (2.30) we have $|k_z v_z(p_0(0))| < |\omega - s\omega_n(p_0(0))|$. If resonance is caused by the central cross-section electrons where $v_z(p_0(0)) = 0$ and $p_0(0) = 0$, then $p_0(k_z) = -k_z[v_z'(0, 0)/s\omega_n''(0)]$. Then for k_z we have $2|\omega_n''(0)[\omega - s\omega_n(0)]| > [v_z'(0, 0)]^2 k_z^2$.

In concluding this section we note that formula (2.23) determines the resonance contribution to the current determined by the skipping electrons. Assuming that $v(p_0, 0)$ is reduced to only a single x-component, we have

$$j_x^{\text{pe3}} = -2e(2\pi\hbar)^{-3}\int dp_z\, G(k, p_z)(\partial^2\sigma/\partial p_z \partial\varphi)_{\varphi=0}[v_x(p_z, 0)/v(p_z, 0)],$$

where σ is the element of the area of the Fermi surface. According to (2.23)

$$-4\pi i\omega c^{-2} j_x^{\text{pe3}} = D[\sqrt{\Delta(k_z)} - A]^{-1} I_{ns}(k, p_0)\int_0^\infty dk' I_{ns}(k', p_0) E_x(k'), \tag{2.31}$$

where

$$D = -\frac{e^2\omega}{\pi\hbar^2 c^2}\left(\frac{\partial^2 \varsigma}{\partial p_z \partial\varphi}\right)_{\varphi=0,\ p_z=p_0} \frac{2v_x^2(p_0, 0)}{v(p_0, 0)\, p_y'(p_0)\,[s\omega_n''(p_0) + k_z v_z''(p_0)]}. \tag{2.32}$$

3. The resonance frequency for a metal with an ellipsoidal Fermi surface (bismuth). Formula for the collisionless damping of surface cyclotron waves in the case of bismuth

Experimental studies [20, 25, 26] have investigated the cyclotron resonance at skipping electrons in bismuth which has an ellipsoidal Fermi surface in sufficient detail. In this case the theory allows significant modification of the general formulae of the preceding section. The energy spectrum of the electrons in the metal with an ellipsoidal Fermi surface is described by the formula

$$\varepsilon = (2m_1)^{-1} p_x^2 + (2m_2)^{-1} p_y^2 + (2m_3)^{-1} p_z^2, \tag{2.33}$$

where m_1, m_2, m_3 are constants having dimensions of the mass, while the Fermi energy ε_F determines the projections of the Fermi momentum and velocity: $p_i^F = (2m_{i'}\varepsilon_F)^{1/2}$; $v_i^F = (2\varepsilon_F/m_{i'})^{1/2}$, where values of the indices $i = x, y, z$ correspond to $i' = 1, 2, 3$. In accordance with (2.33) for the case where the p_z axis of the ellipsoid lies in the direction of the constant magnetic field and $\Omega = |e|B/c\sqrt{m_1 m_2}$, while $p_z'(p_z, 0) = [2m_2(\varepsilon_F - p_z^2/2m_3)]^{1/2}$ the transition frequency entering into formula (2.22)

$$s\omega_n(p_z) = s\,[2\pi^2 e^2 B^2/3\hbar\,(n - 1/4)\, m_1 m_2 c^2]^{1/3}\,(\varepsilon_F - p_z^2/2m_3)^{1/3} \tag{2.34}$$

has the extremum at the central cross-section $p_z = 0$. This allows writing the following solutions for this frequency and its second derivative:

$$s\omega_n = s\,[2\pi^2\varepsilon_F e^2 B^2/3\,(n - 1/4)\,\hbar m_1 m_2 c^2]^{1/3};\quad s\omega_n'' = -s\omega_n/3m_3\varepsilon_F. \tag{2.35}$$

As a result in accordance with formula (2.22) we have

$$\omega = s\omega_n\left\{1 + \frac{2^{1/3}\pi^{2/3} m_3 \alpha^2 L_{ns}^2}{3^{1/3}(n - 1/4)^{4/3}}\left[\frac{\varepsilon_F \hbar^2 e^2 B^2}{m_1 m_2 c^2}\right]^{1/3}\right\}. \tag{2.36}$$

It is useful to represent the quantity α using the parametrization of the function $\Phi(\mathbf{p}, \mathbf{p}')$ introduced in study [58] for an ellipsoidal Fermi surface. Here we must account for the fact that in our case when the p_z axis of the ellipsoid is oriented along the magnetic field, $\Phi(\varepsilon, p_z, \varphi; \varepsilon', p_z', \varphi') = \sqrt{m_1 m_2}\,\Phi(\mathbf{p}, \mathbf{p}')$. In accordance with study [58] we have:

$$\Phi(\mathbf{p}, \mathbf{p}') = \pi^2\hbar^3 (2\varepsilon_F m_1 m_2 m_3)^{-1/2} \sum_{l=0}^{\infty} (2l + 1) A_l P_l(\cos\Theta). \tag{2.37}$$

Here A_l are the parametrization coefficients, P_l is the Legendre polynomial, θ is the angle between the vectors $\hat{\mathbf{w}}$ and $\hat{\mathbf{w}}'$ which are related to $\mathbf{p}$ and $\mathbf{p}'$ by the relations $\mathbf{w} = \hat{T}\mathbf{p}$, $\mathbf{w}' = \hat{T}\mathbf{p}'$. Here the tensor $\hat{T}$ transforms the momentum space p into the space $\mathbf{w}$ in which the surfaces of constant energy are spheres. Assuming in accordance with (2.33) that the axes of the coordinate system lie along the principal axes of the ellipsoid, we have for the matrix T:

$$T = (m_1^2 + m_2^2 + m_3^2)^{1/4} \begin{vmatrix} m_1^{-1/2} & 0 & 0 \\ 0 & m_2^{-1/2} & 0 \\ 0 & 0 & m_3^{-1/2} \end{vmatrix}$$

In accordance with the parametrization (2.37) we write equation (2.3) in the following form:

$$G(\theta, \psi) = G_0(\theta, \psi) + (1/4\pi) \int d\theta_{\mathbf{w}} \sum_{l=0}^{\infty} \sum_{m=-l}^{l} A_l P_l^m(\theta) P_l^m(\theta')(2l+1)\exp[im(\psi - \psi')] \times$$
$$\times G_0(\theta', \psi')(l-m)!/(l+m)!,$$

where θ, ψ and θ', ψ' are the polar and azimuthal angles of the vectors $\mathbf{w}$ and $\mathbf{w}'$, while P_l^m is the associated Legendre polynomial. Bearing in mind $G_0(\theta, \psi) = G(\theta, \psi) - (1/4\pi) \times$

$$\times \int d\theta_{\mathbf{w}} \sum_{l=0}^{\infty} \sum_{m=-l}^{l} (2l+1) \times$$
$$\times [A_l/(1+A_l)] P_l^m(\theta) P_l^m(\theta') \exp[\mathrm{im}(\psi - \psi')] G(\theta', \psi')(l-m)!/(l+m)!$$

and accounting for the fact that $dp_z d\varphi = \sqrt{2m_3\varepsilon_F}\, d\theta_{\mathbf{w}}$, we obtain for the function (2.5) the following expression:

$$\alpha(p_z, \varphi; p_z', \varphi') = (2m_3\varepsilon_F)^{-1/2} \sum_{l=0}^{\infty} (2l+1)[A_l/4\pi(1+A_l)] P_l(\cos\Theta).$$

From here we have the relation relating α and A_l:

$$\alpha = \alpha(0, 0; 0, 0) = (2m_3\varepsilon_F)^{-1/2} \sum_{l=0}^{\infty} (2l+1) A_l/4\pi(1+A_l). \tag{2.38}$$

Relation (2.38) allows representation of formula (2.22) using the constants A_l as

$$\omega = s\omega_n \left\{1 + \frac{\pi^{2/3}}{3^{1/3}2^{14/3}} \left(\frac{\hbar^2 e^2 B^2}{m_1 m_2 c^2 \varepsilon_F}\right)^{1/3} \frac{L_{ns}^2}{(n - 1/4)^{4/3}} \left[\sum_{l=0}^{\infty} \frac{(2l+1)A_l}{1+A_l}\right]^2\right\} - \frac{i}{\tau} =$$
$$= s\omega_n \left\{1 + \frac{\hbar\omega}{32\varepsilon_F} \frac{L_{ns}^2}{s(n - 1/4)} \left[\sum_{l=0}^{\infty} \frac{(2l+1)A_l}{1+A_l}\right]^2\right\} - \frac{i}{\tau}. \tag{2.39}$$

Formula (2.39) will be used in the next section to compare theoretical conclusions to the results from experimental research [20, 25, 26] and to obtain information on the Fermi-liquid interaction constant α.

Using the parametrization (2.37) we may provide here for reference purposes formula (2.19) for Landau damping drafted for the specific case of bismuth. If the energy spectrum of the electrons corresponds to formula (2.33), condition (2.13) is equivalent to the equality

$$\varepsilon_F^{1/2} s\,(n - 1/4)^{-1/3} = (\varepsilon_F - p_1^2\,(m, r)/2m_3)^{1/2} r\,(m - 1/4)^{-1/3}. \tag{2.40}$$

The conventions coincide with those used in (2.13). We find on the basis of (2.40) for bismuth

$$p_1\,(m, r) = (2m_3\varepsilon_F)^{1/2}\,[1 - (s/r)^3\,(m - 1/4)/(n - 1/4)]^{1/2}. \tag{2.41}$$

Substituting (2.35) and (2.41) and accounting for the fact that for bismuth the extremum of $s\omega_n(p_z)$ is achieved at the central cross-section ($p_0 = 0$), we obtain an expression for the decrement of Landau damping as

$$\nu_L\,(n, s)/s\omega_n = 9\,(\pi\hbar)^3\,\alpha N_{ns}\,(m_3/m_2)^{1/2} \times$$
$$\times \sum_{rm} \alpha\,(0, p_1\,(m, r))\,\alpha\,(p_1\,(m, r), 0)\,[N_{ns}^{mr}]^2 \left[\left(\frac{r}{s}\right)^3 \frac{n - 1/4}{m - 1/4} - 1\right]^{-1/2}. \tag{2.42}$$

In obtaining (2.42) we have accounted for the fact that

$$p_y'\,(p_1\,(m, r)) = [2m_2\,(\varepsilon_F - p_1^2\,(m, r)/2m_3)]^{1/2} = (2m_2\varepsilon_F)^{1/2}[(s/r)^3\,(m - 1/4)/(n - 1/4)]^{1/2},$$
$$r\omega_m'\,(p_1\,(m, r)) = -\frac{1}{3}\left(\frac{2}{m_3\varepsilon_F}\right)^{1/2}\left[1 - \left(\frac{s}{r}\right)^3 \frac{m - 1/4}{n - 1/4}\right]^{1/2}\left(\frac{r}{s}\right)^3 \frac{n - 1/4}{m - 1/4}\, s\omega_n.$$

In accordance with formula (2.38)

$$\alpha\,(0, p_1\,(m, r))) = (4\pi)^{-1}\,(2m_3\varepsilon_F)^{-1/2} \sum_{l=0}^{\infty} [(2l + 1)\,A_l/(1 + A_l)] \times$$
$$\times P_l\,([(s/r)^3\,(m - 1/4)/(n - 1/4)]^{1/2}),$$

while for the ratio (2.42) of the decrement of collisionless damping $\nu_L(n, s)$ to the extremal value of the surface transition frequency $\Delta\omega_n$ we obtain

$$\frac{\nu_L\,(n, s)}{s\omega_n} = \frac{9\hbar^3 N_{ns}}{5^{15/2} m_2^{3/2} \varepsilon_F^{3/4}} \sum_{l=0}^{\infty} \frac{(2l + 1)\,A_l}{1 + A_l} \times$$
$$\times \sum_{rm} \left[N_{ns}^{mr} \sum_{i=0}^{\infty} \frac{(2i + 1)\,A_i}{1 + A_i}\, P_i\left(\left(\frac{s}{r}\right)^{3/2}\left(\frac{m - 1/4}{n - 1/4}\right)^{1/2}\right]^2 \left[\left(\frac{r}{s}\right)^3 \frac{n - 1/4}{m - 1/4} - 1\right]^{1/2}.$$

Finally, introducing conventions identical to those used in Chapter 1 of this study and in Appendix 1: $N_{ns} = \pi L_{ns}/2y_n$, $N_{ns}^{mr} = \pi L_{ns}^{mr}/2y_n$, where L_{ns} is determined in (2.21), while L_{ns}^{mr} in accordance with Appendix 1 is equal to

$$L_{ns}^{mr} = \left(\frac{s}{r}\right)^{1/4}\left(\frac{n - 1/4}{m - 1/4}\right)^{1/4} \ln\left[4\left(\left(\frac{s}{r}\right)^{1/2}\left(\frac{n - 1/4}{m - 1/4}\right)^{1/4} - 1\right)^{-1}\right],$$

we finally obtain for collisionless damping in the case of bismuth a formula expressed through the ratio of the Larmor quantum to the Fermi energy and through the $\hbar\omega/\varepsilon_F$ ratio as:

$$\frac{\nu_L(n,s)}{s\omega_n}=\frac{\pi}{2^7}\frac{L_{ns}}{(n-{}^1\!/_4)^2}\frac{\hbar\Omega}{\varepsilon_F}\sum_{l=0}^{\infty}\frac{(2l+1)A_l}{1+A_l}\sum_{rm}\left[\left(\frac{r}{s}\right)^3\frac{n-{}^1\!/_4}{m-{}^1\!/_4}-1\right]^{-1/2}\times$$

$$\times\left[L_{ns}^{mr}\sum_{i=0}^{\infty}\frac{(2i+1)A_i}{1+A_i}p_i\left[\left(\frac{s}{r}\right)^{3/2}\left(\frac{m-{}^1\!/_4}{n-{}^1\!/_4}\right)^{1/2}\right]\right]^2=$$

$$=\sqrt{\frac{3}{2}}\frac{1}{2^7}\left(\frac{\hbar\omega}{\varepsilon_F}\right)^{3/2}\frac{L_{ns}}{[s(n-{}^1\!/_4)]^{3/2}}\sum_{l=0}^{\infty}\frac{(2l+1)A_l}{1+A_l}\sum_{rm}\left[\left(\frac{r}{s}\right)^3\frac{n-{}^1\!/_2}{m-{}^1\!/_4}-1\right]^{-1/2}\times$$

$$\times\left[L_{ns}^{mr}\sum_{i=0}^{\infty}\frac{(2i+1)A_i}{1+A_i}p_i\left[\left(\frac{s}{r}\right)^{3/2}\left(\frac{m-{}^1\!/_4}{n-{}^1\!/_4}\right)^{1/2}\right]\right]^2. \qquad (2.43)$$

Here we provide the estimate of ratio (2.43) for the RF range, when $\omega/2\pi \simeq 3\cdot10^{10}$ Hz, and we select for the sum of the interaction constants in bismuth the value

$$\sum_{l=0}^{\infty}(2l+1)A_l/(1+A_l)=6,$$

which, as we will see below in the next section, derives from an analysis of the experimental data on cyclotron resonance at the skipping electrons in bismuth. Adopting the values: n = 3, s = 2 for the quantum numbers (then L_{32} = 2.12), we obtain

$$\frac{\nu_L(3,2)}{2\omega_3}\simeq 0{,}6\cdot10^{-5}\sum_{r,m}\left[\left(\frac{r}{s}\right)^3\frac{n-{}^1\!/_4}{m-{}^1\!/_4}-1\right]^{-1/2}\times$$

$$\times\left\{\sum_{i=0}^{\infty}\frac{(2i+1)A_i}{1+A_i}p_i\left[\left(\frac{s}{r}\right)^{3/2}\left(\frac{m-{}^1\!/_4}{n-{}^1\!/_4}\right)^{1/2}\right]L_{ns}^{mr}\right\}^2. \qquad (2.44)$$

Even if we consider the quantity in braces in (2.44) to be close to ten, we obtain for the $\nu_L/s\omega_n$ relation in the RF frequency range a negligible quantity of $10^{-3} \ll 1$. Such an estimate confirms the validity of ignoring the collisionless damping quantity in obtaining the spectrum of surface cyclotron waves (2.39) in bismuth.

4. Skipping electron-generated resonance properties of the impedance of a metal

In this section we will demonstrate how accounting for electron interaction produces a qualitatively new possibility for the existence of impedance resonances and surface electromagnetic waves compared to theory [21, 31, 56] ignoring electron interaction. In deriving the impedance expression we take advantage of the fact that the volumetric and surface electrons make an insignificant contribution to the current density which enters into the Maxwell equation. The weak magnetic field has virtually no influence on volumetric electron motion. Hence in order to determine the contribution to total current related to volumetric electrons we must use existing results (see, for example, [34, 59]) obtained for a metal with an anisotropic Fermi surface. If we orient the coordinate axes along the principal axes of the tensor $B_{\alpha\beta}=\int_0^{2\pi} d\Phi n_\alpha n_\beta K^{-1}(\varphi)$, where $K(\varphi)$ is the Gaussian curvature of the Fermi-surface, where $n_i = v_i/v$, we may obtain from Maxwell's equations two independent equations for the electric field components E_x and E_z. If we take the electric field of the surface H-wave to be polarized in the direction of x, we will write the equation for the Fourier-component of the electric field

$$2E'(0)+(k_z^2+k^2-i\delta^{-3}k^{-1})E(k)=4\pi i\omega c^{-2}j^{\text{pes}}(k), \tag{2.45}$$

where $\delta = (c^2\pi\hbar^3/e^2B_{xx}\omega)^{1/3}$ is the depth of penetration of the field into the metal with an anisotropic Fermi surface; $E'(0)$ is the value of the derivative of the electric field at the point $y = 0$.

Noting that when we account for the Fermi-liquid interaction the expression for resonance current (2.31) differs from the corresponding expression obtained in the noninteracting particle theory [55] in the resonance multiplier; we may write the following expression for the surface impedance:

$$Z(k_z)=\rho\frac{4\pi\omega\delta}{3\sqrt{3}c^2}e^{-i\pi/3}+\frac{2\pi^2 i\omega D\alpha^2_{n+s,n}(k)}{c^2}\left\{\sqrt{\Delta(k_z)}-A+\frac{\pi}{2}D\beta_{n+s,n}(k_z)\right\}^{-1}, \tag{2.46}$$

where (compare to [55]):

$$\alpha_{n+s,n}(k_z)=(2/\pi)\int_0^\infty I_{ns}(k,p_0)[k_z^2+k^2-i\delta^{-3}k^{-1}]^{-1}dk,$$

$$\beta_{n+s,n}(k_z)=(2/\pi)\int_0^\infty I^2_{ns}(k,p_0)[k_z^2+k^2-i\delta^{-3}k^{-1}]^{-1}dk, \tag{2.47}$$

In formula (2.46) the first component describes the impedance of the metal in the absence of a magnetic field. With specular reflection of the nonresonance electrons off the surface $\rho = 1$, with diffuse reflection $\rho = 9/8$. The impedance peak corresponding to the vanishing of the denominator of the second component in (2.46) corresponds to the possibility for surface electromagnetic wave excitation. The dispersion equation of such a surface wave is obtained by substituting the solution of (2.45) into the right half of the expression

$$|k_z-\omega/c|^{-1}=(1/\pi)\int_0^\infty dkE(k)/E'(0)\equiv c^2Z(k_z)/4\pi i\omega. \tag{2.48}$$

Before proceeding to a discussion of the possibilities for solving dispersion equation (2.48) for surface cyclotron waves in electron liquid, we will consider in greater detail than before (such as study [55]) the issue of the possibility for propagation of surface waves with small wave vectors k_z near the transition frequencies between the skipping electron levels in an electron gas of noninteracting particles. The need for such an examination is related primarily to the fact that the statements in study [55] regarding the impossibility of the propagation of surface cyclotron waves near the transition frequencies between the levels of the skipping electrons in metals with protrusions, particularly spherical Fermi surfaces were not supported by a consistent solution of the dispersion equation.

The dispersion equation of such surface waves in the electron gas of metals with convex Fermi surfaces may be written based on (2.46) and (2.48) as

$$\frac{1}{|k_z'|} = \frac{4\delta}{3i\sqrt{3}} e^{-i\pi/3} + \frac{\pi D' \alpha^2_{n+s,n}(0)/2}{\sqrt{\Delta + i\gamma} + \beta'}, \tag{2.49}$$

where $k_z' = k_z - \omega/c$. Here we use the same designations as study [48]: k_z is the wave vector of the surface wave propagating in the direction of the constant magnetic field; δ is the depth of the skin-layer in anomalous skin-effect conditions; $D' = D\,|\,\omega_n''/2\omega_n\,|^{1/2}$; $\beta' = \pi D' \beta_{n+s,n}(0)/2$, where $\beta_{n+s,n}(0)$ is the matrix element determined by means of expression (2.47), $s\omega_n$ is the extremal value of the transition frequency $l\omega_n(p_z)$ between surface levels with quantum numbers $n + s$ and n, $s\omega_n''$ is the value of the second derivative $s\omega_n(p_z)$ with respect to momentum p_z at the point corresponding to the extremum of $s\omega_n(p_z)$. The quantity D is determined for metals with an arbitrary Fermi surface by formula (2.32) and, finally,

$$\Delta = \frac{\mathrm{Re}\,\omega - s\omega_n(p_0)}{s\omega_n(p_0)} - \frac{k_z v_z(p_0)}{s\omega_n(p_0)}, \qquad \gamma = \frac{\mathrm{Im}\,\omega}{s\omega_n(p_0)} + \frac{1}{s\omega_n(p_0)\tau}, \tag{2.50}$$

where p_0 is the value of the momentum p_z corresponding to the extremum of the expression $s\omega_n(p_z) + k_z v_z(p_z)$. Thus in the particular case of bismuth when the major axis of one of the three ellipsoids of the electron Fermi surface is parallel to the magnetic field (see study [48])we have

$$-\frac{\omega_n''}{2\omega_n} = (6m_3\varepsilon_F)^{-1}, \qquad D = \frac{6\sqrt{2}\,e^2 m_3 \varepsilon_F^{3/2}}{\pi\hbar^2 c^2 \sqrt{m_1}}, \tag{2.51}$$

$$\Delta = \frac{\mathrm{Re}\,\omega - s\omega_n}{s\omega_n} - \frac{3}{4}\left(\frac{k_z v_z^F}{s\omega_n}\right)^2,$$

where m_1, m_2, m_3 are the values of the effective masses (compare to studies [48, 51]); ε_F is the Fermi energy; $v_z^F = \sqrt{2\varepsilon_F/m_3}$. We will explain in greater detail the dependence of Δ on the wave vector k_z that arises in (2.51). In formula (2.50) the frequency $s\omega_n$ and the velocity v_z depend on the value of the momentum $p_0(k_z)$ that is the solution of the equation $(d/dp_z)[s\omega_n(p_z) + k_z v_z] = 0$. For bismuth when the energy spectrum of the electrons takes the form $\varepsilon(\mathbf{p}) = p_i^2/2m_i$, resonance is caused by the central cross-section electrons and $p_0(k_z) = -k_z[v_z'(0)/s\omega_n''(0)]$. Below we use the conventions $p_0 = p_0(0)$, $s\omega_n(p_0) = s\omega_n$ and $s\omega_n''(0) = s\omega_n''$, and account for the fact that for bismuth in the case of central cross-section resonance $p_0 = 0$, $v_z(p_0)$. In this case

$$\Delta = \frac{\mathrm{Re}\,\omega - s\omega_n(p_0(k_z))}{s\omega_n} - \frac{k_z v_z(p_0(k_z))}{s\omega_n} = \frac{\mathrm{Re}\,\omega - s\omega_n}{s\omega_n} - \frac{1}{2} p_0^2(k_z)\frac{\omega_n''}{\omega_n} +$$
$$+ \frac{k_z^2 v_z'^2}{s^2\omega_n\omega_n''} = \frac{\mathrm{Re}\,\omega - s\omega_n}{s\omega_n} + \frac{k_z^2 v_z'^2}{2s^2\omega_n\omega_n''} = \frac{\mathrm{Re}\,\omega - s\omega_n}{s\omega_n} - \frac{3}{4}\left(\frac{k_z v_z^F}{s\omega_n}\right)^2.$$

We emphasize here that the requirement $|\Delta| \ll 1$ does not infer vanishing of Δ, since in the case $\Delta = 0$ strong collisionless damping of surface waves occurs, and we will attempt to search for solutions of equations (2.50) that correspond to weakly damped waves propagating along the surface (we will use the condition $\gamma \ll \Delta$ in explicit form below).

Following the conventions of study [48] where we designate the depth of occurrence of the nth surface layer of the electron by y_n, we write the matrix elements of (2.46) and (2.47) for a metal with an arbitrary Fermi surface in the following form:

$$\alpha_{n+s,\,n}(0) = \frac{4\delta}{\pi}\int_0^1 dx \cos\pi s x \int_0^\infty du\, \frac{u^4 + iu}{u^6 + 1}\cos\left[u\frac{y_n}{\delta}(1 - x^2)\right], \tag{2.52}$$

$$\beta_{n+s,\,n}(0) = \frac{8\delta}{\pi}\int_0^1 dx\cos\pi s x\int_0^1 dz\cos\pi s z\int_0^\infty du\,\frac{u^4 + iu}{u^6 + 1}\times$$
$$\times\cos\left[u\frac{y_n}{\delta}(1 - x^2)\right]\cos\left[u\frac{y_n}{\delta}(1 - z^2)\right]. \tag{2.53}$$

In deriving expressions (2.52) and (2.53) we used the following quasi-classical expression (compare to (2.12)) for $I_{ns}(k)$

$$I_{ns}(k) = 2\int_0^1 dx\cos\pi s x\cos[k y_n(1 - x^2)]. \tag{2.54}$$

(Here we note that matrix element (2.52) is by definition two times greater than the corresponding matrix element $\psi_{ns}(k)$ given by formula (2.13) in study [55].) As a result our values of $\alpha_{n+s,n}(0)$ and $\beta_{n+s,n}(0)$ are two and four times greater, respectively, than the quantities labeled $\alpha_{ns}(0)$ and β_{ns} in study [55].

We will write out expression (2.49) separately for the real and imaginary parts:

$$\frac{1}{|k_z'|} + \frac{2\delta}{3} = \frac{\pi D'}{2}\times$$
$$\times\frac{\mathrm{Re}\,\alpha_{n+s,\,n}^2(0)\{2^{-1/2}[\Delta + \sqrt{\Delta^2 + \gamma^2}]^{1/2} + \mathrm{Re}\,\beta'\} + \mathrm{Im}\,\alpha_{n+s,\,n}^2(0)\{2^{-1/2}[-\Delta + \sqrt{\Delta^2+\gamma^2}]^{1/2} + \mathrm{Im}\,\beta'\}}{\sqrt{\Delta^2+\gamma^2} + \sqrt{2}\,\mathrm{Re}\,\beta'[\Delta + \sqrt{\Delta^2+\gamma^2}]^{1/2} + (\mathrm{Re}\,\beta')^2 + \sqrt{2}\,\mathrm{Im}\,\beta'[-\Delta + \sqrt{\Delta^2+\gamma^2}]^{1/2} + (\mathrm{Im}\,\beta')^2}, \tag{2.55}$$

$$\frac{2\delta}{3\sqrt{3}} = \frac{\pi D'}{2}\times$$
$$\times\frac{-\mathrm{Re}\,\alpha_{n+s,\,n}^2(0)\{2^{-1/2}[-\Delta + \sqrt{\Delta^2+\gamma^2}]^{1/2} + \mathrm{Im}\,\beta'\} + \mathrm{Im}\,\alpha_{n+s,\,n}^{2(0)}(0)\{2^{-1/2}[\Delta + \sqrt{\Delta^2+\gamma^2}]^{1/2} + \mathrm{Re}\,\beta'\}}{\sqrt{\Delta^2+\gamma^2} + \sqrt{2}\,\mathrm{Re}\,\beta'[\Delta + \sqrt{\Delta^2+\gamma^2}]^{1/2} + (\mathrm{Re}\,\beta')^2 + \sqrt{2}\,\mathrm{Im}\,\beta'[-\Delta + \sqrt{\Delta^2+\gamma^2}]^{1/2} + [\mathrm{Im}\,\beta']^2}. \tag{2.56}$$

Assuming that $\gamma \ll \Delta$ we write equations (2.55) and (2.56) as

$$\frac{1+2\delta|k_z'|}{|k_z'|}=\frac{\pi D'}{2}\frac{\operatorname{Re}\alpha^2_{n+s,n}(0)[\sqrt{\Delta}+\operatorname{Re}\beta']+\operatorname{Im}\alpha^2_{n+s,n}(0)[\gamma/2\sqrt{\Delta}+\operatorname{Im}\beta']}{\Delta[1+\gamma^2/2\Delta^2]+2\operatorname{Re}\beta'\sqrt{\Delta}[1+\gamma^2/8\Delta^2]+(\operatorname{Re}\beta')^2+\operatorname{Im}\beta'\gamma/\sqrt{\Delta}+(\operatorname{Im}\beta)^2}, \tag{2.57}$$

$$\frac{2\delta}{3\sqrt{3}}=\frac{\pi D'}{2}\times$$
$$\times\frac{-\operatorname{Re}\alpha^2_{n+s,n}(0)[\gamma/2\sqrt{\Delta}+\operatorname{Im}\beta']+\operatorname{Im}\alpha^2_{n+s,n}(0)[\sqrt{\Delta}+\operatorname{Re}\beta']}{\Delta[1+\gamma^2/2\Delta^2]+2\operatorname{Re}\beta'\sqrt{\Delta}[1+\gamma^2/8\Delta^2]+(\operatorname{Re}\beta')^2+\operatorname{Im}\beta'\gamma/\sqrt{\Delta}+(\operatorname{Im}\beta')^2}. \tag{2.58}$$

Solving (2.57) and (2.58) we find the following expressions in order to determine the real and imaginary parts of the frequency ω:

$$\frac{\gamma}{2\sqrt{\Delta}}=-\operatorname{Im}\beta'+\frac{\pi D'|k_z|}{2}\times$$
$$\times\frac{(\operatorname{Re}\alpha^2_{n+s,n}(0))^2+(\operatorname{Im}\alpha^2_{n+s,n}(0))^2}{\operatorname{Im}\alpha^2_{n+s,n}(0)[1+(2+{}^2/_3\sqrt{3})\delta|k_z'|]+\operatorname{Re}\alpha^2_{n+s,n}(0)[1+(2-2/\sqrt{3})\delta|k_z'|]},$$
$$\sqrt{\Delta}=-\operatorname{Re}\beta'+\frac{\pi D'}{2}|k_z'|\times$$
$$\times\frac{(\operatorname{Re}\alpha^2_{n+s,n}(0))^2+(\operatorname{Im}\alpha^2_{n+s,n}(0))^2}{\operatorname{Im}\alpha^2_{n+s,n}(0)[1+(2+2/3\sqrt{3})\delta|k_z'|]+\operatorname{Re}\alpha^2_{n+s,n}(0)[1+(2-2/3\sqrt{3})\delta|k_z'|]}, \tag{2.59}$$

$$\frac{\operatorname{Re}\alpha^2_{n+s,n}(0)+2\delta|k_z'|[\operatorname{Re}\alpha^2_{n+s,n}(0)+(2/3\sqrt{3})\operatorname{Im}\alpha^2_{n+s,n}(0)]}{\operatorname{Im}\alpha^2_{n+s,n}(0)+2\delta|k_z'|[\operatorname{Im}\alpha^2_{n+s,n}(0)-(2/3\sqrt{3})\operatorname{Re}\alpha^2_{n+s,n}(0)]}. \tag{2.60}$$

In order to assure clarity we will first write out the dispersion equation for determining the real part of the surface wave frequency in the limit when $k_z' \to 0$ (more precisely $k_z' \ll \omega/c$ while $k_z > \omega/c$). In this case the equation

$$\sqrt{\Delta}=-\operatorname{Re}\beta' \tag{2.61}$$

has no real solutions ($\operatorname{Re}\beta' > 0$) which is in agreement with the conclusions of study [55] regarding the absence of solutions of the dispersion equation of surface oscillations in the model of a metal with a convex Fermi surface in the limit $k_z' \to 0$.

Before proceeding to an analysis of the possibilities for solving dispersion equation (2.60) when $k_z \neq 0$, we will emphasize the differences in the formulation of the problem of the resonances of the surface impedance and the problem of the propagation of surface waves near the transition frequencies between the surface electron levels. The difference is that in investigating the resonance properties of the surface impedance its dependence on the real frequency ω is determined. (Or its dependence on the magnetic field, which is also real.) In the problem of surface wave propagation the real and imaginary parts of the complex frequency ω are found; these describe the frequency spectrum and the decrement of damping of the surface wave. If in conditions of specular electron reflection we write (compare to

(2.46)) the expressions for the surface impedance of the electron gas as

$$Z(\omega, 0) = \frac{16\pi\omega\delta}{3\sqrt{3}c^2} e^{-i\frac{\pi}{3}} + \frac{2\pi^2 i\omega D'\alpha^2_{n+s,n}(0)}{2}\left[\sqrt{\frac{\omega - s\omega_n}{s\omega_n} + \frac{i}{s\omega_n\tau}} + \beta'\right]^{-1} \tag{2.62}$$

The difference between the right half of (2.62) and the right half of (2.49) is primarily that the frequency ω in (2.62) is a purely real quantity. This fact produces resonances of the surface impedance in metals with convex Fermi surfaces as indicated by (2.62) in a metal model that does not account for particle interaction. Such a viewpoint is consistent with that offered by E.A. Kaner and N.M. Makarov in study [55] and was confirmed in study [54] where the gas model was used as the basis for plotting graphs of the oscillating dependencies of the derivative of the real part of the impedance of bismuth as a function of the magnetic field. On the other hand the possibility or impossibility of propagation of a surface cyclotron wave near the transition frequencies $s\omega_n$ is related to the possibility of solving equation (2.60) and equation (2.59).

In order to analyze such a possibility of solving equation (2.60) we write expression for $\alpha_{n+s,n}(0)$ and $\beta_{n+s,n}(0)$ in the simplest case for us: $y_n \ll \delta$. When the condition $y_n \ll \delta$ is satisfied the orbit of a skipping electron is entirely contained within the skin-layer and such an electron may effectively interact with the electromagnetic wave field. In this case we find by direct calculation:

$$\alpha_{n+s,n}(0) = 4y_n(\pi s)^{-2}(-1)^s[1 + 4iy_n/\sqrt{3}(\pi s)^2\delta], \tag{2.63}$$

$$\beta_{n+s,n}(0) = \frac{2y_n}{(\pi s)^2} + iy_n\left(\frac{y_n}{\delta}\right)^3 \frac{32}{(\pi s)^5}\left\{\frac{1}{5} + \left(\frac{3}{2} + \ln\frac{\delta}{y_n}\right)\frac{36}{\pi(\pi s)^3}\right\}. \tag{2.64}$$

We note that in study [55] the imaginary part of $\beta_{n+n,n}(0)$ when $y_n \ll \delta$ is calculated inaccurately. Consistent with [55] (accounting for the difference by a factor of two of the matrix elements $i_{ns}(k)$ and $\psi_{ns}(k)$) we write for $\mathrm{Im}\beta_{n+n,n}(0)$ the expression

$$\mathrm{Im}\,\beta_{n+s,n}(0) = 32[1 - 6/(\pi s)^2]^2 y_n^4 \ln(\delta/y_n)/\pi^5 s^4\delta^3, \tag{2.65}$$

comparing (2.65) to (2.64) and taking $\mathrm{In}(\delta/y_n) \gg 1$ and we see that in study [55] the coefficient with the component $\sim\mathrm{In}(\delta/y_n)$ is incorrect and is differentiated from the corresponding coefficient in (2.64) equal to $32 \cdot 36 : y_n^4/\pi(\pi s)^8\delta^3$ by the factor $[(\pi s)^2/6-1]^2$. Moreover, in study [55] the component equal to $32y_n^4/5\delta^3(\pi s)^4$ that does not contain $\mathrm{In}(\delta/y_n)$ is dropped in the imaginary part of $\beta_{n+n,n}(0)$.

In the opposite limiting case when $(yu/\delta) \gg (\pi s/2) \gg 1$, for $\alpha_{n+n,n}(0)$ and $\beta_{n+n,n}(0)$ we find the expression (compare to study [55]):

$$\alpha_{n+s,n}(0) = (-1)^s\pi s(\delta^3/y_n^2)[(\pi s\delta/2y_n)^3 + i], \tag{2.66}$$

$$\beta_{n+s,\,n}(0) = (2\pi/3\sqrt{3})\,(\delta^2/y_n)\,e^{i\pi/3}. \qquad (2.67)$$

In both limiting cases $y_n \ll \delta$ and $y_n \gg \delta$ the possibility for solving dispersion equation (2.60) may arise only when the second component in the right half of (2.60) is greater than the first. This means that the following inequality will be satisfied:

$$\operatorname{Re}\beta_{n+s,\,n}(0) < k_z' \frac{[\operatorname{Re}\alpha^2_{n+s,\,n}(0)]^2 + [\operatorname{Im}\alpha^2_{n+s,\,n}(0)]^2}{\operatorname{Im}\alpha^2_{n+s,\,n}(0) + \operatorname{Re}\alpha^2_{n+s,\,n}(0)} - \frac{\operatorname{Re}\alpha^2_{n+s,\,n}(0)}{\operatorname{Im}\alpha^2_{n+s,\,n}(0)}. \qquad (2.68)$$

We have taken $k_z'\delta \ll 1$ in deriving (2.68). This is because using the expression $\delta = (2/3)(4v_c^2/3\pi\omega\omega_{\mathrm{Le}}^2)^{1/3}$ for the depth of the skin-layer, we obtain for $k_z' = \omega/v_F$ the value

$$k_z'\delta = (^2/_3)\,(4\omega^2c^2/3\pi\omega_{\mathrm{Le}}^2v_F^2)^{1/3}. \qquad (2.69)$$

At frequencies $\omega \leqslant \omega_{\mathrm{Le}}10^{-2}$ consistent with (2.69) $k_z'\delta \ll 1$. For example, for bismuth $\omega_{\mathrm{Le}} = 10^{14}\ \mathrm{s}^{-1}$ and $k_z'\delta \ll 1$ for both the microwave and the IR spectra. For alkali metals $w_{\mathrm{Le}} \sim 10^{16}\ \mathrm{s}^{-1}$ and $k_z'\delta \ll 1$ in the frequency range $\omega \leqslant 10^{14}\ \mathrm{s}^{-1}$ (for $k_z' \leqslant \omega/v_F$). The values of k_z (or k_z', which is the same thing) will be much less than ω/v_F as indicated by (2.51) in order to avoid the onset of strong collisionless Landau damping of surface waves and so the condition $|\Delta| \ll 1$ is satisfied.

We will now investigate the possibilities for satisfying inequality (2.68). In the case $y_n \ll \delta$ we find (accounting for the fact that $\operatorname{Im}\alpha^2_{n+n,\,n}(0) \ll \operatorname{Re}\alpha^2_{n+n,\,n}(0)$):

$$k_z'\delta > 1/\sqrt{3}. \qquad (2.70)$$

As we have seen above in the microwave and IR regions of the spectrum such an inequality will be satisfied only with large values of $k_z > \omega/v_F$. In this region of k_z we have strong collisionless Landau damping and the wave cannot propagate. With small values of $k_z \geqslant \omega/c$ inequality (2.70) will not be satisfied for frequencies in the microwave and IR ranges and dispersion equation (2.60) has no real solutions and surface waves cannot propagate near the $s\omega_n$ frequencies in the electron gas.

In the case $y_n \gg \delta$ we represent inequality (2.68) as

$$k_z'\delta\,(3\sqrt{3}\pi^5s^6/2^4)\,(\delta/y_n)^9 > 1. \qquad (2.71)$$

Due to the fact that $y_n/\delta > \pi s/2 > 1$, while $k_z\delta < 1$ inequality (2.71) is not satisfied in the microwave and IR-ranges. Propagation of the surface waves with frequencies near the transition frequencies between the surface levels in the electron gas is forbidden in this case as well.

We will now proceed to an examination of the possibilities for the propagation of surface cyclotron waves (near the frequencies $s\omega_n$) in the electron liquid of metals and to a solution of dispersion equation (2.48). When the inequality $k_z\delta \ll 1$ is satisfied we may ignore the contribution related to the volumetric components compared to the contribution of the resonance component in equation (2.48). Bearing in mind values of k_z satisfying the stronger inequality $|k_z v_z(p_0(k_z))| < |\omega - s\omega_n(p_0(k_z))|$ in conditions of the anomalous skin-effect we write the dispersion equation of the surface oscillations as

$$\sqrt{\Delta(k_z)} \equiv \{2[s\omega_n(p_0(k_z)) + k_z v_z(p_0(k_z)) - \omega - i/\tau]/[s\omega_n''(p_0(k_z)) + k_z v_z''(p_0(k_z))]\}^{1/2} = A - \pi D\beta_{n+s,\,n}(k_z)/2. \quad (2.72)$$

In the case $k_z = 0$ with satisfaction of the condition

$$A > \pi D \operatorname{Re} \beta_{n+s,n}(0)/2 \quad (2.73)$$

we obtain from equation (2.72)

$$\operatorname{Re}\omega = s\omega_n - (1/2)\, s\omega_n'' [A - \pi D \operatorname{Re}\beta_{n+s,\,n}(0)/2]^2 + (1/2)\, s\omega_n''[\operatorname{Im}\beta_{n+s,\,n}(0)]^2 (\pi D/2)^2, \quad (2.74)$$

$$\operatorname{Im}\omega = -1/\tau - s\omega_n''[A - \pi D \operatorname{Re}\beta_{n+s,\,n}(0)/2] \operatorname{Im}\beta_{n+s,\,n}(0)\, \pi D/2. \quad (2.75)$$

It follows from (2.75) that two components make a contribution to the imaginary part of ω. First the component equal to τ^{-1} and, second, the component proportional to $\operatorname{Im} \beta_{n+n,n}(0)$. It follows from formulae (2.64) and (2.67) given above that with a depth of the level y_n coinciding with the depth of the skin-layer within an order of magnitude $\operatorname{Im} \beta_{n+n,n}(0) \simeq \delta$.

The imaginary part of the frequency in (2.75) with satisfaction of the inequality

$$\frac{1}{s\omega_n\tau} \gg \left[A - \frac{\pi}{2} D \operatorname{Re}\beta_{n+s,\,n}(0)\right] \operatorname{Im}\beta_{n+s,\,n}(0)\, \pi D \frac{|\omega_n''|}{\omega_n} \simeq$$
$$\simeq \frac{2\alpha\pi^{4/3} e^2 v_x(p_0, 0)\, \delta\, [\hbar\Omega(p_0)]^{1/3} L_{ns}\omega_n(p_0)}{3^{2/3}\hbar^2 c^2 p_y(p_0)\,[p_y(p_0)\, v_y'(p_0)]^{1/3} (n - 1/n)^{2/3} (\omega_n''(p_0))} \left(\frac{\partial^2 \varsigma}{\partial p_z \partial\varphi}\right)_{\varphi=0} \quad (2.76)$$

is determined by the momentum relaxation time.

In the case $k_z \neq 0$ we obtain from equation (2.49) the following frequency spectrum and the damping decrement of the surface cyclotron oscillations:

$$\omega = s\omega_n(p_0)\left\{1 - \frac{s\omega_n''(p_0) + k_z v_z''(p_0)}{2s\omega_n(p_0)}\left[A - \frac{\pi}{2} D \operatorname{Re}\beta_{n+s,\,n}(0)\right]^2 + \frac{k_z v_z(p_0)}{s\omega_n(p_0)}\right\} - \frac{i}{\tau}. \quad (2.77)$$

It is easiest to analyze the possibilities of solving dispersion equation (2.49) in the limiting case of weak collisions $(\omega\tau)^{-1} \to 0$. If in this case $\omega_n'' > 0$, which corresponds to $D \operatorname{Re} \beta_{n+n,n}(0) < 0$, there may be a solution of dispersion equation (2.49) when the following condition is satisfied

$$\frac{\pi}{2} | D \operatorname{Re} \beta_{n+s,n}(0) | > -A. \qquad (2.78)$$

Such a solution exists in the gas model of a metal (here $\omega < s\omega_n$), when $A = 0$ (compare to [55]). If the Fermi-liquid interaction constant has a sufficiently large modulus value and is negative, so that inequality (2.78) is not satisfied, propagation of cyclotron waves near the minimum of the surface transition frequency, including those investigated in study [55] is forbidden. In the opposite case $\omega_n'' < 0$, when $D \operatorname{Re} \beta_{n+n,n}(0) < 0$, consistent with (2.32) the waves investigated in study [55] will not exist. On the other hand solution (2.77) of dispersion equation (2.49) exists in our theory when condition (2.73) is satisfied. The frequency spectrum of (2.79) lies above the surface transition frequency, which serves to suppress the collisionless Landau damping. In the next chapter we will compare the developed theory of surface cyclotron waves and the resonance properties of impedance to available experimental research on bismuth.

Chapter 3

A COMPARISON OF QUASI-CLASSICAL THEORY OF CYCLOTRON RESONANCE AT SKIPPING ORBITS IN ELECTRON LIQUID TO EXPERIMENTAL DATA OBTAINED FOR BISMUTH IN THE MICROWAVE RANGE

1. The possibilities for comparison to experiment

In proceeding to discuss the possibilities for comparing the theory developed above to experiment, we begin by emphasizing that consistent with formula (2.39) the difference between the real part of the resonance frequency and the surface transition frequency is due to electron interaction proportional to the ratio $\hbar L/\varepsilon_F$. Hence we may expect the most explicit manifestation of the influence of electron action on the resonance properties of the surface impedance in metals in which the ratio of the quantum of incident emission $\hbar\omega$ to the Fermi energy ε_F is comparatively large. Bismuth has such a property and $\hbar\omega/\varepsilon_F$ is large compared to other substances due to the low Fermi energy (two orders of magnitude smaller than, for example, the alkali metals or copper). On the other hand bismuth is a convenient semimetal for comparing theoretical and experimental results for at least three reasons. First, the electron Fermi surface has been extensively investigated in bismuth (with better than 1% accuracy) (see, for example, [60, 61]). In the subsequent sections we will determine that good knowledge of the parameters of the Fermi surface of bismuth is one decisive factor that allows us to make quantitative conclusions regarding the magnitude of the Fermi-liquid interaction of electrons. Second, experimenters have succeeded in obtaining sufficiently clean bismuth samples with a high (several millimeters) free path length of the electrons, corresponding to an electron momentum relaxation time of $\tau \gtrsim 10^{-9}$ s [25, 27, 60, 61]. Third, bismuth has been investigated over a broad frequency range [20, 25, 27] and experimental data have been obtained [27] on impedance oscillations related to the cyclotron resonance on surface levels with better than 1% accuracy.

We also emphasize that in accordance with our theory [48] presented in the preceding chapter in the case of rare collisions when $\omega\tau \gg 1$ condition (2.20) for bismuth infers the requirement for a positive sum of the Fermi-liquid interaction constants

$$\mathfrak{A} = \sum_{l=0}^{\infty} \frac{(2l+1)A_l}{1+A_l} > 0. \tag{3.1}$$

Accounting for the electrodynamic effects makes it necessary to satisfy inequality (2.50) which allows us to find a lower estimate for the interelectron interaction value (3.1). Using the values of the parameters in accordance with study [60] for the electron spectrum (2.33) of bismuth: $m_1 = 0.57\cdot10^{-2}m_0$, $m_2 = 1.15\cdot10^{-2}m_0$, $m_3 = 1.27m_0$, where m_0 is the free electron mass and $\varepsilon_F = 2.86\cdot10^{-14}$ ergs in accordance with formulae (2.47) and (2.32) we may write the following inequality that follows from condition (2.50):

$$\mathfrak{A}^+ \geqslant 0{,}3\frac{(n-{}^1/_4)^{4/3}}{s^2 L_{ns} B^{2/3}}, \tag{3.2}$$

where B is the magnetic field strength in oersteds. In deriving this relation it was assumed that the p_z axis of the ellipsoidal Fermi surface of bismuth is oriented in the direction of the constant magnetic field B. Experiments with this orientation of the sample were performed in studies [25, 26]. In deriving inequality (4.2) we utilized the expressions:

$$A = \frac{(3\pi)^{1/3}(m_3\varepsilon_F)^{1/2}L_{ns}}{2^{11/6}(n-{}^1/_4)^{2/3}}\left(\frac{\hbar|e|B}{\sqrt{m_1m_2}\,c\varepsilon_F}\right)^{1/3}\sum_{l=0}^{\infty}\frac{(2l+1)A_l}{1+A_l},$$

$$D = \frac{6\sqrt{2}\,e^2m_3\varepsilon_F^{3/2}}{\pi\hbar^2c^2\sqrt{m_1}},$$

$$\beta_{n+s,\,n}(0) = \frac{R\varphi_n^2}{\pi^2s^2} = \frac{3^{2/3}(n-{}^1/_4)^{2/3}}{2^{1/6}\pi^{4/3}s^2}\left(\frac{\hbar}{\varepsilon_F}\right)^{1/3}\sqrt{\frac{\varepsilon_F}{m_2}}\left(\frac{\sqrt{m_1m_2}\,c}{|e|B}\right)^{1/3}.$$

For the surface impedance resonances of bismuth at the orbits of skipping electrons discovered in study [25] the right half of formula (3.2) is at a maximum for the resonance corresponding to the transition with quantum numbers $n = 3$, $s = 2$ at $B = 2.11$ Oe. The right half of (3.2) is ~0.08 in this case. If we refer to an analysis of the data from study [26] the transition $n = 4$, $s = 2$ in a magnetic field of 1.25 Oe corresponds to the maximum of the right half of formula (3.2) among the impedance resonances that are differentiable in Fig. 19 of this study. These data correspond to a somewhat greater value for the lower estimate of (3.1) since we obtain 0.16 for the right half of formula (3.2). The importance of these estimates lies in the fact that they allow us to determine the magnitude of the parameter $\mathfrak{A}$ where we may ignore electrodynamic effects. For example, if it turns out that $\mathfrak{A} \sim 1$ in processing of experimental data then accounting for electrodynamic effects becomes insignificant if we wish to determine the resonance properties of the impedance of the metal.

It is difficult to use qualitative data for the positions of the impedance resonances of bismuth given in study [25] for detailed comparison to our theory, since the experimental inaccuracy in determining the positions of the resonance peaks of the impedance is comparatively high and equals 2%. On the other hand if we assume that no contradictions to our theory arise within this accuracy then in accordance with formula (2.39) based on analysis of experimental data from study [25] we may state that the value of (3.1) does not exceed 10 for bismuth. The reason for the difficulty in comparing theory to experiment [25] may be found in the poor accuracy in determining the positions of the resonance impedance extrema which is evidently related to the relatively short path length of electrons in study [25] ($\omega\tau \sim 8$). The higher accuracy data obtained in experimental study [27] makes it possible to perform quantitative processing of the measurement results. In the following sections we will perform such an analysis of the data from study [27].

2. The resonance transition frequency between the levels of surface electrons accounting for deviations in the Fermi surface of bismuth from ellipsoidal form

In order to interpret the data from experimental studies [15, 16] on cyclotron resonance at the skipping surface electrons it was sufficient to account for only a single ellipsoid of the electron Fermi surface of bismuth. Qualitatively a different situation exists in the experimental conditions of study [27] when two ellipsoids appear in the cyclotron resonance. In this regard the theory of electron liquid in the conditions of study [27] will account for not only the interaction of electrons of one ellipsoid whose parameters determine cyclotron resonance in studies [20, 25] but will also account for the interaction of electrons of various ellipsoids. In this section based on the theory developed in studies [48, 51, 52], we will identify a method for determining the "interellipsoid" interaction effects. According to study [48] accounting for electron interaction will produce corrections in the frequency spectrum in the order of the ratio of the quantum energy of the incident emission to the Fermi energy. The real deviation of the electron Fermi surface of bismuth from ellipsoidal form may also lead to corrections of comparable magnitude. Hence in order to assure maximum accuracy in our comparison of theory to experiment we will base our examination below on the data from study [62] (see also [63]) describing the Fermi surface of bismuth. Moreover, we will strive to clearly demonstrate how the need to perform more precise measurements in the investigation of cyclotron resonance follows from our theory. We emphasize that for regular consideration of the Fermi electron surface the accuracy of measurements performed in [20, 25, 27] is quite sufficient. On the other hand such measurement accuracy is quite low for determining interelectron interaction.

In the conditions of experimental study [27] the constant magnetic field lying in the plane of the surface of the crystalline bismuth sample was oriented along the C_2 crystallographic symmetry axis. We will orient the z axis in this direction and will plot the y axis perpendicular to the surface of the metal along axis C_3. Based on the data from study [62] (see also [64-67]) the equation of the electron Fermi surface in the coordinate system p_x', p_y', p_z' whose p_x' axis is perpendicular to C_3 and forms the angle $\pi/6$ with the cyrstallographic axis C_1 while the p_y' axis is inclined at $6°20'$ to the C_3 axis, is represented by the formula:

$$B_{00} = (p_x'^2\gamma^2 + p_y'^2\beta^2 + p_z'^2)\,(1 - \overline{\Delta}\,(\mathbf{p}')). \qquad (3.3)$$

Here $\gamma = 13.04$; $\beta = 9.88$; $B_{\infty} = 58.79 \cdot 10^{-2}$ $(\mathrm{g \cdot cm/s})^2$. The quantity entering into (3.3) that is small compared to unity is equal to:

$$\begin{aligned}\overline{\Delta}\,(p') = {} & b_{20}P_2^0(\beta p_y' B_{00}^{-1/2}) + b_{21}P_2^1(\beta p_y' B_{00}^{-1/2})\, p_z'\,(B_{00} - \beta^2 p_y'^2)^{-1/2} + \\ & + b_{22}P_2^2(\beta p_y' B_{00}^{-1/2})\,(\gamma^2 p_x'^2 - p_z'^2)\,(B_{00} - \beta^2 p_y'^2)^{-1} + b_{40}P_4^0(\beta p_y' B_{00}^{-1/2}) + \\ & + b_{41}P_4^1(\beta p_y' B_{00}^{-1/2})\, p_z'\,(B_{00} - \beta^2 p_y'^2)^{-1/2} + b_{42}P_4^2(\beta p_y' B_{00}^{-1/2}) \times \\ & \times (\gamma^2 p_x'^2 - p_z'^2)\,(B_{00} - \beta^2 p_y'^2)^{-1} + \\ & + b_{43}P_4^3(\beta p_y' B_{00}^{-1/2})\, p_z'\,(3\gamma^2 p_x'^2 - p_z'^2)\,(B_{00} - \beta^2 p_y'^2)^{-3/2} + \\ & + b_{44}P_4^4(\beta p_y' B_{00}^{-1/2})\,[(\gamma^2 p_x'^2 - p_z'^2)^2 - 4\,(\gamma p_x' p_z')^2]\,(B_{00} - \beta^2 p_y'^2)^{-2}. \end{aligned} \qquad (3.4)$$

Here P_n^m are the associated Legendre polynomials; the coefficients of the expansion [62] are: $b_{20} = -0.0526$; $b_{21} = -0.0207$; $b_{22} = -0.0346$; $b_{40} = -0.0362$; $b_{41} = -0.0017$; $b_{42} = -0.0036$; $b_{43} = 0.0010$; $b_{44} = -0.0003$. The model of the Fermi surface described by formula (3.3) is called a nonellipsoidal model below. In the first approximation for $\overline{\Delta} = 0$ surface (3.3) is an ellipsoid with the semiaxes $B_{00}^{1/2}/\gamma = 0.588 \cdot 10^{-21}$ g·cm/s, $B_{00}^{1/2}/\beta = 0.0776 \cdot 10^{-21}$ g·cm/s, $B_{00}^{1/2} = 7.667 \cdot 10^{-21}$ g·cm/s. The axes of the coordinate system of $\mathbf{p}'$ lie along the principal axes of this ellipsoid.

If we represent the electron energy spectrum with the formula

$$\varepsilon\,(\mathbf{p}') = [p_x'^2\gamma^2 + p_y'^2\beta^2 + p_z'^2]\,\varepsilon_F B_{00}^{-1}\,[1 - \overline{\Delta}\,(\mathbf{p}')], \qquad (3.5)$$

where ε_F is the Fermi energy, the surface $\varepsilon(\mathbf{p}') = \varepsilon_F$ corresponds to formula (3.3). We will assume it is possible to use formula (3.5) bearing in mind that in the microwave range the ratio of the quantum energy of the incident wave $\hbar\omega$ to energy $\varepsilon_F \sim 10^{-2} \ll 1$.

We set $m_1 = B_{00}/2\gamma^2\varepsilon_F$, $m_2 = B_{00}/2\beta^2\varepsilon_F$, $m_3 = B_{00}/2\varepsilon_F$. Since the masses m_1, m_2, and m_3 are clearly differentiated, in the conditions of Fig. 2 which schematically represents the cross-section of the electron Fermi surface by the plane $p_y = 0$ (p_x, p_y, p_z are the projections of the electron momentum), we may write the following expressions for the energy spectrum of the ellipsoids whose p_x' axes form an angle of $\pm\pi/6$ with the direction of the C_1 axis by means of formula (3.5) (Fig. 2):

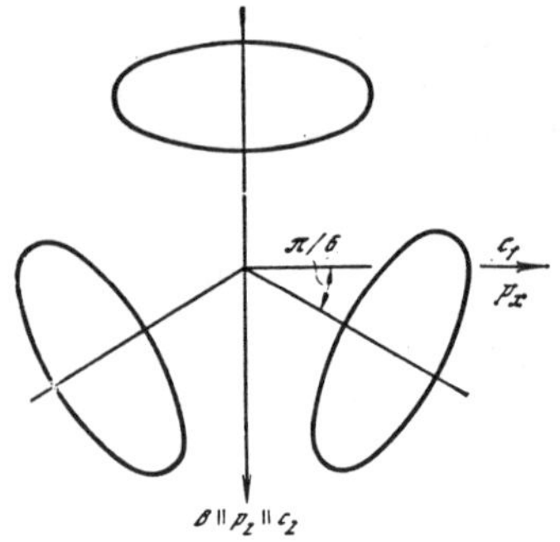

Fig. 2. Configuration of the electron ellipsoids of the Fermi surface of bismuth

$$\varepsilon_{1,2}(\mathbf{p}) = \varepsilon_{1,2}^{(0)}(\mathbf{p})\,[1 - \Delta_{1,2}(\mathbf{p})], \qquad (3.6)$$

$$\varepsilon_{1,2}^{0}(\mathbf{p}) = \frac{3p_x^2}{8m_1}\left[1 + \frac{m_1}{3m_3} + \frac{m_1}{3m_2}\sin^2 6°23'\right] + \frac{p_z^2}{8m_1}\left[1 + \frac{3m_1}{m_3} + \frac{3m_1}{m_2}\sin^2 6°23'\right] +$$
$$+ \frac{p_y'^2}{2m_2}\cos^2 6°23' \mp p_x p_z \frac{\sqrt{3}}{4m_1}\left[1 - \frac{m_1}{m_3} - \frac{m_1}{m_2}\sin^2 6°23'\right] +$$
$$+ \frac{p_y}{4m_2}(\sqrt{3}p_z \mp p_x)\sin 12°46'.$$

For these ellipsoids which we will call the first and second ellipsoids the coordinate system $\mathbf{p}'$ is related to the system $\mathbf{p}$ by rotation about the p_x' axis by 6°23′ and subsequent rotation about the p_y axis by $\pm\pi/6$. Hence the quantities $\Delta_{1,2}(\mathbf{p})$ are obtained from $\Delta(\mathbf{p}')$ by linear coordinate transformation: $p_x' = p_x\sqrt{3}/2 \pm p_z/2$, $p_y' = p_y\cos 6°23' - [p_z\sqrt{3}/2 \pm p_x/2]\sin 6°23'$, $p_z' = (p_x\sqrt{3}/2 \pm p_x/2) \times \cos 6°23' + p_y \sin 6°23'$. Accounting for the smallness of $\bar{\Delta}$ compared to unity, in the zeroeth approximation in the small angle 6°23′ we have:

$$\Delta_{1,2}(p_x, p_y, p_z) = \bar{\Delta}\,(p_x\sqrt{3}/2 \mp p_z/2,\ p_y,\ p_z\sqrt{3}/2 \pm p_x/2).$$

Using formula (3.6) for the frequency of the resonance transition from one surface level (n) to another ($n + s$) in the quasi-classical approximation we have

$$s\left[\frac{\pi^2 e^2 B^2 v_x^2}{3c^2(n - 1/4)}\frac{\partial v_y}{\partial p_y}\right]^{1/3}_{p_y=0} = s\left[\frac{\pi^2 e^2 B^2}{2\hbar c^2 m_1 m_2 (n - 1/4)}\left(\varepsilon_F - \frac{2p_z^2}{3m_3}\right)\right]^{1/2} \times$$
$$\times\,[1 - (2/3)\Delta_{1,2}(p_x^F(p_z), 0, p_z) - (1/3)\,m_2\varepsilon_F(\partial^2/\partial p_y^2)\,\Delta_{1,2}(p_x^F(p_z), 0, p_z)], \qquad (3.7)$$

where $p_x^F(p_z) = \pm(p_z/\sqrt{3}) \pm \sqrt{8m_1\varepsilon_F/3}$; is electron charge; c is the speed of light; B is the field strength of the constant magnetic field. In deriving (3.7) we ignored corrections $m_1/9m_3$ and (1/3) sin 6°23′(~10^{-3}) that are small compared to $\Delta_{1,2}$.

Bearing in mind that $[p_x^F(0)]^2 = (p_x^F)^2 = 8m_1\varepsilon_F/3$, and $\Delta_{1,2}(p_x^F, 0, 0) = -(1/2)b_{20} + 3b_{22} + (3/8)b_{40} - (15/2)b_{42} + 105b_{44}$, $m_2\varepsilon_F\partial^2\Delta(p_x^F, 0, 0)/\partial p_y^2$

$= (3/2)\, b_{20} - 3b_{22} - (15/4)\, b_{40} + 60b_{42} - 210b_{44}$, we obtain the following expression for the extremal transition frequency corresponding to the central cross-section of the Fermi surface:

$$s\omega_n = s\omega_n^{(1)}(0) = s\omega_n^{(2)}(0) = s\left[\frac{\pi^2 e^2 B^2 \varepsilon_F}{2\hbar c^2 m_1 m_2 (n - {}^1/_4)}\right]^{1/2}\left[1 - \frac{b_{20}}{6} - b_{22} + b_{40} - 15b_{42}\right].$$

We will write this formula by expressing its right half through the experimentally observed quantities [61]: S is the area of the extremal cross-section of the Fermi surface and $m^* = (2\pi)^{-1}\partial S/\partial\varepsilon$ is the effective mass.

For this we incorporate the fact that

$$S = 2\sqrt{m_1 m_2 \varepsilon_F}\,\pi\left\{1 + \frac{1}{4}b_{20} + \frac{3}{2}b_{22} + \frac{9}{64}b_{40} + \frac{45}{16}b_{42} + \frac{315}{8}b_{44}\right\}.$$

In accordance with this we have

$$s\omega_n = s\left[\frac{\pi e^2 B^2 S}{4c^2 (n - {}^1/_4)\,\hbar\,(m^*)^3}\right]^{1/2}. \tag{3.8}$$

Here we take account of the fact that in accordance with the values of the coefficients b_{4n} determined from experiment it turns out that

$$1 + \frac{35}{32}b_{40} - \frac{105}{8}b_{42} + \frac{105}{8}b_{44} = 1.$$

Since the resonance transition frequency for the electrons of the third ellipsoid whose major axis lies on p_x is equal to

$$s\omega_n^{(3)}(p_z) = s\left[\frac{2\pi^2 e^2 B^2}{3\hbar c^2 m_2 m_3 (n - {}^1/_4)}\left(\varepsilon_F - \frac{p_z^2}{2m_1}\right)\right]^{1/2},$$

which is significantly less than expression (3.8) with comparable values of the electron energy levels, this allows us in conditions where the numbers n and s do not significantly exceed 10 to limit our examination to incorporating only the contribution of the first two ellipsoids that are symmetrical with respect to the C_2 axis (Fig. 2).

In concluding this section we again emphasize the small difference between the nonellipsoid model and the ellipsoid model. In this regard in describing the comparatively minor contribution of interelectron interaction below we will ignore the squared corrections for these two minor effects.

In accordance with studies [48, 51] we have (i = 1, 2) for the Fourier-components $G_i(k,\ p_z)$ of the perturbations to the distribution functions on the Fermi surface averaged over the angular variable and determining the position of the electron in orbit in the momentum space relating to the (first) two ellipsoids:

$$G_i(k, p_z) = [\hbar/ip_y'(0)]\, I_{ns}(k, 0)\int_0^\infty dk' I_{ns}(k', 0)\,[e\mathbf{v}^{(i)}\mathbf{E}(k') + $$
$$+ i\omega\sum_{j=1,2}\alpha_{ij}\int dp_z' G_j(k', p_z')\,[\omega + (i/\tau) - s\omega_n - ({}^1/_2)\,p_z^2 s\omega_n'']^{-1}. \tag{3.9}$$

In writing this system we ignored the effects of minor collisionless damping and used the following conventions in accordance with [48, 51]: ω and $\mathbf{E}(k)$ is the frequency and Fourier-component of the electric field, $s\omega_n = s\omega_n(p_z = 0)$ is the maximal transition frequency between the levels of the skipping electrons, τ is the free path time of the electron, $\mathrm{v}^{(i)} = \mathrm{v}^{(i)}(p_z = 0)$ is the velocity of the skipping electron at the central ellipsoid cross-section, $s\omega_n'' = (d^2/dp_z^2)\,[s\omega_n(p_z)]$ when $p_z = 0$, $p_y'(0)$ is the derivative of the y-momentum component in the angular variable φ characterizing the position of the electron in orbit, when $\varphi = 0$ corresponding to skipping of the electron along the surface; α_{ij} is the tensor whose diagonal components describe electron interaction of one ellipsoid and whose nondiagonal components describe "interellipsiodal" interaction, where in the case of the geometry of Fig. 2 $\alpha_{11} = \alpha_{22}$, $\alpha_{10} = \alpha_{21}$, finally, $I_{ns}(k, 0)$ is the matrix element equal to, consistent with study [48]:

$$I_{ns}(k, 0) = \int_{-1}^{1} dx\,[1 + 2sx^2/3\,(n - 1/4)]^{-1/4} \cos[\pi sx] \cos[k\Omega\varphi_n^2(0)(1 - x^2)/2v_y'(0)],$$

$$v_y'(0) = p_y'(0)/m_2 = \sqrt{2\varepsilon_F/m_2},\quad \Omega = |e|\,B\sqrt{3}/(2c\sqrt{m_1 m_2}),$$

$$\varphi_n(0) = [3\pi\sqrt{3}\,(n - 1/4)\,\hbar|e|B/4\varepsilon_F c\sqrt{m_1 m_2}]^{1/3}.$$

Equations (3.9) are written for the case where the frequency ω is close to the maximal value of the transition frequency $s\omega_n$. The solution of this system of equations takes the form

$$G_i(k, p_z) = [\hbar/ip_y'(0)]\,[\omega + (i/\tau) - s\omega_n -$$

$$- p_z^2 s\omega_n''/2]^{-1}\Lambda^{-1} I_{ns}(k, 0)\int_0^\infty dk' I_{ns}(k', 0)\, e\mathrm{v}^{(i)}\mathbf{E}(k'),$$

where

$$\Lambda = 1 - \bar{A}\alpha^+/\sqrt{\Delta(0)}.$$

Here $\alpha^+ = \alpha_{11} + \alpha_{12}$, $\Delta(0) = 2\,[s\omega_n - \omega - i/\tau]/s\omega_n''$,

$$\bar{A} = -\frac{2\pi\hbar N_{ns}\omega_n}{p_y'(0)\,\omega_n''} = \frac{3^{3/4}\pi^{4/3}\varepsilon_F^{2/3} L_{ns} m_3}{2^{5/3}(n - 1/4)^{2/3}(m_1 m_2)^{1/6}}\left(\frac{\hbar|e|B}{c}\right)^{1/3}.$$

Here we incorporate the fact that $(\omega_n/\omega_n'') = -9\varepsilon_F m_3/4$ and

$$N_{ns} = \int_0^\infty dk I_{ns}^2(k, 0) = \frac{3^{3/4}\pi^{4/3}\varepsilon_F^{2/3} L_{ns} m_3}{2^{5/3}(n - 1/4)^{2/3}(m_1 m_2)^{1/6}};\qquad L_{ns} \simeq \ln\frac{6(n - 1/4)}{s};\quad n \gg s.$$

The vanishing of the resonance denominator $\Lambda = 0$ in the right half of formula (3.10) corresponding to cyclotron resonance at the skipping electrons produces a sharp peak in the surface impedance of the metal containing the ellipsoidal Fermi surface when the following inequality is satisfied

$$\alpha^+ \gg (-1/\bar{A}^2\omega_n''\tau)^{1/3} > 0.$$

We will call the impedance peak a sharp peak if the inequality

$$|1 - s_1 (n - 1/4)^{2/3}/s (n_1 - 1/4)^{2/3}| \gg (\omega\tau)^{-1},$$

is satisfied, indicating that the width of the peak corresponding to the $n \to n + s$ transition is less than the distance to the neighboring peak corresponding to the transition $n_1 \to n_1 + s_1$. Here the spectrum of resonance frequencies (compare to (2.28)) takes the following form:

$$\omega = s\omega_n - \bar{A}^2 (\alpha^+)^2 s\omega_n''/2 - i/\tau. \tag{3.11}$$

3. Determining the interelectron interaction parameters based on a comparison of quasi-classical theory to experiments in the microwave range

In this section we will address the issue of representing the quantity α^+ by parametrization from study [58] and obtaining quantitative estimates for the interelectron interaction based on a comparison of the developed theory to experimental data from study [27] obtained in the RF range. We first remember that in accordance with the theoretical concepts developed in study [58] it is convenient to make the transformation from the momentum space (**p**) to the space (**w**) in which the ellipsoidal Fermi surface is transformed into a spherical surface $\varepsilon = \mathbf{w}^2$. The Jacobian of such a transformation $D(\,d\rho = Dd\delta d\theta_{\mathbf{w}})$ takes the form$(D = \varepsilon 2m_1 m_2 m_3)^{1/2}$. In this regard consistent with study [22] for the function $\varphi(\mathbf{p}, \mathbf{p}')$ of Fermi-liquid electron interaction we have:

$$2(2\pi\hbar)^{-3} D\varphi_{ij}(\mathbf{p}, \mathbf{p}') = (4\pi)^{-1} \sum_{l=0}^{\infty} (2l+1) A_{ij}^{(l)} P_l(\cos\Theta_{\mathbf{ww}'}).$$

We may easily determine (compare to study [48]) that we have:

$$\alpha^+ = (2\pi)^{-1} (6\varepsilon_F m_3)^{-1/2} \sum_{l=0}^{\infty} (2l+1) A_l^+/(1 + A_l^+), \tag{3.12}$$

where $A_l^+ = A_{11}^{(l)} + A_{12}^{(l)}$. Here we incorporate the fact that $d\varphi\, dp_z = d\Theta_{\mathbf{w}} \left(\frac{3\varepsilon_F m_3}{2}\right)^{1/2}$. Using derived expression (3.12) and in accordance with relation (3.11) we find in the limit $\tau \to 0$ the following formula for the resonance frequency:

$$\omega = s\omega_n \left\{1 + \frac{\pi^{2/3} L_{ns}^2}{2^{16/3} (n - 1/4)^{4/3}} \left(\frac{\hbar e^2 B^2}{\varepsilon_F^2 c^2 m_1 m_2}\right)^{1/3} \left(\sum_{l=0}^{\infty} \frac{(2l+1) A_l^+}{1 + A_l^+}\right)^2\right\}. \tag{3.13}$$

It is quite clear from here that the difference between the experimentally measured values of the resonance frequencies and the resonance transition frequencies of the electrons between the surface levels our theory makes it possible to find:

$$\mathfrak{A}^{+} = \sum_{l=0}^{\infty} \frac{(2l+1)A_l^{+}}{1+A_l^{+}}, \tag{3.14}$$

characterizing interelectron interaction. Specifically, consistent with (3.13) we have

$$\mathfrak{A}^{+} = \sqrt{\frac{\omega}{s\omega_n} - 1}\left(\frac{\varepsilon_F}{\hbar\omega}\right)^{1/2} \frac{[32s(n - 1/4)]^{1/2}}{L_{ns}}. \tag{3.15}$$

We emphasize that expression (3.15) derived by parametrization from study [58] differs from expression (3.1) discussed above due to the existence of the constants $A_{12}^{(l)}$ caused by the "interellipsoidal" electron interaction in bismuth.

The difference $E_{n+s} - E_n$ from the corresponding quasi-classical value $\hbar s\omega_n$ is large for actual experimentally-observed resonances. Hence, for comparison to experiment we will use the following generalization of formula (3.15) below

$$\mathfrak{A}^{+} = (4/L_{ns})[2s(n - 1/4)\varepsilon_F/\hbar\omega]^{1/2}[1 - (3Se^2B_{n+s,n}^2/8c^2(m^*\omega^3)\pi\hbar)^{1/3}(a_{n+s} - a_n)]^{1/2} =$$
$$= (4/L_{ns})[2s(n - 1/4)\varepsilon_F/\hbar\omega]^{1/2}[1 - (B_{n+s,n}/B_{n+s,n}^0)^{2/3}]^{1/2}. \tag{3.16}$$

Here a_n are the zeros of the Airy function ($Ai(-a_n) = 0$) which agrees with $[(3\pi/2)(n - 1/4)]^{23}$, while $B_{n+s,n}$ is the value of the magnetic field where in the case of a given frequency ω resonance is observed corresponding to the transition from the $n + s$ level to the n level. The quantity $B_{n+s,n}^0 = C(\omega)\,[a_{n+s} - a_n]^{-3/2}$ where consistent with (3.8) $C(\omega) = [8\pi\hbar c^2\omega^2(m^*)^3/3Se^2]^{1/2}$ is the resonance value of the magnetic field corresponding to the transition $n + s \rightarrow n$ calculated by the electron theory. Hence, as is obvious from formula (3.16) the difference between the actually-observed values of the resonance magnetic field and the calculated values of one-particle resonance fields determines the interelectron interaction constants.

We will dwell on the possibility for determining the parameters of interelectron interaction provided by formula (3.15). The experimental results of study [27] provide us with the values of the resonance fields $B_{n+s,n}$ measured with very good accuracy. On the one

Table 1

n	$n+s$	$B^0_{n+s,n}$, Oe	$B_{n+s,n}$, Oe	$1-(B_{n+s,n}/B^0_{n+s,n})^{2/3}$
3	5	$3.026\cdot(1\pm0.011)$	$2.92\cdot(1\pm0.007)$	0.023 ± 0.012
4	6	$3.414\cdot(1\pm0.011)$	$3.31\cdot(1\pm0.009)$	0.020 ± 0.013
5	7	$3.762\cdot(1\pm0.011)$	$3.63\cdot(1\pm0.008)$	0.024 ± 0.013
6	8	$4.079\cdot(1\pm0.011)$	$3.98\cdot(1\pm0.010)$	0.016 ± 0.014
7	9	$4.373\cdot(1\pm0.011)$	$4.29\cdot(1\pm0.012)$	0.013 ± 0.015

hand for our purposes we require data on the values of the quantities s and m^* with an equally high degree of accuracy. Such data for bismuth may be found in study [61] which provide: $m^*/m_0 = (0.82 \pm 0.005)\cdot 10^{-2}$, $S = (1.300 \pm 0.003)\cdot 10^{-42}$ $(g\cdot cm/s)^2$. Then, bearing in mind the value of the variable frequency $\omega/2\pi = 3.626\cdot 10^{10}$ Hz used in study [27] we obtain $C(\omega) = 11.41\ (1 \pm 0.011)$ Oe.

Table 1 gives calculated values of $B^0_{n+2,n}$ and resonance values of $B_{n2,n}$ measured in study [27]. We emphasize that in the majority of cases covered here the accuracy of the measured quantities exceeds the accuracy of the calculated quantities. It follows from an analysis of the experimental data and a comparison to data from the theory of resonance transitions of free electrons that high accuracy of the experimentally-measured quantities is required in order to determine the interelectron interaction parameters of interest to us. At the same time it is clear from Table 1 that the use of data from study [27] on the resonance values of the magnetic field and the data from study [61] on the effective mass and cross-sectional area of the Fermi electron surface of bismuth in a number of cases causes the inaccuracy of the experimentally-measured quantities to drop below the quantity characterized by them $1 - [B_{n+s,n}/B^0_{n+s,n}]^{2/3}$. Hence we may use formula (3.16) that takes the following form in the experimental conditions of [27]

$$\mathfrak{A}^+ = 61\sqrt{s\,(n - 1/4)}\, L^{-1}_{ns}\sqrt{1 - [B_{n+s,n}/B^0_{n+s,n}]^{2/3}}.$$

For example, bearing in mind that $L^{-1}_{ns}(n - 1/4)^{1/2}$ adopts the following values, respectively: 0.79 (n = 3), 0.80 (n = 4), 0.82 (n = 5, 0.84 (n = 6), 0.87 (n = 7), we obtain for the quantity $\mathfrak{A}^+$ 12.9-7.47 (n = 3); 12.38-5.50 (n = 4; 13.40-7.05 (n = 5); 12.28-2.89 (n = 6); 12.72-0 (n = 7). It follows from these values and the data provided in Table 1 that the experimental data we used in applying formula (3.16) leads to the conclusion that the value of $\mathfrak{A}^+$ for bismuth lies between 7.5 and 12.3. In this chapter we will not provide results from the processing of data from study [27] on resonances with $s > n$, since at present our theory of accounting for interelectron interaction is formulated in the quasi-classical approximation suitable only when $n \gg s$. In addition we note that we determined frequency (3.11) using the resonance properties of function (3.10) and did not solve electrodynamic equations. In view of the fact that a value greater than unity was obtained for the sum of the interaction constants $\mathfrak{A}^+$, accounting for the electrodynamic effects, as emphasized in the first section of this chapter, is not critical in order to determine the value of $\mathfrak{A}^+$. We note in this context that it is fundamentally possible to determine the interelectron interaction value strictly from experiments on cyclotron resonance at the skipping orbits of surface electrons without incorporating high accuracy data on the parameters characterizing the electrons on the Fermi surface. For this it is sufficient to compare the resonance magnetic fields $B_{n+s,n}$ for various quantum numbers. Such a comparison for the experimental values given in Table 1 makes it possible to obtain $\mathfrak{A}^+ < 25$.

4. A comparison of experimental data on cyclotron skipping electron resonance in bismuth in the microwave range to the electron liquid and electron gas theories accounting for the finite momentum relaxation times of the electrons

In the preceding sections we formulated a quasi-classical theory of cyclotron resonance at the skipping orbits in the electron liquid of metals suitable for interpreting experiments in the microwave range. On the other hand, the study by Prange and Nee [64] as well as extensive subsequent research have formulated the viewpoint that it is possible to interpret Khaykin resonances [30] without any incorporation of interelectron interaction using only concepts of the transitions between the surface electron levels in the electron gas if the product of the electron momentum relaxation time τ and the emission frequency ω is much greater than unity ($\omega\tau \gg 1$). In this section based on results from our study [54] we will compare the corresponding impedance relations obtained in electron liquid theory [40, 51] and electron gas theory [64] to experiments with a specific bismuth sample at $\omega/2\pi$ equal to 36.26 GHz corresponding to the conditions and results from experimental study [27]. We will carry out this comparison accounting for the finite values of the parameter $\omega\tau$ based on data on the Fermi surface of bismuth obtained in study [61] and we will determine the influence of the change in this parameter $\omega\tau$ on the position of the resonance peaks of the magnetic field derivative of the surface impedance. Accounting for the finite values of $\omega\tau$ we also determine the Fermi-liquid interaction parameter that makes theory consistent with experiment.

Assuming, as is the case in study [27] a constant magnetic field B lying in the plane of the sample surface and parallel to the z axis lying on the C_2 crystallographic symmetry axis, we plot the y axis along the C_3 axis perpendicular to the surface of the metal. We again emphasize that the impedance oscillations investigated in study [27] are attributable to the electrons of the two Fermi-surface ellipsoids symmetrical with respect to the C_2 axis.

In the microwave range when using the same frequency as in study [27] $\omega/2\pi$ = 36, 26 GHz, the contribution to the surface impedance from electrons skipping along the surface of the metal is described by quasi-classical theory and consistent with [48] is given by the formula:

$$Z = \sum_{n'n} \frac{2\pi i D\omega\sqrt{2}}{3c^2\sqrt{m_3\varepsilon_F}} \frac{\alpha^2_{n'n}(0)}{[1-(B/B^0_{n'n})^{2/3}+i(\omega\tau)^{-1}]^{1/2}-A^+}. \tag{3.17}$$

In deriving (3.17) we used conventions analogous to those in (3.16). The value of the resonance magnetic field corresponding to the transition between the levels n and n' calculated by gas theory and ignoring electron interaction is equal to $B^0_{n'n} = C(\omega)(a_{n'} - a_n)^{-3/2}$, a_n is n^{th} root of the Airy function $Ai(-a_n) = 0$, $C(\omega) = [8\pi\hbar c^2(\omega m^*)^3/3S|e|]^{1/2}$, where S is the area of the extremal cross-section of the Fermi surface of bismuth equal to $S = (1.300 \pm 0.003)\cdot 10^{-42}$ $(\mathrm{g\cdot cm/s})^2$ according to study [61] while the ratio of the effective mass to the free electron mass is $m^*/m_0 = (0.82 \pm 0.005)\cdot 10^{-2}$, s is the speed of light, ε_F is the Fermi energy and e is electron charge. The constants entering into (3.17) are equal to

$$D = 9\sqrt{3e^2 m_3 \varepsilon_F^{3/2}}/2\sqrt{2\pi\hbar^2 c^2}\sqrt{m_1},$$

$$A^+ = \mathfrak{A}^+ J_{n'n}(a_{n'} - a_n)^{-1/2}\sqrt{\hbar\omega/\varepsilon_F}\sqrt{3}\,\pi/2 = A(\omega)\,J_{n'n}(a_{n'} - a_n)^{-1/2},$$

$$A(\omega) = \mathfrak{A}^+\sqrt{\hbar\omega/\varepsilon_F}\sqrt{3}\,\pi/2, \quad m_1 = 0{,}57\cdot 10^{-2} m_0, \quad m_3 = 1{,}27 m_0,$$

$\mathfrak{A}^+$ is the sum of the Fermi-liquid electron interaction constants. The table of the values of the integral

$$J_{n'n} = \Big[\int_0^\infty dx\,\mathrm{Ai}^2(x - a_n)\,\mathrm{Ai}^2(x - a_{n'})\Big]\Big[\int_0^\infty dx\,\mathrm{Ai}^2(x - a_n)\Big]^{-1} \times$$
$$\times\Big[\int_0^\infty dx\,\mathrm{Ai}^2(x - a_{n'})\Big]^{-1},$$

calculated in our study [54] is provided in Appendix II. In the evaluation

$$\alpha_{n'n}(0) = (2/\pi)\int_0^\infty dk\,[k^2 - i(\delta^3 k)^{-1}]^{-1} I_{n'n}(k),$$

where δ is the depth of the skin-layer in conditions of the anomalous skin-effect we will use the quasi-classical value of the matrix element below (compare to (2.12)):

$$I_{n'n}(k) = 2\int_0^1 dx\cos[\pi(n' - n)x]\cos[ky_n(1 - x^2)],$$

where the depth of occurrence of the nth level is equal to $y_n = \bar{y}_1(n - 1/4)^{2/3}$, $\bar{y}_1 = |3\pi^2\sqrt{3m_1}c\hbar^2/4eBm_2\sqrt{2\varepsilon_p}|^{2/3} \simeq (3\pi/2)(\hbar/2m_2\omega)^{1/2}(a_{n'} - a_n)^{1/2}$, $m_2/m_0 = 1.1\cdot 10^{-2}$. The magnetic field B derivative of expression (3.17) coincides with formula (42) from study [64] if we ignore interelectron interaction, i.e., if we take $A = 0$.

Taking the ratio of the depth of the first level $\bar{y}_n$ to the depth of the skin-layer δ to be small compared to unity $1 \gg \bar{y}_1/\delta \sim [\omega]^{-1/6}$, which is the case for bismuth at frequencies $\omega \gg (\varepsilon_F/\hbar)(\upsilon\hbar\omega_{\mathrm{Le}}/\varkappa\varepsilon_F)^4 \simeq 10^9\ \mathrm{s}^{-1}$, we write expressions for $\alpha_{n'n}(0)$ for different values of the numbers n and n'. If the condition $y_n \gg \delta$ is satisfied or, the equivalent

$$n' - 1/4 \ll (n - 1/4)[1 + \beta^{-2}(3\pi/2)^{-2/3}(n - 1/4)^{-2}]^{3/2}, \tag{3.18}$$

then (compare to study [55]) for $\alpha_{n'n}(0)$ we have the following expression

$$\alpha_{n'n}(0) = 4y_n\,[\pi(n'-n)]^{-2}(-1)^{n'-n}\,[1 + 4y_n/\sqrt{3}\delta\pi^2(n'-n)^2]. \tag{3.19}$$

The quantity β entering into (3.18) that is small compared to unity is equal to

$$\beta = (\bar{y}_1/\delta)(a_{n'} - a_n)^{-1/2} = (3\pi/2)^{2/3}\,(\hbar/2m_2\omega)^{1/2}/\delta.$$

Inequality (3.18) is violated for large values of n and n'. In this case the condition $y_n \gg \delta$ is satisfied, and we write the following formula for $\alpha_{n'n}(0)$ (compare to study [55]):

$$\alpha_{n'n}(0) = i\pi(n'-n)\,\delta^3\,(-1)^{n'-n}/y_n^2. \tag{3.20}$$

Below we will compare the conclusions of formula (3.17) to experiment. For this accounting for (3.19) and (3.20) we will write the expression for the magnetic field derivative of the real part of the impedance as

$$\frac{\partial R}{\partial B} = \left(\frac{3\pi}{2}\right)^{1/3} \frac{3^2 2^2 e^2 \varepsilon_F \sqrt{3m_3}}{\pi^3 \hbar c^4 m_2 \sqrt{m_1}}\,\mathrm{Re}\left(\sum_{\substack{nn' \\ y_n<\delta}} \frac{(n-1/4)^{4/3}(a_{n'}-a_n)}{(n'-n)^4}\,\times\right.$$
$$\times\left\{-\frac{8\beta(n'-n)^{2/3}(a_{n'}-a_n)^{1/2}}{\sqrt{3}\,\pi^2(n'-n)^2} + i\left[1 - \frac{16\beta^2(n-1/4)^{4/3}(a_{n'}-a_n)}{3\pi^4(n'-n)^4}\right]\right\} -$$
$$\left.-\sum_{\substack{nn' \\ y_n>\delta}} \frac{i\pi^6(n'-n)^2}{2^4\beta^6(a_{n'}-a_n)^2(n-1/4)^{8/3}}\right)\frac{\partial}{\partial B}\left\{\left[1-\left(\frac{B}{C(\omega)}\right)^{2/3}(a_{n'}-a_n)+\frac{i}{\omega\tau}\right]^{1/2} -\right.$$
$$\left.-A(\omega)\,\frac{J_{n'n}}{\sqrt{a_{n'}-a_n}}\right\}^{-1}. \tag{3.21}$$

We will now estimate the relative magnitude of the contribution of each sum in the right half of (3.21). In deriving this estimate we ignore the dependence of the resonance factor in (3.21) containing the magnetic field on the numbers n and n'. We find for the first sum corresponding to condition (3.18) $y_n \ll \delta\,(x = (n'-1/4)/(n-1/4))$,

$$\sum_{n-1/4<\beta^{-2}} \left(n-\frac{1}{4}\right)^{-2} \sum_{1+\left(n-\frac{1}{4}\right)^{-1} \leqslant x \ll [1+(n-1/4)^{-2}\beta^{-2}(2/3\pi)^{2/3}]^{3/2}} \frac{x^{2/3}-1}{(x-1)^4} \simeq \sum_{n-1/4<\beta^{-2}} \simeq \beta^{-2}. \tag{3.22}$$

For the second sum in the right half of (3.21) we obtain an upper estimation of:

$$\frac{\pi^6}{2^4}\left(\frac{2}{3\pi}\right)^2\beta^{-6} \sum_{n-1/4>\beta^{-2}}\ \sum_{x-1\ll\beta^2} \frac{(x-1)^2}{(x^{2/3}-1)^2} \leqslant 1{,}5\beta^{-4} \sum_{n-1/4>\beta^{-2}} \left(n-\frac{1}{4}\right)^{-2} \leqslant 1{,}5\beta^{-2}. \tag{3.23}$$

The smallness of the ratio of the left half of (3.23) to the left half of (3.22) allows us to ignore the contribution to the sum over n and n' of the components for which condition (3.18) is violated.

In the experimental situations examined above the resonance factor in formula (3.21) dependent on the magnetic field which we ignored in deriving (3.23) has no influence on the size of the second sum in the right half of (3.21). This is because the sharp peaks in the derivative of the impedance appearing in the range of magnetic fields shown in Figs. 4, 6-10 from study [27] are related only to the smallness of the magnetic field-dependent denominator in the first sum of formula (3.21). This first sum, as discussed above, corresponds to small values of the numbers n and n' (satisfying condition (3.18)). On the other hand with large values of n and n' corresponding to the second sum in the right half of (3.21) the denominator in (3.21) is not small compared to unity across the entire range of magnetic fields investigated experimentally in study [27] $(1 - (B/B^0_{n'n})^{2/3} \sim 1)$. Hence in this range of magnetic fields no sharp changes appear in the impedance derivative related to the second sum in (3.21) whose contribution to $\partial R/\partial B$ is insignificant compared to the contribution of the first sum.

For comparison of theory to experiment at $\omega/2\pi$ = 36.26 GHz we will use the following expression below for the magnetic field derivative of the real part of the impedance:

$$\frac{\partial R}{\partial B}=\left(\frac{3\pi}{2}\right)^{1/3}\frac{(6e)^2\varepsilon_F\sqrt{3m_3}}{\hbar\pi^3c^4m_2\sqrt{m_1}}\sum_{nn'}\frac{(n-1/4)^{4/3}(a_{n'}-a_n)}{(n'-n)^4}\,\mathrm{Re}\,\frac{\partial}{\partial B}\times$$
$$\times\left[\left\{-\frac{8\beta(n'-n)^{2/3}(a_{n'}-a_n)^{1/2}}{\sqrt{3}\,\pi^2(n'-n)^2}+i\left[1-\frac{16\beta^2(n-1/4)^{4/3}(a_{n'}-a_n)}{3\pi^4(n'-n)^4}\right]\right\}\left\{\left[1-\right.\right.\right.$$
$$\left.\left.\left.-\left(\frac{B}{C(\omega)}\right)^{2/3}(a_{n'}-a_n)+\frac{i}{\omega\tau}\right]^{1/2}-A(\omega)\frac{J_{n'n}}{\sqrt{a_{n'}-a_n}}\right\}^{-1}\right]. \qquad (3.24)$$

Before we compare the experimental curves $\partial R/\partial B$ and graphs of relation (3.24) drafted with various values of the parameters $A(\omega)$ and $(\omega\tau)^{-1}$ we will discuss the problem of the number of components that must be included in the sum in the right half of (3.24). In order to answer this question we note that the frequency $\omega/2\pi$ = 32.26 GHz corresponds to β = 0.14 and consistent with inequality (3.18) $n - 1/4 \ll (3/2)^{1/3}\beta^{-2}\pi^{-2/3}$ $\simeq$ 20. For comparison we shown in Fig. 3a a graph of relation (3.24) where summation is carried out over n from 1 to 5 and over n' from n + 1 to 15, $A(\omega)$ = 0, $(\omega\tau)^{-1}$ = 0.05, while in Fig. 3b the graph of the same (3.24) relation with identical values of $A(\omega)$ = 0 and $(\omega\tau)^{-1}$ = 0.05, although here summation is carried out over n from 1 to 10 and over n' from n + 1 to 25.

We see in comparing Fig. 3 a and b that with an increase in the number of components in the sum of (3.24) that first the positions of the extrema on the X-axis do not change and, second, the amplitude of the resonance peaks remains constant. On this basis in drafting graphs 3 c-7 we will limit summation over n from 1 to 5 and over n' from n + 1 to 15 in formula (3.24) below.

In comparing the graphs of relation (3.24) for different values of the parameters $A(\omega)$ and $(\omega\tau)^{-1}$ we note that the experimental values of the resonance magnetic fields were provided to us in a personal communication from R. Doezma, one of the authors of study [27]. Below

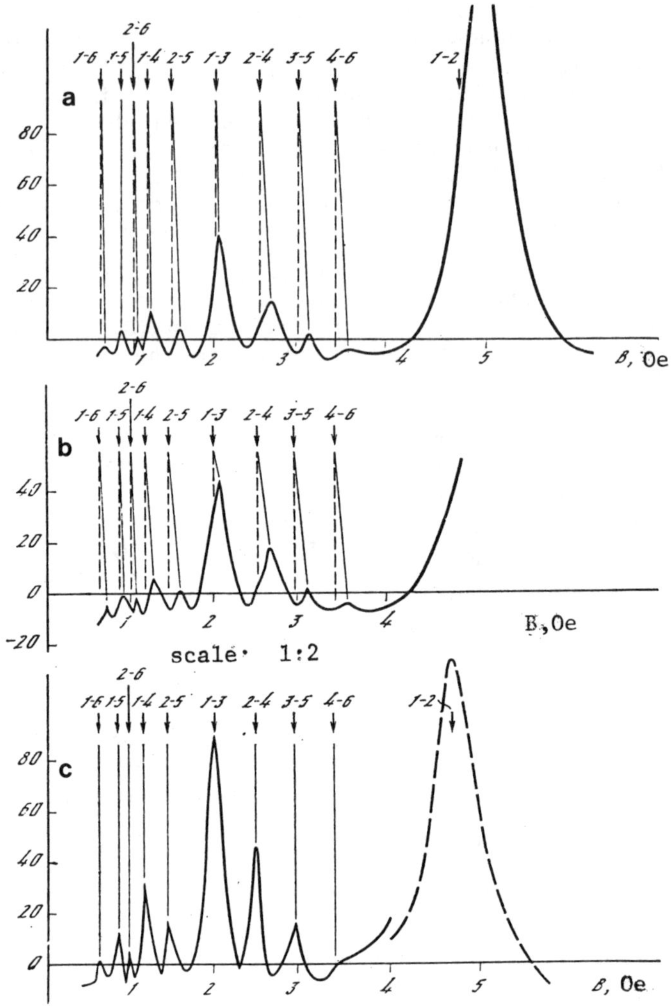

Fig. 3. Graph plotting the derivative of the real part of the impedance as a function of the magnetic field
a – $\omega\tau$ = 20, $A(\omega)$ = 0; b – $\omega\tau$ – 20 $A(\omega)$ = 0; c = $\omega\tau$ = 20, $A(\omega)$ = 2.5. $\partial R/\partial B \cdot 1.3 \cdot 10^{-16}$ s/cm·Oe is plotted on the y-axis. Frequency $\omega/2\pi$ = 36.26 GHz, β = 0.14.

we compare graphs of the (3.24) relation drafted for $A(\omega)$ = 0 and $A(\omega)$ = 2.5 corresponding to electron gas and electron liquid theories, respectively (Fig. 3a and c). These graphs are drafted with β = 0.14 corresponding to a frequency of $\omega/2\pi$ = 36.26 GHz and values of $(\omega\tau)^{-1}$ = 0.05 corresponding to those provided in study [27]. The values of $J_{n'n}$ were calculated numerically on a computer. An examination of Fig. 3 a, c shows that with both $A(\omega) \neq 0$ and with $A(\omega)$ = 0 we observe oscillations in curve (3.24), while the number of peaks in both dia-

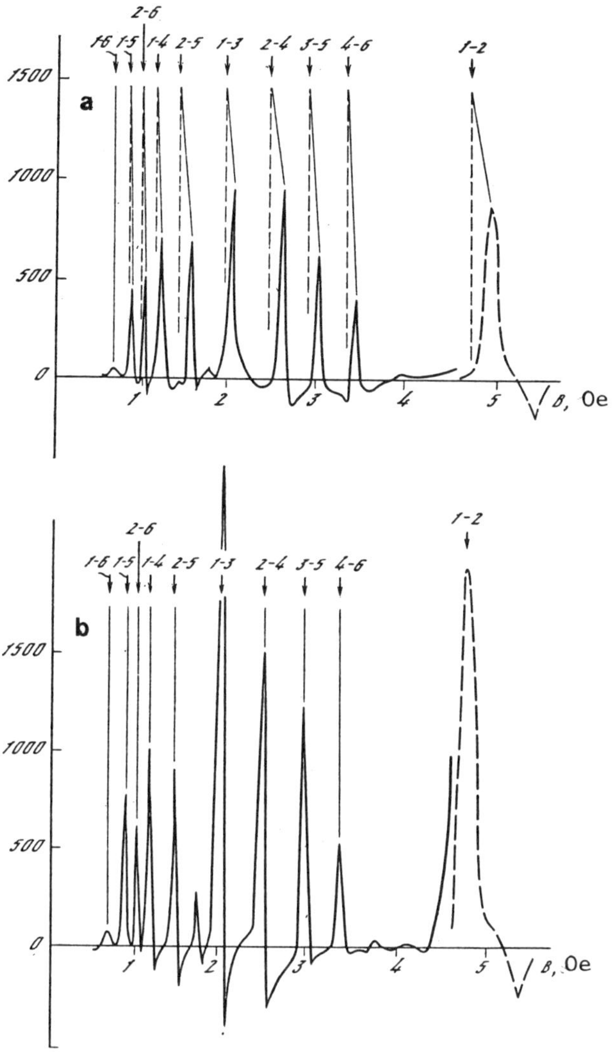

Fig. 4. The impedance derivative plotted as a function of magnetic field (scale 1:10)
a - $\omega\tau$ = 100, $A(\omega)$ = 0; b - $\omega\tau$ = 100, $A(\omega)$ = 2.5. $\partial R/\partial B \cdot 1.3 \cdot 10^{-16}$ s/cm·Oe is plotted on the Y axis. Frequency $\omega/2\pi$ = 33.26 GHz, β = 0.14.

grams coincides. The amplitude of the peaks of the impedance derivative when $A(\omega) \neq 0$ is several times greater than the amplitude of the peaks when $A(\omega)$ = 0. We may therefore assume that measurement of the amplitude of the impedance derivative peaks in absolute units rather than the arbitrary units as performed in study [27] could point to an additional possibility for determining the constant $A(\omega)$. In our view it is significant that the position of the maxima of curve (3.24)

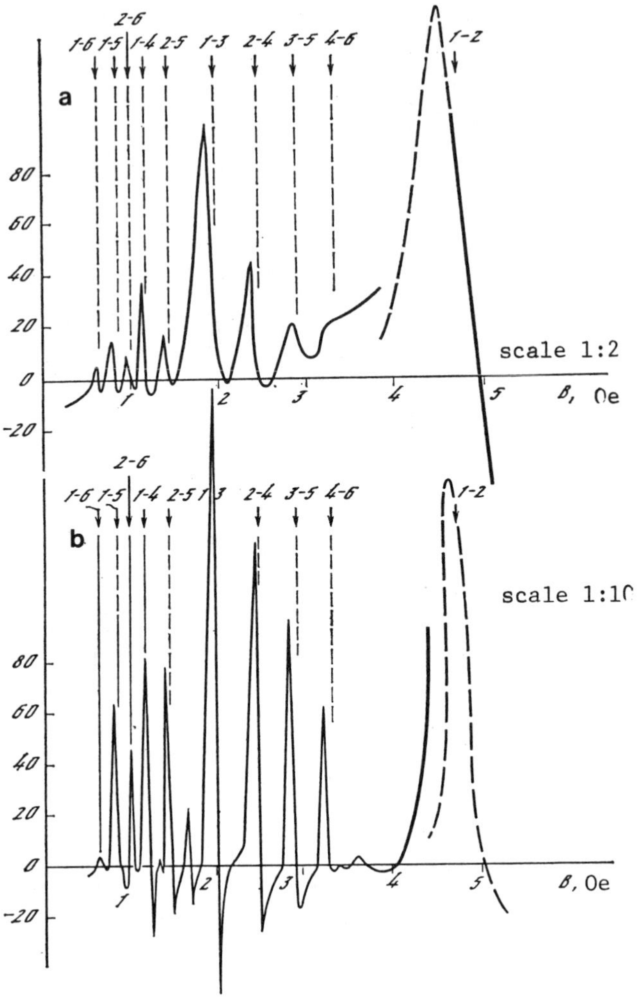

Fig. 5. The impedance derivative plotted as a function of magnetic field
a – $\omega\tau = 20$, $A(\omega) = 3$; b = $\omega\tau = 100$, $A(\omega) = 3$. $\partial R/\partial B \cdot 1.3 \cdot 10^{-16}$ s/cm·Oe is plotted on the Y axis. Frequency $\omega/2\pi = 36.26$ GHz, $\beta = 0.14$.

$\partial R/\partial B$ is in good agreement with the experimentally measured values from study [27] only when we account for $A(\omega) \neq 0$. On the other hand, the maxima of the curve in our Fig. 3a corresponding to electron gas theory shift towards higher fields $A(\omega) = 0$ and $A(\omega) = 2.5$, which contradict experiment. (The arrows in the graphs represent the values of the experimentally-measured resonance magnetic fields $B_{n'n}$ with the dotted lines perpendicular to the X-axis, while the dotted lines are

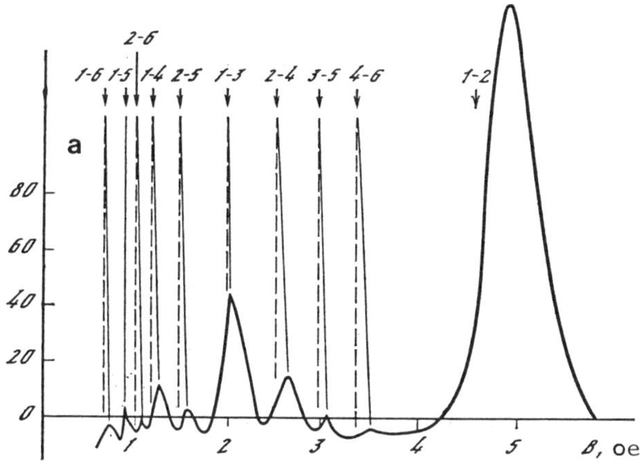

Fig. 6. Graph plotting the impedance derivative as a function of magnetic field
$\omega\tau = 20$, $A(\omega) = 0.15$. $\partial R/\partial B \cdot 1.2 \cdot 10^{-16}$ s/cm·Oe is plotted on the Y-axis. Frequency. $\omega/2\pi = 36.26$ GHz, $\beta = 0.14$.

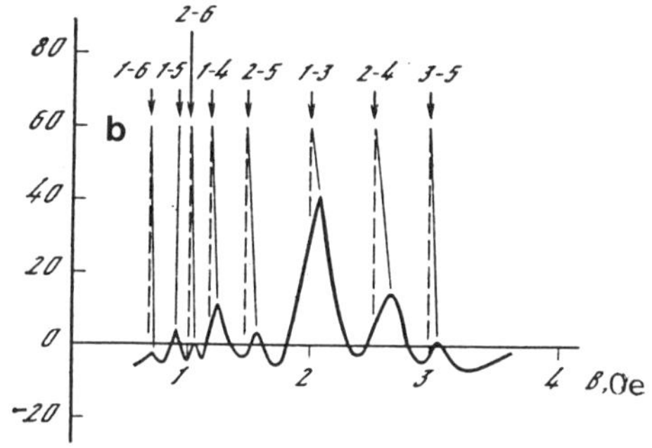

Fig. 7. Graph plotting the impedance derivative as a function of magnetic field
$\omega\tau = 20$, $A(\omega) = 0$. $\partial R/\partial B \cdot 1.2 \cdot 10^{-16}$ s/cm·Oe is plotted on the Y-axis. Frequency. $\omega/2\pi = 36.26$ GHz, $\beta = 0.11$

not shown for the extrema of $(\omega\tau)^{-1} = 0.05$ coinciding with $b_{n'n}$). Accounting for electron interaction ($A(\omega) \neq 0$) corresponding to the curve shown in Fig. 3c eliminates this contradiction by shifting the maxima of the impedance derivative in the direction of smaller magnetic fields away from $B^0_{n'n}$.

We should note that the change in the parameter $(\omega\tau)^{-1}$ does not shift the position of the extrema system shown in Figs. 3a and c. If, for example, we take a value of 0.01 for $(\omega\tau)^{-1}$, even in the case of large values of $\omega\tau = 100$ we cannot eliminate the discrepancy between the positions of the maxima of $\partial R/\partial B$ measured in experiment and those calculated by formula (3.24) when $A(\omega) = 0$. The curves given in Fig. 4a, b corresponding to $A(\omega) = 0$ and $A(\omega) = 2.5$ for $(\omega\tau)^{-1} = 0.01$ and $\beta = 0.14$ reveal that with reduction of $(\omega\tau)^{-1}$ from 0.05 to 0.01 the width of the separate peaks drops, their amplitude grows and the position of the extrema on the X-axis changes insignificantly compared to Fig. 3a, c. This justifies our processing of experimental data based on formulae obtained in the limit $(\omega\tau)^{-1} \to 0$ in the preceding section.

In order to demonstrate the influence of a change in the parameter $A(\omega)$ on the position of the extrema of the curve corresponding to formula (3.24) we have provided graphs of $\partial R/\partial B$ in Fig. 5a, b plotted for $A(\omega) = 3$, $\beta = 0.14$ and for two values of $(\omega\tau)^{-1}$: 0.05 and 0.01. It follows from an examination of these graphs and their comparison to graphs 3c and 4b that when increasing the value of $A(\omega)$ from 2.5 to 3 there is, first, an even greater shift to smaller magnetic fields from values of $B^0_{n'n}$, and second, the amplitude of the resonance peaks grows. With growth of $A(\omega)$ from 2.5 to 3 the agreement with experiment deteriorates. When $A(\omega) = 3$ as in the preceding cases (Fig. 3a, 4b) the reduction of the parameter $(\omega\tau)^{-1}$ from 0.05 to 0.01 does not change the position of the $\partial R/\partial B$ extrema on the X-axis. Comparing Fig. 5a and b we see that a reduction in $(\omega\tau)^{-1}$ serves to increase the amplitude of the $\partial R/\partial B$ extrema, reduce the width of the peaks and provide better resolution of the separate resonances, although the number of peaks remains the same. Fig. 6 gives the $\partial R/\partial B$ graph corresponding to values of $(\omega\tau)^{-1} = 0.05$, $\beta = 0.14$, $A(\omega) = 0.15$. In comparing Fig. 6 and Fig. 3a, we see that the small,(compared to 2.5) value of $A(\omega)$ will not cause the shift towards smaller fields from values of $B^0_{n'n}$ in the system of extrema shown in Fig. 3a when $A(\omega) = 0$ which would satisfy experimental observation.

A change in the parameter β has only a weak influence on the position of the resonance peaks of the $\partial R/\partial B$ curve. In Fig. 7 we provide a portion of the $\partial R/\partial B$ graph drafted for the case $A(\omega) = 0$, $(\omega\tau)^{-1} = 0.05$ and $\beta = 0.11$. Such a change compared to Fig. 3a where $A(\omega) = 0$, $\beta = 0.14$ in the parameter β from 0.14 to 0.11 would correspond to a threefold change in the frequency. As we see from Fig. 7 in spite of this change Figs. 7 and 3a virtually coincide which reveals a weak dependence of the position of the $\partial R/\partial B$ peaks on the value of the parameter β.

The comparison of theoretical curves corresponding to the magnetic field derivative of the real part of the impedance carried out in this section to experimental data from study [27] in the microwave range shows that it is possible to provide a sufficiently comprehensive and noncontradictory interpretation of experimental results only by comprehensive incorporation of interelectron interaction. The advisability of performing experimental measurements of the absolute impedance or its derivative in the microwave range also follows from the results presented in this section. Such measurements would reveal an additional possibility for determining the magnitude of Fermi-liquid electron interaction.

Chapter 4

QUANTUM THEORY OF CYCLOTRON RESONANCE AT SKIPPING ORBITS NEAR THE TRANSITION FREQUENCY BETWEEN SURFACE ELECTRON LEVELS IN A DEGENERATE ELECTRON LIQUID

1. Solution of a quantum kinetic equation for the density matrix of skipping electrons. Incorporation of quantum corrections

In the preceding chapter we formulated a quasi-classical theory of cyclotron resonance at the skipping electrons in the electron liquid of metals and semimetals. The use of results from this theory to process the experimental results from studies [25-27] obtained for the microwave range made it possible to conclude that the experimental data indicate strong interelectron interaction in bismuth [48, 51, 54]. On the other hand we must emphasize that there exists among the experimental data certain other data that is difficult to interpret by quasi-classical theory [48, 51]. Specifically, the results of study [27] contain data obtained in the infrared range and for transitions from n' and n for which the condition $n' - n \ll n$ is violated. The electron gas theory [21, 25, 55, 64] easily overcame such difficulties since the eigenvalues of energy needed for such a theory were, first, known exactly and, secondly, the exact eigenvalues of the energy levels are near identical to the corresponding results from quasi-classical quantization in virtually all cases. The situation is more complex in the case of electron liquid theory where in order to determine the resonance values of the magnetic fields it is necessary to know, in addition to the eigenvalues of the energy of the skipping electrons, the corresponding matrix elements. In this section we will formulate the quantum theory of cyclotron resonance at skipping electrons presented in our study [53]; this theory accounts for: a) interelectron interaction of electrons, b) the existence of critical momentum values on the levels before and after electron transition and, finally, c) the theory makes it possible to consider all possible (and not just those satisfying the quasi-classicality condition) numbers of the initial and final states of the skipping electrons determining cyclotron resonance.

In order to describe the weakly excited states of the electron liquid it is productive to use the quantum kinetic equation for the minor deviation $\delta\rho$ of the density matrix from the equilibrium Fermi distribution ρ^0. Below we will use an approximation of Fermi-liquid interaction by means of a single constant (compare to [60, 69]). We will orient the z axis in the direction of the constant magnetic field B lying on the plane of the surface of the sample and will plot the y axis perpendicular to this surface and we take the eigenvalues of the energy of the electron on the surface level $E_n(p_z, p_x)$ as functions of the number of the level n and p_z, and the p_x-projections of the quasi momentum of the electron in the plane of the metallic surface. Assuming a time dependence of $\sim\exp(-i\omega t)$ and that the metal is in the half space $y > 0$, for the case of a surface wave with wave vector k_z propagating in the direction of the constant magnetic field we write the following quantum kinetic equation (compare to study [69]):

$$[\hbar(\omega + i/\tau) - E_{n'}(p_z + \hbar k_z, p_x) + E_n(p_z, p_x)]\,\delta\rho(n', n, p_z p_x, k_z) +$$
$$+ \frac{\rho^0(E_{n'}(p_z + \hbar k_z, p_x)) - \rho^0(E_n(p_z, p_x))}{E_{n'}(p_z + \hbar k_z, p_x) - E_n(p_z, p_x)} \Big\{ \sum_{q_y} i\hbar \mathbf{j}_{n'n}(q_y)\,\delta\mathbf{E}(q_y) +$$
$$+ [E_{n'}(p_z + \hbar k_z, p_x) - E_n(p_z, p_x)] \times$$
$$\times V_0 \sum_{n_1' n_1 p_{1z} p_{1x}} \sum_{q_y} I_{n'n}(q_y) I_{n_1' n_1}(q_y)\,\delta\rho(n_1', n_1, p_{1z}, p_{1x}, k_z)\Big\} = 0. \qquad (4.1)$$

Here τ is the electron momentum relaxation time; V_0 is the amplitude of the electron interaction potential; ω and $\delta\mathbf{E}(q_y)$ is the frequency and Fourier component of the nonequilibrium electric field; $I_{n'n}(q_y) = \int_0^\infty dy\,\psi_{n'}(y) \exp(iq_y y)\,\psi_n(y)$ — is the matrix element in which $\psi_n(y)$ is the coordinate-dependent part of the wave function of the surface state $(2\pi\hbar)^{-1}\exp\{ip_x x/\hbar + ip_z z/\hbar\}\psi_n(y)$. The component of the Fourier-operator of the current density $\mathbf{j}_{n'n}(q_y)$ obeys the equation

$$\mathbf{j}_{n'n}(q_y) = \mathbf{j}^0_{n'n}(q_y) + V_0 \sum_q I_{n'n}(q) \sum_{n_1' n_1 p_{1z} p_{1x}} I_{n_1' n_1}(q)\,\mathbf{j}_{n_1' n_1}(q_y),$$

where ignoring paramagnetic effects $\mathbf{j}^0_{n'n}(q_y)$ is the component of the Fourier-operator of the current density of an electron that does not interact with other electrons. Equation (4.1) leads us to the following inhomogeneous integrodifferential equation:

$$S(0, k_y, k_z) + \sum_{q_y} Q(k_y, q_y)\, S(0, q_y, k_z) =$$
$$= - \sum_{n_1' n_1} I_{n_1' n_1}(k_y) \frac{i\hbar}{\hbar\omega' - E_{n_1'}(p_z + \hbar k_z, p_x) + E_{n_1}(p_z, p_x)} \times$$
$$\times \frac{\rho^0(E_{n_1'}(p_z + \hbar k_z, p_x)) - \rho^0(E_{n_1}(p_z, p_x))}{E_{n_1'}(p_z + \hbar k_z, p_x) - E_{n_1}(p_z, p_x)} \sum_{q_y} \mathbf{j}_{n_1' n_1}(q_y)\,\delta\mathbf{E}(q_y), \qquad (4.2)$$

where we use the conventions: $S(0, k_y, k_z) = \sum_{n_1' n_1 p_z p_x} I_{n_1' n_1}(k_y)\,\delta\rho(n_1', n_1, p_z p_x, k_z)$, $\omega' = \omega + i/\tau$, while the kernel of equation (4.2) takes the form (compare to formula (1.10)):

$$Q(k_y, q_y) = V_0 \sum_{n_1' n_1} I_{n_1' n_1}(q_y) I_{n_1' n_1}(k_y) \frac{\rho^0(E_{n_1'}(p_z + \hbar k_z, p_x)) - \rho^0(E_{n_1}(p_z, p_x))}{\hbar\omega' - E_{n_1'}(p_z + \hbar k_z, p_x) + E_{n_1}(p_z, p_x)}.$$

In the kernel $Q(k_y, q_y)$ we will isolate the sum of volumetric components, the sum of nonresonance surface components and the resonance component with quantum numbers n' and n for which the extremal transition frequency between the surface levels n' and n is close to the value of ω. Here

$$Q(k_y, q_y) = Q_{\text{об}}(k_y, q_y) + Q_{\text{нр}}(k_y, q_y) + Q_{\text{р}}(k_y, q_y),$$

$$Q_{\text{р}}(k_y, q_y) = V_0 \sum_{p_x p_y} I_{n'n}(q_y) I_{n'n}(k_y) \frac{\rho^0(E_{n'}(p_z + \hbar k_z, p_x)) - \rho^0(E_n(p_z, p_x))}{\hbar\omega' - E_{n'}(p_z + \hbar k_z, p_x) + E_n(p_z, p_x)},$$

$$Q_{\text{нр}}(k_y, q_y) = V_0 \sum_{p_z p_x r' \neq n'\ r \neq n} I_{r'r}(q_y) I_{r'r}(k_y) \frac{\rho^0(E_{r'}(p_z + \hbar k_z, p_x)) - \rho^0(E_r(p_z, p_x))}{\hbar\omega' - E_{r'}(p_z + \hbar k_z, p_x) + E_r(p_z, p_x)},$$

$$Q_{\text{об}}(k_y, q_y) = V_0 \sum_{p_z p_x r' r} J_{|r'-r|}(k_y \sqrt{2c\hbar(r + 1/2)/|e|B}) \times$$
$$\times J_{|r'-r|}(q_y \sqrt{2c\hbar(r + 1/2)/|e|B}) \times$$
$$\times \frac{\rho^0(E_{r'}(p_z + \hbar k_z, p_x)) - \rho^0(E_r(p_z, p_x))}{\omega' - (r' - r)\Omega(p_z) - k_z v_z} \simeq$$
$$\simeq V_0 \delta_{k_y q_y} \sum_{p_z p_x r} \frac{\partial\rho^0(E_r)}{\partial E_r} \frac{k_y v_r(p_z) + k_z v_z}{\omega' - k_y v_r(p_z) - k_z v_z} =$$
$$= A_0 \delta_{k_y q_y} \left\{1 - \left(\sum_{p_z p_x r} \frac{\partial\rho^0}{\partial E_r}\right)^{-1} \sum_{p_z p_x r} \frac{\omega'(\partial\rho^0/\partial E_r)}{\omega' - k_y v_r(p_z) - k_z v_z}\right\},$$
$$v_r(p_z) = \Omega(p_z)[2(r + 1/2)c\hbar/|e|B]^{1/2}.$$

Here $J_m(x)$ is the Bessel function; $A_0 = -V_0 \sum_{p_z p_x r} \frac{\partial\rho^0}{\partial E_r}$; $\Omega(p_z)$ is the cyclotron frequency. The characteristic scale of the variation in k_y is the inverse depth of penetration of the electromagnetic field into the sample δ^{-1}. In the case of the anomalous skin-effect in the temperature range $\omega \ll \omega_{Le} v/c$ (v is the Fermi velocity, ω_{Le} is the Langmuir electron frequency) the depth of the skin-layer is small compared to the distance traveled by the electron during the change in the field $\delta \ll v/\omega$. Since we are interested in such a case we ignore in the last expression for $Q_{\text{об}}(k_y, q_y)$ the second component in braces. As demonstrated in study [48] for the case of moderate values of n' and n (less than 10) the resonance contribution to $Q(k_y, q_y)$ is much greater than the contribution of the nonresonance components. In this case we find the solution of equation (4.2) as

$$\delta\rho(n', n, p_z, p_x, k_z) = -\frac{i\hbar}{\hbar\omega' - E_{n'}(p_z + \hbar k_z, p_x) + E_n(p_z, p_x)} \times$$
$$\times \frac{\rho^0(E_{n'}(p_z + \hbar k_z, p_x)) - \rho^0(E_n(p_z, p_x))}{E_{n'}(p_z + \hbar k_z, p_x) - E_n(p_z, p_x)} \times$$
$$\times \left\{1 + \bar{N}_{n'n} V_0 \sum_{p_z p_x} \frac{\rho^0(E_{n'}(p_z + \hbar k_z, p_x)) - \rho^0(E_n(p_z, p_x))}{\hbar\omega' - E_{n'}(p_z + \hbar k_z, p_x) + E_n(p_z, p_x)}\right\}^{-1} \sum_{q_y} \mathbf{j}_{n'n}(q_y)\, \delta\mathbf{E}(q_y). \tag{4.3}$$

Here we set:

$$\bar{N}_{n'n} = \frac{1}{1 + A_0} \int_0^\infty \frac{dk_y}{\pi} I^2_{n'n}(k_y) = \int_0^\infty \frac{dy}{1 + A_0} \psi^2_{n'}(y) \psi^2_n(y), \tag{4.4}$$

where $\psi_n(y)$ is the y-dependent part of the wave function of the electron surface state. In order to determine the spectrum of surface oscillations in electron liquid (compare to (2.6)) we will solve the equation corresponding to the vanishing of the resonance denominator in the right half of (4.3),

$$1 - \tilde{N}_{n'n} V_0 \Pi_{n'n}(\omega, k_z) = 0, \tag{4.5}$$

where

$$\Pi_{n'n}(\omega, k_z) = \sum_{p_z p_x} \frac{\rho^0(E_n(p_z, p_x)) - \rho^0(E_{n'}(p_z + \hbar k_z, p_x))}{\hbar\omega' - E_{n'}(p_z + \hbar k_z, p_x) + E_n(p_z, p_x)}. \tag{4.6}$$

Incorporating the electrodynamic effects investigated in study [55] as demonstrated in the third and fourth chapters of this study produces negligible corrections to the spectra of surface waves, if the inter-electron interaction constant $\mathfrak{A}^+ = A_0/(1 + A_0)$ is not small compared to unity. In the specific case of bismuth of interest to us the experimental data from studies [25-26] make it possible to indeed ignore the electrodynamic effects.

Below we will use equation (4.5) to analyze the experimental data from study [27] requiring for its understanding quantum rather than quasi-classical theory. Here the quantum approach is necessary with small quantum numbers n and n' for calculating matrix element (4.4) for which we may write the following formula:

$$\tilde{N}_{n'n} = \frac{1}{1 + A_0}\left[\frac{2v_x eB}{c\hbar^2}\left(\frac{\partial^2\varepsilon}{\partial p_y^2}\right)^{-1}\right]^{1/3} J_{n'n}. \tag{4.7}$$

here $\varepsilon(\mathbf{p})$ is electron energy as a function of momentum, the values $v_x = \partial\varepsilon/\partial p_x$ and $\partial^2\varepsilon/\partial p_y^2$ are taken for $p_y = 0$, while

$$J_{n'n} = \left[\int_0^\infty dx\, \mathrm{Ai}^2(x - a_{n'})\, \mathrm{Ai}^2(x - a_n)\right] \times$$

$$\times \left[\int_0^\infty dx\, \mathrm{Ai}^2(x - a_n)\right]^{-1} \left[\int_0^\infty dx\, \mathrm{Ai}^2(x - a_{n'})\right]^{-1}, \tag{4.8}$$

(which coincides with the conventions from the preceding chapter), where $\mathrm{Ai}(x)$ is the Airy function, and a_n and a_n' are the zeros of the Airy function. In the quasi-classical approximation where in accordance with study [51]:

$$J_{n'n} \simeq J^0_{n'n} = 4^{-1}(2/3\pi)^{2/3}(n - 1/4)^{-2/3} \ln[6(n - 1/4)/(n' - n)].$$

In Appendix II Table 2 gives the values of integral (4.8) obtained from a computer and the corresponding quasi-classical values of $J^0_{n'n}$ are given for comparison.

A quantum approach that accounts for corrections of $\sim\hbar\omega/\varepsilon_F$ is also necessary for evaluating integral (4.6).

In order to incorporate corrections of $\sim\hbar\omega/\varepsilon_F$ in evaluating (4.6) in the case of an arbitrary Fermi-surface we write (see, for example, [48, 51]) expressions for the eigenvalues of electron energy and the transition frequency between the surface levels with quantum numbers n' and n in the form

$$E_n(p_z, p_x) = \varepsilon(p_x, 0, p_z) + a_n\left[\frac{1}{2}\frac{\partial^2}{\partial p_y^2}\varepsilon(p_x, p_y, p_z)\right]^{1/3}_{p_z=0} \times$$
$$\times\left[\frac{eB}{c}\frac{\partial}{\partial p_x}\varepsilon(p_x, 0, p_z)\right]^{2/3},$$
$$\omega_{n'n}(p_z, p_x) = \frac{1}{\hbar}\left[E_{n'}(p_z, p_x) - E_n(p_z, p_x)\right] =$$
$$= \frac{1}{\hbar}\left[\frac{1}{2}\frac{\partial^2}{\partial p_y^2}\varepsilon(p_x, p_y, p_z)\right]^{1/3}_{p_y=0}\left[\frac{eB}{c}\frac{\partial}{\partial p_x}\varepsilon(p_x, 0, p_z)\right]^{2/3}(a_{n'} - a_n). \quad (4.9)$$

In evaluating $\Pi_{n'n}(\omega, k_z)$ we must account for the fact that $E_n(p_z, p_x) \leqslant \varepsilon_F$, while $E_{n'}(p_z + \hbar k_z, p_x) \geqslant \varepsilon_F$. This process produces the following constraint on the range of integration with respect to p_x in formula (4.6)

$$p_x(n', p_z + \hbar k_z) \leqslant p_x \leqslant p_x(n, p_z).$$

Here

$$p_x(n, p_z) = p_x^F(p_z)[1 - \Delta(n, p_z)] \quad (4.10)$$

is the solution of the equation $\varepsilon(p_x, 0, p_z) + a_n[\partial^2\varepsilon(p_x, p_y, p_z)/2\partial p_y^2]^{1/3}_{p_y=0} \times \times [eB\partial\varepsilon(p_x, 0, p_z)/\partial p_x \cdot c]^{2/3} = \varepsilon_F$, while $p_x^F(p_z)$ satisfies the equation $\varepsilon(p_x^F(p_z), 0, p_z) = \varepsilon_F$ determined by equality (4.10) is small compared to unity due to the smallness of the second component in the right half of (4.9) compared to the first. The corrections $\Delta(n, p_z)$ and $\Delta(n', {}_z)$ are proportional to $\hbar\omega/\varepsilon_F$.

Incorporating formulae (4.9) and (4.10) in the longwave approximation $\hbar k_z \ll p_x^f$, we may write integral (4.6) in the following form ($\varepsilon(0, 0, p_z^F = \varepsilon_F)$):

$$\Pi_{n'n}(\omega, k_z) = \int\limits_{-p_z^F}^{p_z^F}\frac{dp_z}{(2\pi\hbar)^2}\left[\frac{\partial\omega_{n'n}(p_z, p_x^F(p_z))}{\partial p_x^F(p_z)}\right]^{-1}\times$$
$$\times\ln\frac{\omega' - k_z v_z - \omega_{n'n}(p_z, p_x^F(p_z)) + p_x^F(p_z)\Delta(n', p_z)\partial\omega_{n'n}(p_z, p_x^F(p_z))/\partial p_x^F(p_z)}{\omega' - k_z v_z - \omega_{n'n}(p_z, p_x^F(p_z)) + p_x^F(p_z)\Delta(n, p_z)\partial\omega_{n'n}(p_z, p_x^F(p_z))/\partial p_x^F(p_z)}. \quad (4.11)$$

When frequency ω is close to the extremal value of the expression $\omega_{n'n}(p_z, p_x^F(p_z)) + k_z v_z$ may be advanced further. Now expanding the integrand in (4.11) in powers of $p_z - \bar{p}_z(k_z)$ at the point $\bar{p}_z(k_z)$ near the extremum of $(d/dp_z)[\omega_{n'n}(p_z, p_x^F(p_z)) + k_z v_z]_{p_z=\bar{p}_z(k_z)} = 0$, we obtain

$$\Pi_{n'n}(\omega, k_z) = \frac{p_x^F(p_z)}{2\pi^2\hbar^2\omega}\frac{\Delta(n', p_z) - \Delta(n, p_z)}{[-\omega^{-1}(d^2/2dp_z^2)(\omega_{n'n}(p_z, p_x^F(p_z)) + k_z v_z)]^{1/2}}\times$$

$$\times\left\{\left[\frac{\omega'-\omega_{n'n}(p_z, p_x^F(p_z))-k_z v_z}{\omega}+\Delta(n', p_z)\frac{p_x^F(p_z)}{\omega}\frac{\partial\omega_{n'n}(p_z, p_x^F(p_z))}{\partial p_x^F(p_z)}\right]^{1/2}+\right.$$

$$+\left[\frac{\omega'-\omega_{n'n}(p_z, p_x^F(p_z))-k_z v_z}{\omega}+\Delta(n, p_z)\frac{p_x^F(p_z)}{\omega}\times\right.$$

$$\left.\left.\times\frac{\partial\omega_{n'n}(p_z, p_x^F(p_z))}{\partial p_x^F(p_z)}\right]^{1/2}\right\}_{p_z=\bar{p}_z(k_z)}. \qquad (4.12)$$

This formula is suitable near resonance when $|\omega'-\omega_{n'n}(p_z, p_x^F(p_z))-$ $-k_z v_z|_{p_z=\bar{p}_z(k_z)}\ll\omega$. It was obtained for an arbitrary Fermi surface. If as is the case, for example, for bismuth, resonance is assumed to occur near the central cross-section of the Fermi surface so that $p_z^F\gg$ $\bar{p}_z(k_z)\sim k_z$, while the vector k_z satisfies the condition

$$|k_z v_z|\ll|\omega-\omega_{n'n}(0, p_x^F)|,$$

where $p_x^F = p_x^F(0)$, formula (4.12) is simplified

$$\Pi_{n'n}(\omega, k_z)=\frac{p_x^F}{2(\pi\hbar)^2\omega}\frac{\Delta(n', 0)-\Delta(n, 0)}{[-\omega^{-1}d^2\omega_{n'n}(p_z, p_x^F(p_z))/2dp_z^2]^{1/2}_{p_z=0}}\times$$

$$\times\left\{\left[\left[\frac{\omega'-\omega_{n'n}}{\omega}+\Delta(n', 0)\frac{p_x^F}{\omega}\frac{\partial\omega_{n'n}(0, p_x^F)}{\partial p_x^F}\right]^{1/2}+\right.\right.$$

$$\left.\left.+\left[\frac{\omega'-\omega_{n'n}}{\omega}+\Delta(n, 0)\frac{p_x^F}{\omega}\frac{\partial\omega_{n'n}(0, p_x^F)}{\partial p_x^F}\right]^{1/2}\right]^{-1}+\frac{k_z v_z}{\sqrt{\omega(\omega'-\omega_{n'n})}}\right\}.$$

Here $\omega_{n'n}\equiv\omega_{n'n}(0, p_x^F)$. Formula (4.12) in conjunction with Table 2 from Appendix II for integral (4.8) determining matrix element (4.7) make it possible to speak of a quantum theory of the spectrum determined by dispersion equation (4.5) of the surface waves in an electron liquid with frequencies close to the transition frequencies between the skipping electron levels.

2. Interpretation of experimental results for bismuth in the IR and determination of interelectron interaction parameters

In this section we use the quantum theory of cyclotron resonance at skipping electrons developed above as the basis for interpreting the experimental results of study [27] obtained from irradiation of bismuth samples in the IR for the case $\omega/2\pi$ = 890.7, 1362.5, 1747.1, 2526.6 GHz. As in the case of study [27] we take the constant magnetic field B (see Fig. 2) to be parallel to the crystallographic symmetry axis C_2 and assume the y axis lies along C_3. For electrons belonging to one of the two ellipsoids of the Fermi-surface of bismuth, symmetric with respect to C_2 the dependence of the energy on the quantum numbers is (compare to study [52]):

$$E_n(p_z, p_x)=\frac{2p_z^2}{3m_3}+\frac{3}{8m_1}\left[p_x-p_z\frac{1}{\sqrt{3}}\right]^2+$$

$$+\left[\frac{3\pi\sqrt{3}}{4}\hbar\Omega\right]^{2/3}\left(n-\frac{1}{4}\right)^{2/3}\left[\frac{3(p_x-p_z/\sqrt{3})}{8m_1}\right]^{1/3}. \qquad (4.13)$$

Here m_1, m_2, m_3 are constants having the dimensions of mass, $\Omega = |e|B/c\sqrt{m_1 m_2}$ is the cyclotron frequency. Study [27] experimentally measured the values of the magnetic fields corresponding to the resonances of the magnetic field derivative of the real part of the impedance. Here we will determine which cross-sections of the Fermi surface correspond to the results from study [27]. For this we account for the fact that, consistent with (4.13) for bismuth in the geometry of Fig. 2 $p_z^F = (8m_1\varepsilon_F)^{1/2}$, $p_x^F(p_z) = [8m_1(\varepsilon_F - 2p_z^2/3m_3)/3]^{1/2} + p_z/\sqrt{3}$, $v_z = -\sqrt{3}p_x^F(p_z)/4m_1$, $\bar{p}_z(k_z) \simeq -9k_z m_3\varepsilon_F/16m_1\omega$. Taking $k_z \sim \omega/c$ where $\omega \approx 10^{13}$ s^{-1} according to study [27] we find that resonance occurs near the central cross-section of the Fermi surface at values of p_z equal in absolute value to $\sim 9m_3\varepsilon_F/16m_1c = 1.1\cdot 10^{-22}$ g·cm/s $\ll p_z^F = 10^{-21}$ g·cm/s. In this case the difference between the resonance value of the transition frequency $\omega_{n'n}(\bar{p}_z, p_x^F(\bar{p}_z))$ and the value of the transition frequency $\omega_{n'n}(0, p_x^F)$ corresponding to the central cross-section $p_z = 0$

$$1-\omega_{n'n}(\bar{p}_z, p_x^F(\bar{p}_z))/\omega_{n'n}(0, p_x^F) = 2\bar{p}_z^2/9m_3\varepsilon_F = 9m_3\varepsilon_F/128m_1^2c^2 \simeq 0{,}8\cdot 10^{-4}$$

is negligible compared to corrections in the order of 1% to the resonant frequency related to electron interaction. The quantity $|k_z v_z|/\omega$ is equal to $\sim(\varepsilon_F/2m_1)^{1/2}$ s^{-1} $= 1.7\cdot 10^{-3}$ and by virtue of smallness compared to $|1 - \omega_{n'n}/\omega|$ (see (4.16) below) it plays no role in determining the critical value of the resonance frequency of the surface oscillations. On this basis if we take the resonance to occur at the central cross-section $p_z = 0$ we solve dispersion equation (4.5) by substituting the value $\Pi_{n'n}(\omega, 0)$ into this equation; this was calculated for the specific case for the dependence (4.13) of energy on the quantum numbers. Here formula (4.12) is written as

$$\Pi_{n'n}(\omega, 0)\frac{\sqrt{3m_1m_2}}{2\pi\hbar}\left\{\left[\frac{\omega'}{\omega_{n'n}} - 1 + \frac{\hbar\omega}{3\varepsilon_F}\frac{a_{n'}}{a_{n'}-a_n}\right]^{1/2} + \left[\frac{\omega'}{\omega_{n'n}} - 1 + \frac{\hbar\omega}{3\varepsilon_F}\frac{a_n}{a_{n'}-a_n}\right]^{1/2}\right\}^{-1}. \qquad (4.14)$$

In deriving (4.14) we incorporated, consistent with (4.13) $p_x^F = p_x^F(0) = (8m_1\varepsilon_F/3)^{1/2}$, $\Delta(n', 0) = (\sqrt{3}\hbar\Omega/2\varepsilon_F)^{2/3} a_{n'}/2 \simeq \hbar\omega a_{n'}/2\varepsilon_F(a_{n'} - a_n)$, $\Delta(n, 0) = (\sqrt{3}\hbar\Omega/2\varepsilon_F)^{2/3} a_n/2 \simeq \hbar\omega a_n/2\varepsilon_F(a_{n'} - a_n)$, $\partial\omega_{n'n}(0, p_x^F)/\partial p_x^F = 2\omega_{n'n}/3p_x^F$, $d^2\omega_{n'n}(p_z, p_x^F(p_z))/dp_z^2|_{p_z=0} = -4\omega_{n'n}/9m_3\varepsilon_F$. The extremal frequency $\omega_{n'n}$ entering into (4.14) in the case of bismuth is equal to $\omega_{n'n} = [\sqrt{3}\Omega/2]^{2/3}(\varepsilon_F/\hbar)^{1/3}[a_{n'} - a_n]$. Substituting the derived values of (4.7), (4.14) into dispersion equation (4.15) and accounting for $v_x = (3\varepsilon_F/2m_1)^{1/2}$, while $\partial v_y/ddp_y = m_2^{-1}$, we obtain

$$\left[\frac{\omega'}{\omega_{n'n}}-1+\frac{\hbar\omega}{3\varepsilon_F}\frac{a_{n'}}{a_{n'}-a_n}\right]^{3/2}+\left[\frac{\omega'}{\omega_{n'n}}-1+\frac{\hbar\omega}{3\varepsilon_F}\frac{a_n}{a_{n'}-a_n}\right]^{3/2}=$$

$$=\mathfrak{A}^+J_{n'n}\frac{\sqrt{3}\,\pi}{(a_{n'}-a_n)^{1/2}}\sqrt{\frac{\hbar\omega}{\varepsilon_F}}\,. \qquad (4.15)$$

When $(\omega\tau)^{-1}\ll\mathfrak{A}^{+2}3\pi^2J_{n'n}\,\hbar\omega/4\varepsilon_F$, solving (4.15) we find the critical frequency of the spectrum of surface oscillations:

$$\omega=\omega_{n'n}\left\{1+\frac{\hbar\omega}{\varepsilon_F}\left[\frac{3\pi^2}{4}\frac{J^2_{n'n}}{a_{n'}-a_n}\mathfrak{A}^{+2}-\frac{1}{6}\frac{a_{n'}+a_n}{a_{n'}-a_n}+\frac{a_{n'}-a_n}{108\pi^2J^2_{n'n}\mathfrak{A}^{+2}}\right]\right\}-\frac{i}{\tau}. \qquad (4.16)$$

Formula (4.16) is valid if $\mathfrak{A}^+>[a_{n'}-a_n]^{1/2}(3\pi J_{n'n})^{-1}$ (the values of $\mathfrak{A}^+$ obtained below satisfy such an inequality). Expression (4.16) makes it possible to compare the developed theory and experimental results from study [27] and to determine the value of the constant $\mathfrak{A}^+$ using such a comparison. In section 3 of Chapter 4 we determine the constant $\mathfrak{A}^+$ based on the processing of experimental data obtained in the microwave range for bismuth using values of the effective mass and area of the extremal cross-section of the Fermi surface. This same chapter indicates the determination method that involves comparing different values of the resonance magnetic fields without requiring high accuracy of data on the parameters of the Fermi-surface of bismuth. At the same time the accuracy of experimental measurements in the microwave range [27] using this approach makes it possible to determine only the upper boundary for the quantity $\mathfrak{A}^+$. The situation is different in the IR range due to the significantly higher value of the ratio $\hbar\omega/\varepsilon_F$ compared to the microwave range. The accuracy of only single measurements in the IR range in study [27] is such that it makes it possible to identify the lower and upper boundaries for $\mathfrak{A}^+$ based on a comparison of the experimental data contained therein. After writing formula (4.16) for the two values of the resonance magnetic fields $B_{n'n}$ and $B_{r'r}$ and dividing these values by one another, we obtain for $\omega\tau\to\infty$ the following relation which is important for processing experimental data:

$$\left(\frac{B_{r'r}}{B_{n'n}}\right)^{2/3}=\frac{a_{n'}-a_n}{a_{r'}-a_r}\left\{1+\frac{3\pi^2\hbar\omega}{4\varepsilon_F}\mathfrak{A}^{+2}\left[\frac{J^2_{n'n}}{a_{n'}-a_n}-\frac{J^2_{r'r}}{a_{r'}-a_r}\right]-\right.$$

$$\left.-\frac{\hbar\omega}{6\varepsilon_F}\left[\frac{a_{n'}+a_n}{a_{n'}-a_n}-\frac{a_{r'}+a_r}{a_{r'}-a_r}\right]+\frac{1}{108\pi^2\mathfrak{A}^{+2}}\frac{\hbar\omega}{\varepsilon_F}\times\left[\frac{a_{n'}-a_n}{J^2_{n'n}}-\frac{a_{r'}-a_r}{J^2_{r'r}}\right]\right\}.$$

This relation makes it possible to write the following formula for the experimental determination of $\mathfrak{A}^+$:

$$\mathfrak{A}^{+2}=(P^{r'r}_{n'n}/2)\,[1+\sqrt{1+4q^{r'r}_{n'n}/(P^{r'r}_{n'n})^2}], \qquad (4.17)$$

where

$$P^{r'r}_{n'n}=\frac{4\varepsilon_F}{3\pi^2\hbar\omega}\frac{\mathscr{F}_{r'r,\,n'n}}{J^2_{n'n}/(a_{n'}-a_n)-J^2_{r'r}/(a_{r'}-a_r)}\,,$$

$$\tilde{\mathcal{F}}_{r'r,\,n'n} = \left(\frac{B_{r'r}}{B_{n'n}}\right)^{2/3} \frac{a_{r'} - a_r}{a_{n'} - a_n} - 1 + \frac{\hbar\omega}{\varepsilon_F}\left[\frac{a_{n'} + a_n}{a_{n'} - a_n} - \frac{a_{r'} + a_r}{a_{r'} - a_r}\right],$$

$$q^{r'r}_{n'n} = \frac{1}{81\pi^4} \frac{(a_{n'} - a_n)(a_{r'} - a_r)}{J^2_{n'n} J^2_{r'r}}.$$

The results from processing data from study [27] obtained for four IR frequencies $\omega/2\pi$ are equal to 890.7; 1362.5; 1747.1; 2526.6 GHz and those calculated by formula (4.17) are given in Table 3 of Appendix III. Here the lower boundary of $\mathfrak{A}^+$ for these four frequencies lies within 0-7.46, 0-6.50, 0-5.75, 0-6.17, respectively. In a number of cases of processing the values of the $B_{r'r}/B_{n'n}$ ratio obtained in study [27] we obtained a lower boundary of $\mathfrak{A}^+ = 0$. For the upper boundary of $\mathfrak{A}^+$ we obtained: 4.67-18.13; 5.11-10.26; 4.94-7.19; 4.27-7.28. We should emphasize that these results do not contradict the values of $\mathfrak{A}^+$ obtained in this chapter (see also [51]) by processing experimental data for the microwave range. In addition we mention that when using the value of the relative measurement error in measurements of the resonance magnetic fields of 1.5% quoted in study [27], we obtain by using formula (4.17) at each frequency a set of values of the constants $\mathfrak{A}^+$ in which the greatest lower boundary of $\mathfrak{A}^+$ is greater than the greatest upper boundary.

Doezma, one of the authors of study [27], has reported to us in a private communication that they also have an absolute measurement error of the resonance magnetic fields equal to 1%. If we process the data using formula (4.17) bearing this in mind, then, compared to the preceding case when such an absolute error was ignored, the greatest lower boundary of $\mathfrak{A}^+$ drops, and the lowest upper boundary grows. Here for the lower boundary of $\mathfrak{A}^+$ we obtain values of 0-6.47; 0-5.88; 0-5.59; 0-5.80, and for the upper boundary we obtained: 6.27-14.32; 5.50-11.65; 5.21-7.32; 4.67-7.63. As in the preceding case of processing data for the majority of values of $B_{r'r}/B_{n'n}$, the lower boundary is greater than a few units. The position of the resonance peaks with such quantum numbers cannot be explained without incorporating electron interaction. The results are given in the second column of Table 3 in Appendix III.

The data from study [27] obtained for the IR range also allow determination of $\mathfrak{A}^+$ using the area of the extremal cross-section of the Fermi surface $S = 1.300(I \pm 0.003)\cdot 10^{-42}$ $(\text{g}\cdot\text{cm/s})^2$ and the effective mass $m^* = (0.82 \pm 0.005)\cdot 10^{-2} m_0$, where m_0 is the mass of a free electron (compare to study [22]) using the formula

$$\mathfrak{A}^+ = (a_{n'} - a_n)^{1/2} J^{-1}_{n'n} \frac{1}{\sqrt{3}\,\pi} \sqrt{\frac{\varepsilon_F}{\hbar\omega}} \left\{\left[1 - \left(\frac{B_{n'n}}{B^0_{n'n}}\right)^{2/3} + \frac{\hbar\omega}{3\varepsilon_F}\frac{a_{n'}}{a_{n'} - a_n}\right]^{1/2} + \right.$$
$$\left. + \left[1 - \left(\frac{B_{n'n}}{B^0_{n'n}}\right)^{2/3} + \frac{\hbar\omega}{3\varepsilon_F}\frac{a_n}{a_{n'} - a_n}\right]^{1/2}\right\}, \qquad (4.18)$$

where

$$B^0_{n'n} = [8\pi\hbar c^2/3Se^2]^{1/2} (\omega m^*)^{3/2} [a_{n'} - a_n]^{-3/2}.$$

In the quasi-classical approximation for $n' - n \ll n$ and $\hbar\omega \ll \varepsilon_F$ this formula coincides with (3.16) which is used in the third chapter of this study to determine $\mathfrak{A}^+$. As a result of this interpretation of the data from study [27] using (4.18) we find for the lower boundary of $\mathfrak{A}^+$ at four IR frequencies the following values: 6.28-12.05; 5.72-9.63; 5.99-10.59; 5.41-9.73; for the upper boundary we obtain: 7.39-16.82; 6.47-11.43; 6.28-11.85; 5.84-10.66. The values of the $\mathfrak{A}^+$ interaction constants obtained from processing data from study [27] using formula (4.18) are given in Table 4 of Appendix III. These values are greater than the corresponding values of $\mathfrak{A}^+$ obtained by means of (4.17).

The primary conclusion we may draw based on a comparison of our quantum theory of cyclotron resonance at skipping orbits and specific experimental data obtained for bismuth in the IR is that, first, our theory correctly explains the positioning of the resonance extrema of the impedance derivative and, second, makes it possible to determine the values of the Fermi-liquid electron interaction constant in bismuth. On the other hand, in view of the fact that our theory is valid for the electron liquid of metals with an arbitrary Fermi surface, we may hope that the use of such a theory to interpret highly-accurate experimental data on cyclotron resonance at skipping orbits in both bismuth and other metals will make it possible to implement a program of determining the parameters of Fermi-liquid interaction for a wide range of metals.

3. Comparison of experimental data on cyclotron resonance in bismuth at the surface levels in the IR to the electron liquid and electron gas theories incorporating the finite values of the electron momentum relaxation time

In Chapter 3, section 4 we compared theoretical and experimental values of $\partial R/\partial B$ in the microwave range. We will carry out a somewhat similar analysis in this section, although for experimental data in the infrared range. In this section we will use conventions identical to those employed in the preceding chapter and based on the results of our study [54] we will interpret the results obtained in study [27] for frequencies of 890.7, 1362.5, 1747.1, and 2526.6 GHz.

In proceeding to a discussion of the dependence of the impedance on the magnetic field in the IR where the quasi-classical approximation is not valid we must, first, incorporate resonance corrections proportional to $\hbar\omega/\varepsilon_F$, and second, write a formula for $\partial R/\partial B$ that is valid for arbitrary, including small, values of n' and n. In accordance with study [51] we will write the following expression for the real part of the magnetic field derivative of the impedance:

$$
\begin{aligned}
\frac{\partial R}{\partial B} = {} & \left(\frac{3\pi}{2}\right)^{1/3} \frac{(6e)^2 \varepsilon_F (3m_3)^{1/2}}{\hbar \pi^3 c^4 m_2 m_1^{1/2}} \sum_{n'n} \frac{(n - 1/4)^{4/3} (a_{n'} - a_n)}{(n' - n)^4} \operatorname{Re} \frac{\partial}{\partial B} \times \\
& \times \left[\left\{ - \frac{8\beta (n' - n)^{2/3} (a_{n'} - a_n)^{1/2}}{\sqrt{3}\,\pi^2 (n' - n)^2} + i \left[1 - \frac{16\beta^2 (n - 1/4)^{2/3} (a_{n'} - a_n)}{3\pi^4 (n' - n)^4} \right] \right\} \times \right. \\
& \times \left\{ \left[1 - \left(\frac{B}{C(\omega)}\right)^{2/3} (a_{n'} - a_n) + \frac{\hbar\omega}{3\varepsilon_F} \frac{a_{n'}}{a_{n'} - a_n} + \frac{i}{\omega\tau} \right]^{1/2} + \right. \\
& + \left[1 - \left(\frac{B}{C(\omega)}\right)^{2/3} (a_{n'} - a_n) + \frac{\hbar\omega}{3\varepsilon_F} \frac{a_n}{a_{n'} - a_n} + \frac{i}{\omega\tau} \right]^{1/2} - \\
& \left. \left. - \frac{2A(\omega)}{\sqrt{a_{n'} - a_n}} J_{n'n} \right\}^{-1} \right].
\end{aligned} \tag{4.19}
$$

We have ignored terms of the sum over n and n' for which inequality (3.18) ($y_n \ll \delta$) is not satisfied similar to the process earlier in the microwave range based on a comparison of formulae (3.22) and (3.23). We will carry out summation in (4.19) over the following values of n' and n: n = 2, n' = 5, 6, 7, 8, 9, 10, 11; n = 3, n' = 6, 7. The values of the resonance magnetic fields in Table 1 of study [27] are provided for transitions with such quantum numbers. It is not necessary to account for the large number of components in (4.19) for a comparison of theory to the data in Table 1 from study [27], since this process produces negligible changes in the $\partial R/\partial B$ curve, as we observed previously in the microwave range from a comparison of Fig. 3a, b. Graphs of relation (4.19) corresponding to $\omega/2\pi$ = 890.7 GHz, $A(\omega)$ = 0 and $A(\omega)$ = 6 for $(\omega\tau)^{-1}$ = 0.5 and β = 0.08 are given in Fig. 8a, b. An examination of these graphs shows that when $A(\omega)$ = 0 incorporating only quantum corrections does not produce a match in the positions of the maxim of curve (4.19) and the experimental values of the resonance magnetic fields $B_{n'n}$. As in the case of the microwave range the amplitude of the peaks of the impedance derivative for $A(\omega) \neq 0$ is several times greater than when $A(\omega)$ = 0. The maxima of relation (4.19) in Fig. 6a when $A(\omega)$ = 0 are shifted towards higher field values from the experimentally observed values of $B_{n'n}$, which is not in agreement with experiment. Incorporating electron interaction ($A(\omega) \neq 0$) produces the best match between the maxima of impedance derivative (4.19) and the values of $B_{n'n}$. Fig. 8c which shows the graph of the (4.19) relation with $A(\omega)$ = 0, β = 0.08, $(\omega\tau)^{-1}$ = 0.01 reveals that a reduction in the value of the parameter $(\omega\tau)^{-1}$ of 0.05 to 0.01 does not make it possible to achieve a match between gas theory and experiment in the case of weak collisions when $\omega\tau$ = 100. The position of the maxima of curve (4.11) is similar to that in the microwave range and is virtually identical to the position of the maxima in Fig. 8a for $(\omega\tau)^{-1}$ = 0.05.

A comparison of our Fig. 8a-c to Fig. 7 from study [27] shows that even the use of a comparatively simple model in which summation is carried out over a finite number $n'$and n, and quasi-classical asymptotic formula (3.19) is used allows us, by incorporating electron interaction ($A(\omega) \neq$) to achieve better agreement with experiment than when using gas theory formulae in our calculation, where $A(\omega)$ = 0. A comparison between Fig. 8a and c shows that the amplitude of the peaks

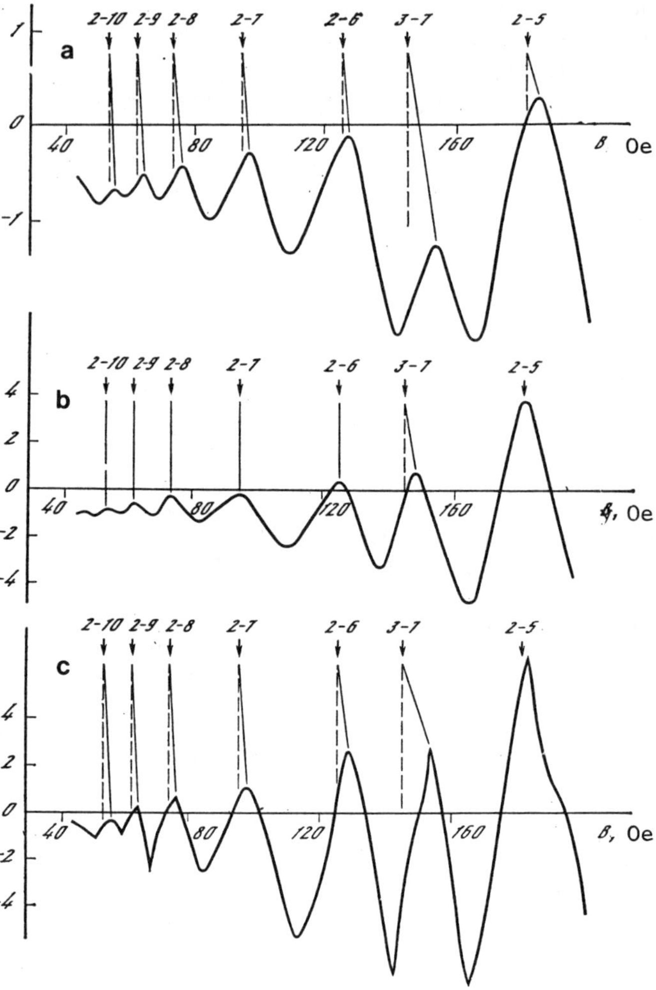

Fig. 8. The magnetic field dependence of the impedance derivative a = $\omega\tau$ = 20, $A(\omega)$ = 0; b = $\omega\tau$ = 20, $A(\omega)$ = 6; c - $\omega\tau$ = 100, $A(\omega)$ = 0. $\partial R/\partial B \cdot 2.1\cdot 10^{-17}$ s/cm·Oe is plotted on the Y-axis. Frequency $\omega/2\pi$ = 890.7 GHz, β = 0.08.

of the curve in Fig. 8a may increase for two reasons. First, when incorporating electron interaction when $A(\omega) \neq 0$ and, second, with a reduction in the parameter $(\omega\tau)^{-1}$. The difference is that when incorporating $A(\omega) \neq 0$ in addition to an increase in the amplitude of the maxima they also shift towards lower fields, while with a reduction of $(\omega\tau)^{-1}$ there is no such shift towards lower fields. On the other hand in comparing Fig. 8a-c we may identify a significant shift of the maxima towards larger fields with reduction of $(\omega\tau)^{-1}$ from 0.05 to 0.01. In Fig. 8a-c the resonance curve is located below the Y-axis on the lower field side over a comparatively broad range, and on the higher

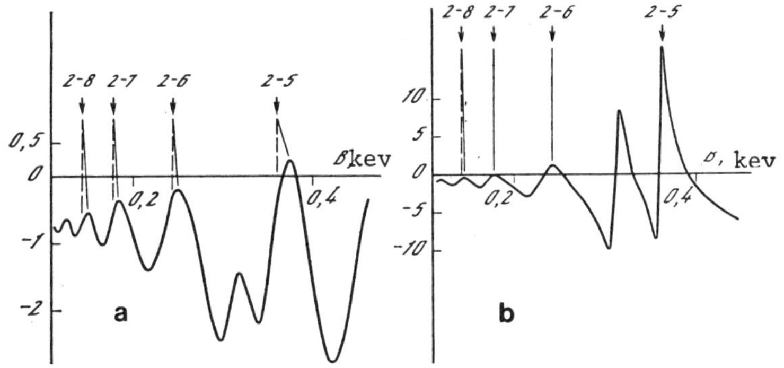

Fig. 9. The real part of the impedance plotted against the magnetic field
a - $\omega\tau$ = 20, $A(\omega)$ = 0; b - $\omega\tau$ = 20, $A(\omega)$ = 7.5. $\partial R/\partial B \cdot 1.1 \cdot 10^{-17}$ s/cm·Oe is plotted on the Y-axis. Frequency $\omega/2\pi$ = 1362.5 GHz, β 0.075.

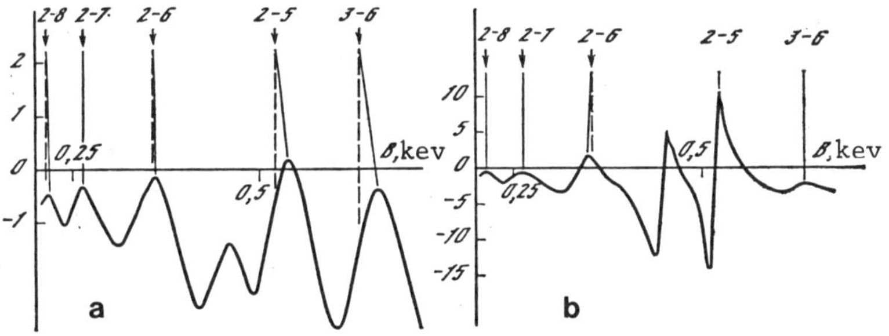

Fig. 10. Graph of the derivative of the real part of the impedance
a - $\omega\tau$ = 20, $A(\omega)$ = 0; b - $\omega\tau$ = 20, $A(\omega)$ = 8.4. $\partial R/\partial B \cdot 7.8 \cdot 10^{-18}$ s/cm·Oe is plotted on the Y-axis. Frequency $\omega/2\pi$ = 1747.1 GHz, β = 0.07.

field side the function $\partial R/\partial B$ adopts both positive and negative values.

Graphs of the $\partial R/\partial B$ relation corresponding to frequencies of $\omega/2\pi$ equal to 1362.5, 1747.1, and 2526.6 GHz and parameter values of $(\omega\tau)^{-1}$ = 0.05, $A(\omega)$ = 0 and $A(\omega) \neq 0$ are given in Fig. 9a-11b. (Each figure indicates the values of the parameters corresponding to the given graph.) An examination of these graphs reveals conclusions analogous to those obtained earlier; specifically, inclusion of interaction serves to increase the amplitude of the resonance peaks of $\partial R/\partial B$ and causes a shift in the system of maxima towards smaller fields in accordance with experiment. We note, that, as we see from Fig. 11b, incorporation of electron interaction in the case of $\omega/2\pi$ =

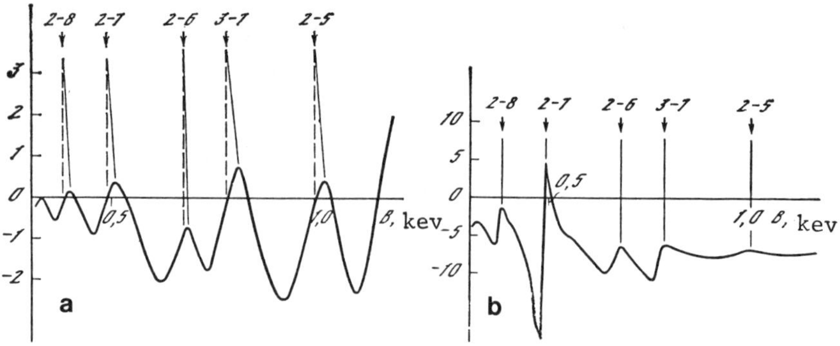

Fig. 11. Graph of the magnetic field derivative of the impedance a - $\omega\tau$ = 20, $A(\omega)$ = 0; b = $\omega\tau$ = 20, $A(\omega)$ = 10. $\partial R/\partial B \cdot 4.5 \cdot 10^{-18}$ s/cm·Oe is plotted on the Y-axis. Frequency $\omega/2\pi$ = 2526.6 GHz, β = 0.065

2526.6 GHz produced a broad region of negative values of $\partial R/\partial B$ with a constant sign on the high field side. This fact is in agreement with the data from Fig. 10 of study [22] and contradicts the data from our Fig. 11a obtained in an idealized electron gas model.

This examination reveals that it is impossible, without comprehensive incorporation of interelectron interaction, to provide a comprehensive interpretation of experimental data in the infrared range on cyclotron resonance at skipping electrons.

We note that the discussion above identifies the advisability of further experimental investigation of the phenomenon of cyclotron resonance at electrons skipping along the surface of a metal, particularly for obtaining results from absolute measurements of impedance oscillations for its magnetic field derivative.

CONCLUSION

A theory of surface quantum spin waves near the transition frequencies between the levels of skipping electrons in the electron liquid of metals has been developed. The frequency spectra and decrements of the damping of surface spin waves have been derived within the scope of this theory for metals with both an isotropic and anisotropic electron dispersion law.

A quasi-classical and quantum theory of cyclotron resonance at the skipping orbits in electron liquid is given. Such a theory which is suitable for interpreting experiments over a broad range of frequencies and for arbitrary quantum numbers of the skipping electron levels is developed in greatest detail for the specific case of bismuth. A comparison of the formulae of this theory to experimental data on surface impedance oscillations in the microwave and infrared regions has made it possible to determine the value of the Fermi-liquid interaction constant in bismuth. Our estimates of the inter-electron interaction constants for bismuth significantly exceed the corresponding values for the alkali metals. Specifically we believe it is timely to draw the attention of experimenters to the importance of obtaining new experimental results on cyclotron resonance at skipping surface electrons. Here it is desirable to improve measurement accuracy of the resonance values of the magnetic field by an order of magnitude compared to existing measurements. On the basis of the developed theoretical concepts we compare experimentally-measured and theoretically-calculated curves corresponding to the derivative of the real part of the surface impedance of bismuth for various frequency values. Here it is demonstrated that only by incorporating inter-electron interaction is it possible to correctly interpret results from the experimental determination of the positions of the resonance maxima of the surface impedance of bismuth. New experimental possibilities are noted (measuring the absolute values of the amplitudes of the resonances of the impedance derivative) for determining the electron interaction parameters in conductors.

In formulating the conclusions of this study we may identify the following concepts as the primary conclusions:

1. Incorporation of Fermi-liquid electron interaction which always exists in normal conductors results in the possibility for surface quantum spin wave propagation in metals near the transition frequencies between the skipping electron levels. A positive value of the interelectron interaction constant is a necessary condition for implementing such a possibility in metals with a convex Fermi surface. Specifically, for the surface quantum spin waves to propagate near the minimum of the surface transition frequency the Fermi-liquid interaction constant as a function of one of the quasi-momentum projections must be negative.

2. We demonstrated that the collisionless Landau damping of surface spin waves that arises in metals with a noncylindrical Fermi surface is insignificant in the transition frequency range between the skipping surface electron levels with quantum numbers n' and n within 10. On the other hand, near transition frequencies between levels with larger numbers n' and n (greater than 10) the surface quantum spin waves will not propagate due to powerful collisionless Landau damping.

3. Investigating the resonance properties of the distribution function of skipping surface electrons obtained by solving a quasi-classical kinetic equation subject to the boundary condition makes it possible to formulate a dispersion equation for the surface cyclotron waves propagating in the electron liquid of metals and semimetals. A particularly simple form of the dispersion equation is obtained when the Fermi-liquid electron interaction constant is not small compared to unity. If this is the case incorporating the electrodynamic effects

in the dispersion equation of surface cyclotron waves is not important.

4. The difference between the critical value of the frequency of a surface cyclotron wave (with small wave vectors) and the extremal value of the transition frequency $\omega_{n'n}$ between the levels of skipping electrons with numbers n' and n is determined by the Fermi-liquid electron interaction constant. On the one hand the existence of such a frequency shift of the surface wave from the value of $\omega_{n'n}$ makes it possible for a surface cyclotron wave to propagate in conditions of insignificant collisionless damping. On the other hand the difference between experimentally-measured values of the resonance magnetic fields $B_{n'n}$ corresponding to the extrema of the surface impedance or its derivative from the magnetic field values $B^0_{n'n}$ calculated by gas theory and corresponding to transition frequencies $\omega_{n'n}$ may be used to determine the magnitude of interelectron interaction.

5. Incorporation of the difference of the shape of the bismuth Fermi surface from an ellipsoidal form was not critical for determining the interaction constant $\mathfrak{A}^+$ based on experimental data on surface impedance oscillations in the microwave and IR frequency ranges.

Corrections to the frequency spectrum of surface cyclotron waves proportional to $\hbar\omega/\varepsilon_F$ and generated by the difference in the critical values of the electron momentum on the levels prior to and after transition become significant in the IR range. Based on a solution of the dispersion equation we find expressions for the frequency spectra of the surface cyclotron waves incorporating the corrections proportional to $\hbar\omega/\varepsilon_F$. Here the values of the matrix element $j_{n'n}$ determining (with $\mathfrak{A}^+$) the shift of the resonance frequency from $\omega_{n'n}$ were calculated numerically, which assured suitability of the derived formulae for the arbitrary values of the numbers n' and n that do not satisfy only the quasi-classical limit $n' - n \ll n$.

7. It is demonstrated using bismuth as an example that by simply incorporating the values of the Fermi-liquid interaction constant $\mathfrak{A}^+$ that are not small compared to unity it is possible to match the positions of the extrema of a theoretically-calculated curve corresponding to the magnetic field derivative of the real part of the impedance for various frequencies $\omega/2\pi$ to experimentally measured values of the resonance magnetic fields.

APPENDIX I

ON THE STRUCTURE OF THE MATRIX ELEMENTS

In order to calculate the quasi-classical matrix elements $\langle n + s|\cos ky|n\rangle$ we will use results from the theory of quasi-classical quantization of orbits of skipping electrons for which we have, by virtue of the smallness of y and p_y, the following expression for the Hamiltonian of the skipping electrons

$$\hat{H}\left(p_x + \frac{e}{c} By, p_y, p_z\right) \simeq \varepsilon(p_x, 0, p_z) + \frac{e}{c} Bv_x y + \frac{1}{2}\frac{\partial v_y}{\partial p_y} p_y^2,$$

where v_x and $\partial v_y/\partial p_y$ are taken for $p_y = 0$. Here motion along the x and y axes is free, and the dependence of the wave function on the coordinate y is described by the Airy function $\mathrm{Ai}(-\xi - a_n)$, where

$$\xi + a_n = [-y + (E_n - \varepsilon(p_x, 0, p_z))\, c/v_x \,|e|\, B]\, |2v_x eB/c\hbar^2 (\partial v_y/\partial p_y)|^{1/3}.$$

Here E_n is the eigenvalue of the Schrodinger equation

$$\hat{H}\psi_n = E_n\psi_n.$$

Here quantization of motion is determined by the condition $\mathrm{Ai}(-a_n) = 0$ for $y = -\xi|2v_x eB/c\hbar^2(\partial v_y/\partial p_y)|^{1/2} = 0$. This quantization condition infers that the quantity

$$a_n = y_n\left|\frac{2v_x eB}{c\hbar^2 \partial v_y/\partial p_y}\right|^{1/3},$$

where $y_n = \dfrac{E_n - \varepsilon(p_x, 0, p_z)}{v_x |e| B} c$, is a root of the Airy function. A table of the roots a_n of the Airy function may be found, for example, in study [43]. With large numbers n a good approximation is provided by the asymptotic formula $a_n = [(3\pi/2)(n - 1/4)]^{2/3}$. For the eigenvalues of the energy of a skipping electron we obtain

$$E_n(p_z, p_x) = \varepsilon(p_x, 0, p_z) + |e| Bv_x y_n/c,$$

where the maximal value of the coordinate y on the skipping orbit is equal to

$$y_n = a_n\left|\frac{2v_x eB}{c\hbar^2(\partial v_y/\partial p_y)}\right|^{-1/3} \simeq \left(n - \frac{1}{4}\right)\left|\frac{9\pi^2 c\hbar^2}{8v_x eB}\frac{\partial v_y}{\partial p_y}\right|^{1/3} \equiv$$
$$\equiv \left(n - \frac{1}{4}\right)\frac{(3\pi)^{2/3}}{2}\left(\frac{\hbar\Omega}{p_y' v_y'}\right)^{2/3} R(p_z).$$

Here we account for $|e|Bv_x/c = \Omega p_y'$ and $\partial v_y/\partial p_y = v_y'/p_y'$ where the prime infers the φ derivative when $\varphi = 0$. For the eigenvalues of the energy of a skipping electron we may now write an expression that is valid for an arbitrary Fermi surface,

$$E_n(p_z, p_x) = \varepsilon(p_x, 0, p_z) + a_n\left[\frac{\hbar v_x eB}{c}\right]^{2/3}\left[\frac{\partial v_y}{2\partial p_y}\right]^{1/3}.$$

We will write the wave function of a skipping electron in the quasi-classical approximation using an asymptotic representation of the Airy functions as

$$\psi_n = [3\pi(n - 1/4)/2]^{1/6} y_n^{-1/2}\, |\xi + a_n|^{-1/4} \sin(2[\xi + a_n]^{3/2}/3 + \pi/4),$$

satisfying boundary condition (2.4).

Since we may now write the following quantization formula for the electron/metal surface collision angles

$$\varphi_n^2 (p_z) = 2y_n/R (p_z),$$

then, bearing in mind integration with respect to φ in formula (3.10) we make the transformation to the argument in the wave function using the following quantum generalization of the formula following from (3.6),

$$y = y_n - R (p_z) \varphi^2/2 = R (p_z)[\varphi_n^2 (p_z) - \varphi^2]/2.$$

Then, finally, bearing in mind that $s \ll n$ and ignoring everywhere s where this does not produce singularities or cause vanishing of the matrix element, we obtain:

$$\langle n + s | \cos ky | n\rangle \simeq 2 \int_0^{\varphi_n(p_z)} d\varphi \varphi_n^{-1} (p_z) \cos [\pi s \varphi \varphi_n^{-1} (p_z)] \times$$

$$\times \cos \{kR (p_z) [\varphi_n^2 (p_z) - \varphi^2]/2\} [1 + 2s\varphi_n^2 (p_z)/3 (n - {}^1/_4) \varphi^2]^{-1/4}.$$

Obviously this expression produces $\psi_{ns}^{KB}(k, p_z, \varphi)$ and the integral $I_{ns}(k, p_F)$ in the form of (3.12).

We provide here for reference purposes and based on the formulae derived above calculations of the matrix element entering into (2.21):

$$N_{n+s,\, n} = \int_0^\infty dk I_{n+s,\, n}^2 (k, p_F) = 2 \int_0^1 dx\, [1 + 2s/3 (n - {}^1/_4) x^2]^{-1/4} \cos \pi s x \times$$

$$\times \int_0^1 dy\, (1 + 2s/3 (n - {}^1/_4) y^2)^{-1/4} \cos \pi s y\, (\pi/y_n) [\delta (x^2 - y^2) + \delta (2 - x^2 - y^2)] =$$

$$= (\pi/y_n) \int_0^1 dx\, (x^2 + 2s/3 (n - {}^1/_4))^{-1/2} \cos^2 \pi s x =$$

$$= (2\pi/y_n) \ln [\sqrt{3 (n - {}^1/_4)/2s} + \sqrt{1 + 3 (n - {}^1/_4)/2s}]^2 \simeq$$

$$\simeq (2\pi/y_n) \ln [6 (n - {}^1/_4)/s].$$

For the matrix elements $N_{n'n}^{r'r}$ we may write the expression ($s = n' - n$, $m = r' - r$):

$$N_{n'n}^{r'r} = 2 \int_0^\infty dk \int_0^1 dx\, [1 + 2s/3 (n - {}^1/_4) x^2]^{-1/4} \cos \pi s x \times$$

$$\times \cos [k y_n (1 - x^2)] \int_0^1 dz\, [1 + 2m/3 (r - {}^1/_4) z^2]^{-1/4} \cos \pi m z \times$$

$$\times \cos [k y_r (1 - z^2)] = (2\pi/y_n) \int_0^1 dx\, [1 + 2s/3 (n - {}^1/_4) x^2]^{-1/4} \times$$

$$\times \cos \pi s x \int_0^1 dz\, [1 + 2m/3 (r - {}^1/_4) z^2]^{-1/4} \cos \pi m z \left\{\delta (1 - x^2 - y_r (1 - z^2)/y_n) + \right.$$

$$\left. + \delta \left(1 - x^2 + \frac{y_r}{y_n} (1 - z^2)\right)\right\} = \frac{2\pi}{y_n} \int_0^1 dz\, \frac{\cos \pi s z}{[1 + 2m/3 (r - {}^1/_4) z^2]^{1/4}} \times$$

$$\times \cos \pi s z \sqrt{1-\frac{y_r}{y_n}+\frac{y_r}{y_n} z^2}\left[1-\frac{y_r}{y_n}+\frac{y_r}{y_n} z^2\right]^{-1/4} \times$$

$$\times\left[1-\frac{y_r}{y_n}+\frac{y_r}{y_n} z^2+\frac{2s}{3(n-1/4)}\right]^{-1/4}$$

If we ignore the components $m/(r-1/4)$ and $s/(n-1/4)$ compared to unity in this expression, we may obtain

$$N_{n'n}^{r'r}=\frac{2\pi}{y_n}\int_0^1 dz \cos \pi s z \cos\left[\pi s \sqrt{1-\frac{y_r}{y_n}(1-z^2)}\right]\left[1-\frac{y_r}{y_n}(1-z^2)\right]^{1/2}=$$

$$=\frac{\pi}{\sqrt{y_n y_r}}\left\{\ln\left[1+\sqrt{\frac{y_n}{y_r}}\right]-\ln\sqrt{\frac{y_n}{y_r}-1}\right\} \simeq \frac{\pi}{2y_n} L_{ns}^{rm},$$

where

$$L_{ns}^{rm}=(y_n/y_r)^{1/2} \ln [4/(y_n/y_r-1)].$$

The quantity y_n entering into this expression is determined above and in the particle case of a metal with isotropic dispersion law (2.9) is equal to

$$y_n=(n-1/4)^{2/3}(3\pi\hbar/2)^{2/3}[c/2|e| B\sqrt{2m\varepsilon_F}]^{1/3}.$$

Carrying out calculations similar to those above we obtain the following expression for matrix element $M_{n'n}^{r'r}$:

$$M_{n'n}^{r'r}=\int_0^\infty dk I_{n'n}(-k, p_F) I_{r'r}(k, p_F[1-(\omega_{n'n}/\omega_{r'r})^3]^{1/2})=$$

$$=\frac{\pi}{2y_n}\left[\frac{n-1/4}{r-1/4}\right]^{1/2} \ln \frac{4}{(n-1/4)/(r-1/4)-1}.$$

APPENDIX II

INTEGRAL VALUE

Here we provide the computer calculated values of the integral

$$J_{n'n}=\left[\int_0^\infty dx \operatorname{Ai}^2(x-a_{n'}) \operatorname{Ai}^2(x-a_n)\right]\left[\int_0^\infty dx \operatorname{Ai}^2(x-a_n)\right]^{-1} \times$$

$$\times\left[\int_0^\infty dx \operatorname{Ai}^2(x-a_{n'})\right]^{-1},$$

determining matrix element (4.7) where $\operatorname{Ai}(x)$ is the Airy function; a_n is the nth root of the Airy function. The last column in Table 2 contains values calculated by the quasi-classical asymptotic formula for $J_{n'n}$, in which according to study [51]

$$J_{n'n}^0=(2/3\pi)^{2/3} 4^{-1}(n-1/4)^{-2/3} \ln [6(n-1/4)/(n'-n)].$$

Table 2

n	n'	$J_{n'n}$	$J^0_{n'n}$
1	2	0.164	0.162
	3	0.113	0.088
	4	0.085	0.044
	5	0.070	0.013
	6	0.061	−0.011
	7	0.055	−0.031
	8	0.049	−0.048
	9	0.045	−0.062
	10	0.041	−0.075
	11	0.039	−0.086
2	3	0.135	0.144
	4	0.098	0.102
	5	0.078	0.077
	6	0.067	0.059
	7	0.058	0.045
	8	0.052	0.034
	9	0.048	0.025
	10	0.044	0.017
	11	0.041	0.009
3	4	0.120	0.127
	5	0.091	0.096
	6	0.074	0.077
	7	0.064	0.064
	8	0.057	0.054
	9	0.051	0.046
	10	0.047	0.039
	11	0.043	0.033

n	n'	$J_{n'n}$	$J^0_{n'n}$
4	5	0.109	0.115
	6	0.084	0.089
	7	0.070	0.074
	8	0.061	0.064
	9	0.055	0.055
	10	0.052	0.049
	11	0.045	0.043
5	6	0.101	0.105
	7	0.079	0.084
	8	0.067	0.071
	9	0.059	0.062
	10	0.053	0.055
	11	0.048	0.049
6	7	0.094	0.098
	8	0.075	0.079
	9	0.064	0.068
	10	0.057	0.060
	11	0.050	0.054
7	8	0.089	0.092
	9	0.072	0.075
	10	0.062	0.065
	11	0.055	0.058

APPENDIX III

VALUES OF THE $\mathfrak{A}^+$ CONSTANTS

Table III gives the values of the $\mathfrak{A}^+$ constants obtained from processing data from study [27] from four infrared frequencies $\omega/2\pi$ equal to 890.7, 1362.5, 1747.1, and 2526.6 GHz by formula (4.17). The experimental values of the measured resonance magnetic fields $B_{n'n}$ were selected from Table 1 in study [27]. The next to the last column in our Table 3 gives the values of $\mathfrak{A}^+$ calculated assuming that the measurement error for measurement of $B_{n'n}$ in study [27] is 1.5%. In the last column we provide values of $\mathfrak{A}^+$ for transitions where the lower boundary of $\mathfrak{A}^+$ calculated assuming a measurement error of $B_{n'n}$ equal to 2.5% is determined.

Table 4 contains results from processing of experimental data by formula (4.18).

Table 3

ω/2π, GHZ	n'	n	r'	r	𝔄+	𝔄+	ω/2π, GHz	n'	n	r'	r	𝔄+	𝔄+
890.7	8	2	7	2	0—8.05		1362,5	6	3	8	2	5.87—8.86	5.53—7.17
			6	2	0—7.03					7	2	5.28—6.53	4.84—6.91
			5	2	4.17—5.77	0—6.27				6	2	6.50—8.16	5.88—8.64
	7	2	6	2	0—8.31			8	2	7	2	4.98—10.26	0—11,65
			5	2	3,92—5.93	0—6.53				6	2	0—5.98	
	6	2	5	2	0—6,01					5	2	4.07—5.11	0—5.50
	9	2	8	2	0—14,22			7	2	6	2	0—6.42	
			7	2	0—9,29					5	2	0—7.05	
			6	2	5.20—7,65	0—8.40		6	2	5	2	0,16—5.35	0—6,00
			5	2	4.88—6.24	4.45—6.66	1747,1	6	3	8	2	5.29—6.10	5.01—6.35
	10	2	9	2	0—16.20					7	2	5.50—6.47	5.16—6.77
			8	2	6.60—12.69	0—14.32				6	2	5,75—7.19	5.59—7.32
			7	2	5.88—9.46	0—10.54		8	2	7	2	0—6.58	
			6	2	5,90—7.97	5.25—8,61				6	2	0—5.60	
			5	2	5,34—6.55	4.95—6.93				5	2	4.13—4.94	3.89—5.21
	11	2	10	2	0—18,13			7	2	6	2	0—6.35	
			9	2	0—18,74					5	2	4.03—5,07	3.70—5.40
			8	2	7.46—8.15	0—13.39		6	2	5	2	3,34—5.07	0—5.60
			7	2	6.56—9.53	0—10.44	2526,6	7	3	8	2	6.17—7.28	5.80—7.63
			6	2	6.29—8.11	5.72—8.68				7	2	5.28—7.09	4.66—7,62
			5	2	5.64—6.73	5,29—7.08				5	2	3.09—4,27	0—4.67
	7	3	6	3	5.15—7,95	4.01—8.69		6	3	8	2	4,49—5,06	4.30—5.25
			11	2	7.07—8.86	6.47—9,40				7	2	4.27—5.01	4.04—5.24
			10	2	6.73—8.75	6.05—9.36				6	2	4.00—5.23	3.60—5.59
			9	2	6.25—8.62	5.44—9.31		8	2	7	2	0—6.57	
			8	2	5.17—8.29	0—9.17				6	2	0—5.36	
			7	2	5.37—9,90	0—11,04				5	2	4.01—4.56	3.85—4.75
			5	2	0—4.67			7	2	6	2	0—5.51	
	6	3	11	2	6,96—8.02	6.06—8.34				5	2	3.72—4.43	3.50—5.67
			10	2	6.79—7.93	6,04—8.28		6	2	5	2	3.14—4.34	0—4.73
			9	2	6.51—7.77	6.07—8.15							
			8	2	6,05—7.51	5.52—7.94							
			7	2	6,31—8.00	5.69—8.49							
			6	2	6.60—9.01	5.63—9.67							

Table 4

ω/2π GHz	n'	n	𝔄+	ω/2π, GHz	n'	n	𝔄+
890,7	2	5	6.38—7.39	1747.1	2	5	5.99—6,28
	3	6	7,70—8.56		2	6	7,29—8.08
	2	6	7.67—9.12		2	7	8.66—9.70
	3	7	8.44—9.78		2	8	10,59—11.85
	2	7	8.77—10,79		3	6	6.89—7,38
	2	8	10,33—12.80	2526.6	2	5	5.41—5.84
	2	9	10,59—13.86		2	6	6.75—7,35
	2	10	11.09—15.20		2	7	8,13—8.92
	2	11	12.05—16.82		2	8	9,73—8.66
1362,5	2	5	5,72—6.47		3	6	5.56—5.98
	2	6	6.87—7.97		3	7	7,85—8.37
	2	7	9.03—10.36				
	2	8	9.63—11.43				
	2	9	7.07—7.69				

APPENDIX 4

THE POSSIBILITY FOR DETERMINING THE MAGNITUDE OF FERMI-LIQUID INTERACTION IN COPPER BASED ON EXPERIMENTAL DATA ON SURFACE IMPEDANCE OSCILLATIONS IN WEAK MAGNETIC FIELDS

Study [22] contains experimental data on the cyclotron resonance at skipping electrons in tin, indium, and aluminum obtained in RF range in weak magnetic fields less than 100 Oe. Extensive experimental data on surface impedance oscillations in copper in a 0-250 Oe magnetic field range at 36 GHz may be found in studies [24, 70]. The resonance values of the magnetic fields obtained in experiment [24] are given in this study in the form of tables that permit theoretical processing similar to that performed in the third chapter of our study for bismuth. Data on surface impedance oscillations in aluminum at 36 GHz may be found in study [71] although these data are presented in a less convenient form than that of studies [24, 20] for copper.

Here we will consider the issue of the possibility for a theoretical interpretation of data from [24] on surface impedance oscillations in copper attributable to the transitions between the quantum levels of skipping electrons and determination of the magnitude of interelectron interaction in copper. The Fermi surface of copper has been extensively studied (see, for example, [72-74]). Such a surface is described by the formula from the study by Halse [75] accurate to 0.2%

$$\begin{aligned}\varepsilon_F = \varepsilon(\mathrm{p}) = C_0'\{&3-\sum\cos(ap_x/2)\cos(ap_y/2)+C_{200}\left[3-\sum\cos ap_x\right]+\\&+C_{211}\left[3-\sum\cos ap_x\cos(ap_y/2)\cos(ap_z/2)\right]+C_{220}\left[3-\sum\cos ap_y\cos ap_z\right]+\\&+C_{310}\left[6-\sum\cos(3ap_x/2)\cos(ap_y/2)-\sum\cos(3ap_y/2)\cos(ap_x/2)\right]+\\&+C_{222}\left[1-\cos ap_x\cos ap_y\cos ap_z\right]+\\&+C_{321}\left[6-\sum\cos(3ap_x/2)\cos ap_y\cos(ap_z/2)-\sum\cos(3ap_z/2)\times\right.\\&\left.\times\cos ap_y\cos(ap_x/2)\right]\},\end{aligned} \tag{5.1}$$

where the values of the coefficients c_{ijk} of the expansion are given in Table 6 of study [75]. The Fermi electron energy in copper is designated ε_F while the value of the ratio $\varepsilon_F/C_0' = C_0$ is given in Table 6 of study [75] and the letter a represents the lattice constant of copper divided by $\hbar$. According to study [75][$a = (3.418 + 0.0004)\cdot 10^{19}$ cm/erg·s, while the quasi-momentum components of the electron are designated p_i. The sign of the sum in (5.1) represents the cyclic commutation of p_x, p_y, p_z.

Study [24] provides experimentally-measured values of the resonance magnetic field B_{21} corresponding to the transition between surface levels with quantum numbers 1 and 2 for various orientations of the constant magnetic field B with respect to the crystallographic axes of the sample. In accordance with study [48] we will consider the coordinate y axis to be perpendicular to the sample surface extending horizontally into the metal and will take the magnetic field parallel to the sample surface along the z axis, with that parallel to the metallic axis on the x axis.

According to theory [48] interelectron interaction will produce a shift in the resonance value of the magnetic field B_{21} from the B^0_{21} value calculated by theory which does not account for Fermi-liquid electron interaction. The relative value of this shift is equal to:

$$\left(\frac{B^0_{21}}{B_{21}}\right)^{2/3} - 1 = -\frac{\omega_{21}}{\omega''_{21}} \frac{2^{11/3}3^{1/3}\pi^4\alpha^2}{a_2 - a_1} \frac{\hbar\omega J^2_{21}}{(eB_{21}/c\Omega)^2 v_x^2 (\partial v_y/\partial p_y)} . \qquad (5.2)$$

Here e is electron charge; c is the speed of light; ω_{21} and ω''_{21} is the transition frequency between the quantum levels of the surface electrons with numbers 2 and 1 and its second derivative; α is the interaction constant (compare to (3.18)); Ω is the cyclotron frequency; v_x and $\partial v2y/\partial p_y$ is the electron velocity component and the velocity derivative on the Fermi surface; J_{21} is the matrix element equal to J_{21} = 0.160 according to study [27]; a_2 and a_1 are the roots of the Airy function ($a_2 - a_1 = 1.7498$). The value of B_{21} according to study [53] calculated ignoring electron interaction is equal to $B^0_{21} = C(\omega)(a_2 - a_1)^{-3/2}$, where $(a_2 - a_1)^{-3/2} = 0.432019$, while $C(\omega) = (\hbar\omega)^{3/2}\ c\sqrt{2}/\hbar|e|\Upsilon_x(\partial\Upsilon_y/\partial P_y)^{1/2}$.

In proceeding to an interpretation of the experimental data from study [24] we will examine the simplest case where the magnetic field B lies on the (100) crystallographic plane and is parallel to the <100> crystallograhic axis. Here in formula (5.1) the p_z axis lies along <100>, while the p_y axis is perpendicular to the (100) plane. In this case the resonance value of the magnetic field B_{21} = 22.7 Oe found in study [24] corresponds to resonance electrons at point a of the Fermi surface in Fig. 6 from study [24]. We obtain the cross-section of the Fermi surface shown in this diagram if we take $p_y = 0$ in formula (5.1) then point a in Fig. 6 from study [24] corresponds to $p_z = 0$. Direct calculation of velocity v_x by (5.1) yields

$$v_x = C'_0 a\,\{\sin(ap^F_x/2)(1 + C_{211} + C_{310} + C_{321}) + \sin ap^F_x (C_{200} + C_{211} + 2C_{220} + C_{222} + 2C_{321}) + \sin(3ap^F_x/2)(3C_{310} + C_{321})\}. \qquad (5.3)$$

We will find the value of $ap^F_x/2$ taking $p_y = p_z = 0$ in (5.1) and solving the cubic equation for $\cos(ap^F_x/2)$,

$$C_0 = 2 - 2\cos(ap^F_x/2) + C_{200}(1 - \cos ap^F_x) + C_{211}(3 - \cos ap^F_x - 2\cos(ap^F_x/2)) + C_{220}(2 - 2\cos ap^F_x) + C_{310}(4 - 2\cos(3ap^F_x/2) - 2\cos(ap^F_x/2)) + C_{222}(1 - \cos ap^F_x) + C_{321}(6 - 2\cos(3ap^F_x/2) - 2\cos ap^F_x - 2\cos(ap^F_x/2)). \qquad (5.4)$$

For example, by accounting for the five nonzero coefficients of the expansion in formula (5.1)(such a model is called the Cu5+ model [75]) the solution of equation (5.4) yields $ap^F_x/2$ = 2.6182. Here the velocity value on the Fermi surface consistent with (5.4) is

$$v_x = C'_0 a\,[0{,}53535 \sin(ap^F_x/2) - 0{,}45811 \sin ap^F_x - 0{,}11730 \sin(3ap^F_x/2)] =$$

$$= C_0' a \cdot 0{,}54692. \qquad (5.5)$$

By comparison to the experimentally measured velocity on the Fermi surface we find the constant C_0'. We note that there is some discrepancy in the experimental data on the determination of v_x. For example, study [75] provides a value of $v_x = 1.04 \cdot (1 \pm 0.05) \cdot 10^8$ cm/s calculated using de Haas-van Alphen data. The use of this value accounting for the fact that $a = 3.4184 \cdot (1 \pm 0.0001) \cdot 10^{19}$ cm/erg·s, yields for C_0' a quantity of $C_0' = 5.56 \cdot (1 \pm 0.05) \cdot 10^{-12}$ ergs. In subsequent studies Lee [76, 77] focused on determining the Fermi energies and the velocity from an analysis from de Haas-van Alphen data and cyclotron resonance (at volumetric electrons) and provided a velocity value of $v_x = 1.111 \cdot (1 \pm 0.017) \cdot 10^8$ cm/s. Accounting for such a value of v_x produced a value of $C_0' = 5.945(1 \pm 0.017) \cdot 10^{-12}$ ergs.

We find the value of the second derivative $\partial v_y / \partial p_y$ by the formula

$$\frac{\partial v_y}{\partial p_y} = C_0' a^2 \left\{ \frac{1}{4} + C_{200} + C_{220} + \frac{10}{4} C_{310} + \right.$$
$$+ \cos \frac{a p_x^F}{2} \left[\frac{1}{4} + \frac{5}{4} C_{211} + \frac{9}{4} C_{310} + \frac{13}{4} C_{321} \right] +$$
$$+ \cos a p_x^F \left[\frac{1}{4} C_{211} + C_{220} + C_{222} + \frac{10}{4} C_{321} \right] +$$
$$\left. + \cos \frac{3 a p_x^F}{2} \left[\frac{1}{4} C_{310} + \frac{5}{4} C_{321} \right] \right\} = C_0' a^2 \cdot 0{,}39670. \qquad (5.6)$$

Using the results from derived formulae (5.5) and (5.6) and accounting for the fact that $\omega/2\pi = 35.9$ GHz $C(\omega)(a_2 - a_1)^{-3/2} = 1.32721 \cdot 10^{23}$ $\text{erg}^{1/2}$ s/cm, we find for copper in the case of resonance attributable to the electrons on the "spherical" part of the Fermi surface the following values for B_{21}^0: $B_{21}^0 = 24.60(1 \pm 0.08)$ Oe (if we take $C_0' = 5.56 \cdot (1 \pm 0.05) \cdot 10^{-12}$ erg), and $B_{21}^0 = 22.745 \cdot (1 \pm 0.025)$ Oe (if we assume that $C_0' = 5.945 \cdot (1 \pm 0.017) \cdot 10^{-12}$ erg). For both these values of B_{21}^0, as was the case previously for bismuth, the value of the left half in formula (5.2) is positive, which corresponds to the maximal value of the transition frequency at the central cross-section of the "sphere", or, in other words, corresponds to a positive value of $\frac{\omega_{21}''}{\omega_{21}}$ for a convex Fermi surface. According to study [24] $B_{21} = 22.7 \cdot (1 \pm 0.007)$ Oe, and hence the quantity characterizing Fermi-liquid interaction is equal to

$$\left(\frac{B_{21}^0}{B_{21}} \right)^{2/3} - 1 = \begin{cases} 0{,}050 \pm 0{,}053 & \text{(based on data from [57])} \quad (5.7) \\ 0{,}002 \pm 0{,}017 & \text{(based on data from [76])} \quad (5.8) \end{cases}$$

The values in (5.7) and (5.8) obtained for the "spherical" part of the Fermi surface of copper may be used to determine the Fermi-liquid interaction constant by the formula:

$$\alpha^2 = -\frac{\omega''_{21}}{\omega_{21}}(a_2 - a_1)\frac{(eB/c\Omega)^2 v_x^2 (\partial v_y/\partial p_y)}{(2\pi)^4 \hbar\omega J_{21}^2}\left[\left(\frac{B_{21}^0}{B_{21}}\right)^{2/3} - 1\right]. \tag{5.9}$$

We may determine the quantity $\omega''_{21}/\omega_{21}$ entering into (5.9) by the formula

$$\frac{\omega''_{21}}{\omega_{21}} = \left[\frac{\partial^2}{\partial p_z^2}\left(v_x^{2/3}\left(\frac{\partial v_y}{\partial p_y}\right)^{1/3}\right)\right]\left[v_x^{2/3}\left(\frac{\partial v_y}{\partial p_y}\right)^{1/3}\right]^{-1},$$

while the value of the effective mass $|e|B/c\Omega$ coincides with that given in study [75] and is equal to 1.370 $m_0(1 \pm 0.005)$, where m_0 is the free electron mass. Direct calculation using relation (5.1) yields:

$$\frac{\partial^2}{\partial p_z^2}\left[v_x^{2/3}\left(\frac{\partial v_y}{\partial p_y}\right)^{1/3}\right] = \frac{2}{3}v_x^{-1/3}\frac{\partial^2 v_x}{\partial p_z^2}\left(\frac{\partial v_y}{\partial p_y}\right)^{1/3} +$$
$$+ v_x^{2/3}\left(\frac{\partial v_y}{\partial p_y}\right)^{-2/3}\frac{\partial^2}{3\partial p_z^2}\frac{\partial v_y}{\partial p_y} = -C_0' a^{10/3}\cdot 0{,}21635.$$

We obtained $\partial^2 v_x/\partial p_z^2 = -C_0' a^3\cdot 0{,}00669$, $(\partial^2/\partial p_z^2)(\partial v_y/\partial p_y) = -C_0' a^4\cdot 0{,}16980$, with the p_z derivatives taken with $p_z = 0$. For the ratio of the second derivative of the transition frequency with numbers 1 and 2 we have:

$$\omega''_{21}/\omega_{21} = -a^2\cdot 0{,}44027. \tag{5.10}$$

Substituting the values of (5.5)-(5.8) and (5.10) into formula (5.9) we obtain an upper estimate for the Fermi-liquid interaction constant α in the case of resonance at the central cross-section of the "spherical" part of the Fermi surface of copper:

$$0 \leqslant \alpha_{Cu} \leqslant \begin{cases} 0{,}68\cdot 10^{21} & \text{s/g·cm (based on data from [75])} \\ 0{,}29\cdot 10^{21} & \text{s/g·cm (based on data from [76])} \end{cases} \tag{5.11}$$

We derive the value of the constant $\mathfrak{A}^+$ = 7.5-12.3 by processing data corresponding to the microwave range previously in Chapter 4 (see also study [51]) for bismuth. In the case of bismuth the $\mathfrak{A}^+$ constant is related to α by the relation:

$$\alpha^{Bi} = (4\pi)^{-1}(2m_3\varepsilon_F)^{-1/2}\mathfrak{A}^+.$$

Substituting the constants $m_3 = 1.27\ m_0$, $\varepsilon_F = 2.86\cdot 10^{-14}$ erg into this relation we find for bismuth $0.08\cdot 10^{21}$ s/g·cm $\leqslant \alpha_{Bi} \leqslant$ 0.12·1021 s/g·cm.

By comparison to the values of α_{Cu} obtained above (5.11) we find that within an order of magnitude the ratio of the upper boundaries of α_{Cu} and α_{Bi} is equal to unity. If we introduce the quantity $\mathfrak{A}$ ($\mathfrak{A}^+$ in bismuth) using the formula

$$\mathfrak{A} = \left(-\frac{2\omega_{21}}{3\omega''_{21}}\right)^{1/2} 4\pi\alpha,$$

we may state consistent with the values of (5.10) and (5.11) obtained above that for copper the value of $\mathfrak{A}$ lies in the range

$$0 \leqslant \mathfrak{A} \leqslant \begin{cases} 3\cdot 10^2 & \text{(based on data from [75])} \\ 1,3\cdot 10^2 & \text{(based on data from [76])} \end{cases}$$

We emphasize here that such a high value of the derived upper boundary on $\mathfrak{A}$ is attributable primarily to the high inaccuracy (≥100%) in determining $(B^0_{21}/B_{21})^{2/3} - 1$. Hence the upper boundary of $\mathfrak{A}$ is in fact determined in this case by the measurement error of $(B^0_{21}/B_{21})^{2/3} - 1$.

In our view the path to a more exact determination of the Fermi-liquid interaction constants in copper based on experimental data on resonances at skipping surface electrons is as follows. First it is necessary to obtain data on the positions of the resonance peaks with numbers other than $n' = 2$ and $n = 1$. Such information makes it possible to avoid using data on the Fermi velocity (or energy) and to determine the constants directly from data on the resonances at the magnetic surface levels. Second, it would be desirable to raise the frequency $\omega/2\pi$ used in the experiments in study [24] by one or two orders of magnitude (i.e., to jump to the IR range). Such an increase in frequency would make it possible to increase the accuracy of determination of $\mathfrak{A}$ and to identify a range of values of $\mathfrak{A}$ for copper and not simply estimate the lower boundary as was obtained from an interpretation of experimental data from study [24] obtained in the microwave range.

BIBLIOGRAPHY

1. Silin, V.P. "Theory of degenerate electron liquid and electromagnetic waves" FMM, 1970,, V. 29, p. 681-734.

2. Silin, V.P. "A theory of collective description of electron interaction in solids" FMM, 1956, V. 3, p. 193-203.

3. Silin, V.P. "Degenerate electron liquid oscillations" ZhETF, 1958, V. 35, p. 1243-1250.

4. Akhiezer, A.I., Bar'yakhtar, V.G., Peletminskiy, S.B. "Spinovye volny" [Spin waves] Moscow: Nauka, 1967, 368 p. Addendum to this book: Silin, V.P. Spinovye volny v neferromagnitnykh metallakh [Spin waves in nonferromagnetic metals].

5. Silin, V.P. "Spin waves in a degenerate electron liquid" ZhETF, 1968, V. 55, p. 697-703.

6. Shultz, S., Dunifer, G. "Observation of spin waves in sodium and potassium" PHYS. REV. LETT., 1967, V. 18, p. 283-287.

7. Platzman, P.M., Wolff, P.A. "Spin wave excitation in non-ferromagnetic metals" PHYS. REV. LETT., 1967, V. 18, p. 280-283.

8. Dunifer, G., Schultz, S., Schmidt, P.H. "Spin waves in sodium and potassium" J. APPL. PHYS., 1968, V. 39, p. 397-402.

9. Ying, S.C., Quinn, J.J., "Spin independent oscillations of a degenerate electron liquid" PHYS. REV., 1969, V. 180, p. 193-217.

10. Ying, S.C., Quinn, J.J. "Spin waves in a degenerate electron liquid" PHYS. REV., 1969, V. 180, p. 218-224.

11. Silin, V.P. "Theory of the anomalous spin-effect in metals" ZhETF, 1958, V. 33, p. 1282-1285.

12. Silin, V.P. "Theory of a degenerate electron liquid" ZhETF, 1957, V. 33, p. 495-500.

13. Zyryanov, P.S., Okulov, V.I., Silin, V.P. "Quantum waves in the degenerate electron liquid of a metal" ZhETF, 1970, V. 58, p. 1295-1309.

14. Okulov, V.I., Pamyatnykh, Ye.A. "The spectrum of quantum waves in the electron liquid of a metal" FMM, 1974, V. 38, p. 279-288.

15. Zyryanova, N.P., Okulov, V.I., Silin, V.P. "Problems in solid state physics" 1975, No. 31, p. 38-86.

16. Zyryanov, P.S., Okulov, V.I., Silin, V.P. "Quantum spin waves" PIS'MA I ZhETF, 1968, V. 8, p. 489-492.

17. Zyryanov, P.S., Okyalov, V.I., Silin, V.P. "Quantum electron spin-acoustic waves" PIS'MA I ZhETF, 1969, V. 9, p. 371-374.

18. Butikov, Ye.I., Kondrat'ev, A.S., Kuchma, A.Ye. "Collective excitations in thin metallic films" FMM, 1973, V. 36, p. 479-492.

19. Kondrat'ev, A.S., Kychma, A.Ye., Meylanov, R.P. "Quantum wave theory in thin films" FMM, 1974, V. 37, p. 1138-1145.

20. Khaykin, M.S. "The oscillatory dependence of the surface resistance of a metal on a weak magnetic field" ZhETF, 1960, V. 39, p. 212-214; "Magnetic surface levels" UFN, 1968, V. 96, p. 409-440.

21. Nee, T.W., Prange, R.E. "Quantum spectroscopy of the low field oscillations of the surface impedance" PHYS. LETT. A, 1967, V. 25, p. 582-583.

22. Koch, J.F., Kuo, C.C. "Surface-impedance oscillations in a weak magnetic field" PHYS. REV., 1966, V. 143, p. 470.

23. Koch, J.F. Technical Report N. 898, Wash., Univ. MD., 1968.

24. Doezema, R.E., Koch, J.F. "Magnetic surface levels in Cu-observation and analysis of microwave resonances and determination of Fermi velocities" PHYS. REV. B, 1972, V. 5, p. 3866-3882.

25. Koch, J.F., Jensen, J.D. "Magnetic field induced surface states in bismuth" PHYS. REV., 1969, V. 184, p. 643-655.

26. Edel'man, V.S. "Electrons in bismuth" ADV. PHYS., 1976, V. 25, p. 555-613.

27. Wanner, M., Doezema, R.E., Strom, D. "Far infrared surface-Landau-level spectroscopy in Bi" PHYS. REV. B, 1975, V. 12, p. 2883-2892.

28. Strom, U., Kamgar, A., Koch, J.F. "Quantum aspects and electrodynamics of high-frequency cyclotron resonances in bismuth" PHYS. REV. B, 1973, V. 7, p. 2435-2450.

29. Takaoka, S., Kawamura, H., Murase, K., Takano, S. "Electron band model of bismuth by magnetic surface resonances" PHYS. REV. B, 1976, V. 13, p. 1428-1433.

30. Herrmann, R. "Oszillationen der Oberflachenimpedanz von Wolfram in schwachen Magnetfeldern" PHYS. STATUS SOLIDI, 1967, V. 21, p. 703-707.

31. Fishbeck, H.J., Mertsdung, J. "The theory of low-field oscillations of the surface impedance in metals" PHYS. STATUS SOLIDI, 1968, V. 27, p. 345-357.

32. Fishbeck, H.J. "Low-field magentoconductivity of metals in the half-space" PHYS. STATUS SOLIDI (b), 1971, V. 45, p. 221-233.

33. Fal'kovskiy, L.A. "The density and damping of surface magnetic states" ZhETF, 1970, V. 58, p. 1830-1842.

34. Abrikosov, A.A. "Vvedenie v teoriyu normal'nykh metallov" [An introduction to normal metal theory] Moscow: Nauka, 1972, 288 p.

35. Andreev, A.F. "Interaction of conduction electrons with the surface of a metal" UFN, 1971, V. 105, p. 113-124.

36. Kaner, E.A., Makarov, N.M. "Surface electromagnetic waves in metals in a magnetic field" PIS'MA I ZhETF, 1969, V. 10, p. 253-257.

37. Kaner, E.A., Makarov, N.M., Fuks, I.M. "The spectrum and damming of surface electron states in a magnetic field" ZhETF, 1968, V. 55, p. 931-941.

38. Kaner, E.A., Makarov, N.M. "The anomalous skin-effect in metals in a magnetic field" ZhETF, 1969, V. 57, p. 1435-1444.

39. Kaner, E.A., Makarov, N.M. Fal'ko, V.L., Yampol'skii, V.A. "Theory of cyclotron resonance in metals with a near-specular boundary" ZhETF, 1977, V. 73, p. 1400-1406.

40. Grishin, A.M., Kaner, E.A., Tarasov, Yu.V. "Geometrical resonance of the Rayleigh sound wave absorption as a method for direct measurements of the electron specular reflection coefficient from the metal surface" Solid State Communs, 1975, V. 16, p. 425-429.

41. Grishin, A.M., Tarasov, Yu.V. "Propagation of surface acoustic waves in metals in a magnetic field" ZhETF, 1973, V. 65, p. 1571-1582.

42. Grishin, A.M., Kaner, E.A., Lyubimov, O.I., Makarov, N.M. "Rayleigh and electromagnetic waves in metals in a weak magnetic field" ZhETF, 1970, V. 59, p. 629-640.

43. Fok, V.A. "Problemy difraktsii i rasprostraneniya radiovoln" [Problems of radio wave diffraction and propagation] Moscow: Sov. radio, 1970, 517 p.

44. Grishin, A.M., Kaner, E.A., Tarasov, Yu.V. "Magnetoacoustic relaxational effects in the absorption of Rayleigh acoustic waves in metals" ZhETF, 1976, V. 70, p. 196-213.

45. Bilenkin, A.V., Kaner, E.A., Fuks, I.M. "On the width of resonance on magnetic surface levels" ZhETF, 1972, V. 63, p. 315-328.

46. Silin, V.P., Tolkachev, O.M. "Resonances at skipping orbits in an electron liquid" PIS'MA I ZhETF, 1975, V. 21, p. 791-721.

47. Silin, V.P., Tolkachev, O.M. "Surface quantum spin waves in the degenerate electron liquid of metals in a weak magnetic field" ZhETF, 1976, V. 70, p. 639-656.

48. Silin, V.P., Tolkachev, O.M. "Cyclotron resonance at skipping orbits in an electron liquid" ZhETF, 1978, V. 74, p. 2138-2153.

49. Silin, V.P., Tolkachev, O.M. "On the electron resonance at skipping orbits in electron liquid" PHYS. LETT. A, V. 57, p. 267-269.

50. Silin, V.P., Tolkachev, O.M. "Cyclotron resonance and surface waves at skipping orbits in the electron liquid of metals" In: "193 Vsesoyuz. sobeshch. po fizike nizkikh temperatur" [The 19th All Union Conference on Low Temperature Physics] 14-18 September, 1976. Topic papers. Minsk, 1976, p. 74-75.

51. Silin, V.P., Tolkachev, O.M. "Cyclotron resonance at skipping electrons and determination of interelectron interaction parameters in bismuth" FTT, 1979, V. 21, p. 1300-1306.

52. Silin, V.P., Tolkachev, O.M. "Determination of Fermi-liquid interaction in bismuth by means of experiments on cyclotron resonance at skipping electrons" In: "20-3 Vsesoyuz. sobeshch. po fizike nizkikh temperatur [The 20th All-Union Conference on Low Temperature Physics] (23-26 January, 1979. Moscow, 1979, p. 57-58.

53. Silin, V.P., Tolkachev, O.M. "Quantum theory of cyclotron resonance at skipping orbits near the transition frequencies between surface electron levels in a degenerate electron liquid" FTT, 1980, V. 22, p. 374-382.

54. Prinkhalter, A., Silin, V.P., Tolkachev, O.M. "Cyclotron resonance at skipping electrons in the IR" FMM, 1981, V. 51, p. 510-518.

55. Kener, E.A., Makarov, N.M. "Theory of surface electromagnetic waves in metals in a weak magnetic field" ZhETF, 1970, V. 58, p. 1972-1986.

56. Nee, T.W., Koch, J.E., Prange, R.E. "Surface quantum states and impedance oscillations in a weak magnetic field numerical aspects" PHYS. REV., 1968, V. 174, p. 758-766.

57. Meyerovich, B.E. "The influence of a sample surface on cyclotron resonance in metals" ZhETF, 1970, V. 58, p. 324-336.

58. Alodzhants, G.P. "Theory of cyclotron and spin waves in an anisotropic electron liquid" ZhETF, 1970, V. 59, p. 1429-1439.

59. Silin, V.P., Rukhadze, A.A. "Electromagnetic properties of a plasma and plasma-like media" Moscow: Atomizdat, 1961, 230 p.

60. Edel'man, V.S. "Magnetoplasma waves in bismuth" UFN, 1970, V. 102, p. 55-85.

61. Edel'man, V.S. "Properties of electrons in bismuth" UFN, 1977, V. 123, p. 257-287.

62. Edel'man, V.S. "The shape of the electron Fermi surface of bismuth" ZhETF, 1973, V. 64, p. 1734-1745.

63. Edel'man, V.S. "Investigation of bismuth in a quantizing field" ZhETF, 1975, V. 68, p. 257-271.

64. Ketterson, J.B., Windmiller, J.R. "Inversion of de Haas-van Alphen data on nearly ellipsoidal surface: Application to As and Symmetrical basis" PHYS. REV., 1970, V. 1, p. 463-470.

65. Cohen, M.N. "Energy bands in the bismuth structure" PHYS. REV., 1961, V. 121, p. 387-395.

66. Takano, S., Kawamura, H. "The quantum variation of Fermi energy in

bismuth under high magnetic fields" J. PHYS. SOC. J. APPL. PHYS.,., 1970, V. 28, p. 348-359.

67. Mueller, F.M. "New inversion scheme for obtaining Fermi-surface radii from de Haas-van Alphen areas" PHYS. REV., 1966, V. 148, p. 636-637.

68. Eyryanov, P.S., Okulov, V.I., Silin, V.P. "Coupled spiral and quantum spin-acoustic waves in the electron liquid of a metal" FMM, 1970, V. 30, p. 1093-1095.

69. Kondrat'ev, A.S., Kuchma, A.Ye. "Elektronnaya zhidkost' normal'nykh metallov" [Electron liquid of normal metals] Leningrad: 1980, 200 p.

70. Doezema, R.E., Koch, J.F. "Magnetic surface levels in Cu-determination of electron-phonon scattering rates" PHYS. REV. B, 1972, V. 6, p. 2071-2077.

71. Wegehaupt, T., Doezema, R.E. "Measurement of the anisotropic Fermi velocity in Al" PHYS. REV. B, 1977, V. 16, p. 2515-2525.

72. Garcia-Moliner, F. "On the Fermi surface of copper" PHILOS. MAGNETIC., 1958, V. 3, p. 207.

73. Krenell, A., Uong, K. "Poverkhnost' Fermi" [The Fermi surface] Moscow: Atomizdat, 1978, 350 p.

74. Roaf, D.J., "The Fermi surfaces of copper, silver, and gold" PHILOS. TRANS. ROY. SOC. LONDON A, 1963, V. A255, p. 135-152.

75. Haise, M.H. "Fermi surface of noble metals" PHILOS. TRANS. ROY. SOC. LONDON A, 1969, V. A265, p. 507-532.

76. Lee, M. "Dynamical properties of quasi particle excitations in metallic copper" PHYS. REV. B, 1970, V. 2, p. 250-263.

77. Lee, M. "Phase-shift analysis of the Fermi surface of copper" PHYS. REV., 1969, V. 187, p. 901-911.

Theory of the Surface Impedance of Normal Metals

A.V. Kobelev, V.P. Silin

Abstract: A method is developed for formulating a theory of the surface impedance of a metal in a magnetic field perpendicular to the surface within the scope of degenerate electron liquid theory. General formulae expressing the impedance through the transverse conductivity of the metal are found in a model where the correlation function is approximated by two parameters A_1 and A_2 for the case of specular reflection and diffuse scattering of electrons off the surface. In the anomalous skin-effect regime the dependence of the impedance on the frequency and the field strength of the external magnetic field is analyzed in the cyclotron and cyclotron-wave resonance regions. A method is proposed and the necessary conditions are noted for experimental determination of the magnitude and sign of the Fermi-liquid interaction parameter A_2 in experiments employing bulk samples.

INTRODUCTION

In examining many effects in solid state physics a critical role is played by the incorporation of interelectron interaction, specifically the interaction between conduction electrons in normal metals. Often the influence of this interaction is more significant than the periodic potential generated by the ion lattice. In this case commonly-used approximations such as the approach to quasi particles with a renormalized effective mass are not sufficient for a theoretical description of the effects attributable to the interaction between quasi particles, which may be rather strong.

One of the theories that is valid for an arbitrary interaction value is the Landau-Silin theory of a degenerate electron liquid. A useful concept introduced in this theory is the assumption that the small change in quasi particle energy is determined by the deviation from equilibrium of the one-particle distribution function of the quasi particles, which may be expressed by means of a momentum- and coordinate-nonlocal integral operator. The kernel of this operator describes both the long-range Coulomb interaction and the close correlation, where the dependence of the close correlation on the coordinates manifests a delta nature. The momentum dependence of this close electron correlation function in the case of isotropic metals is reduced to a dependence on a single quantity: the angle between two vectors, since their modulus value is equal to the Fermi momentum due to degeneration. Consequently the correlation function may be represented as a series expansion in terms of Legendre polynomials of this angle with coefficients A_l and B_l, whose complete set determines the Fermi-liquid interelectron interaction.

Since the function characterizing electron correlations (or, the set of phenomenological Fermi-liquid parameters, which is equivalent) cannot be sufficiently strictly calculated, it must be experimentally determined which may be achieved in principle by investigating the high-frequency properties of the metal in a magnetic field. Thus, we may reliably determine the first two parameters (B_0 and B_1) of the spin-dependent part of the correlation function in a number of alkali metals based on observation of spin waves in experiments on signal propagation through the sample. The parameters A_l (l = 1, 2, ...) related to the orbital degrees of freedom are accessible to investigation by observation of cyclotron waves.

The orientation of the external constant magnetic field perpendicular to the surface of the metal when the cyclotron waves propagate in the direction of the magnetic field is of particular interest. In the gas model of the metal the propagation of such waves is forbidden due to strong collisionless Landau damping and hence their existence is entirely attributed to interelectron interaction such as the nonzero value of the parameter A_2.

In investigating cyclotron waves the experimentally observed quantity is the real part of the surface impedance which is proportional to the absorbed power in experiments on bulk samples or the magnetic field derivative of the impedance in thin wafers. Hence for an actual comparison of theoretical and experimental results we need to develop a theoretical description in the electron Fermi-liquid model of an experimentally observed quantity such as the surface impedance.

At present experimental data do not agree with theoretical predictions of the impedance in the vicinity of cyclotron-wave resonance frequencies. This is attributable to the absence of a strict theory of the frequency dependence of the impedance in the cyclotron and cyclotron-wave resonance regions, particularly in the case of diffuse electron scattering on the metal surface. The formulation of such a theory which accounts for Fermi-liquid interaction and conduction scattering simultaneously even in the simplest approximation of the diffusion parameter involves significant mathematical difficulties. For this reason to date no analytical expressions have been derived for the electron liquid impedance in diffuse scattering, which has slowed the further development of the theory. Consequently in subsequent theoretical studies the resonance additions to the impedance in diffuse scattering attributable to Fermi-liquid interaction were investigated by an approximate variational method.

This study provides a survey and coverage of the results from the application of a general method of calculating the surface impedance of a half-bounded metal in the electron liquid model accounting for the constants A_1 and A_2 in a magnetic field perpendicular to the surface with both specular and diffuse electron scattering by the surface. When accounting for diffuse scattering we obtain general relations that relate the impedance to the conduction of the metal dependent on A_1 or A_2. New results include asymptotic expansions of the impedance in terms of the inverse powers of the parameter characterizing the anomalous nature of the skin-effect with term accuracy which largely determine the resonance frequency dependence attributable to Fermi-liquid interaction. The behavior of the impedance in the vicinity of cyclotron and cyclotron-wave resonances is investigated in detail and asymptotic formulae are given that characterize the resonance singularities in the collisionless limit. A method is proposed for determining the magnitude and sign of the constant A_2 in experiments using bulk samples together with experimental conditions.

The following are the primary results of this study:

1. A comprehensive theoretical description of the impedance singularity attributable to cyclotron-wave resonance and the identification of magnification of this singularity in the case of diffuse scattering.

2. Determination of the nonresonance nature of corrections in the impedance and of the field passed through the wafer attributable to the constant A_1; these are nonetheless comparable in magnitude to the primary component in diffuse scattering.

3. The identification of smoothing of the impedance singularities in the vicinity of cyclotron resonance with specular reflection and the enhancement of the singularity in the case of diffuse scattering.

4. Analysis of the frequency dependence of the impedance and the relative position of the singularities in the vicinity of cyclotron and cyclotron-wave resonances in diffuse scattering.

The first chapter discusses the primary concepts used in further calculations, formulates the initial system of integral equations for the moments of the distribution function in the electron liquid in the case of specular and diffuse electron scattering at the boundary. An expression is derived for the transverse conduction of electron liquid and is used as the basis to examine the characteristic properties of the cyclotron wave spectrum in the simplest model with a single Fermi-liquid parameter A_2.

In the second chapter we calculate the impedance with specular electron reflection at the boundary in the model with two nonzero Fermi-liquid interaction parameters A_1 and A_2. The position and relative magnitude of the impedance singularity in the vicinity of the cyclotron-wave resonance frequency are investigated and these are compared to experimental data.

Calculation of the impedance with diffuse electron scattering is much more complex and hence the influence of the Fermi-liquid constants A_1 and A_2 in diffuse scattering are accounted for separately, which makes it possible to significantly simplify the solution of the problem in each case.

In the third chapter for the case of diffuse electron scattering we formulate a theory of surface impedance in a model with a single nonzero Fermi-liquid parameter A_1. The possibility for the existence of resonance singularities in the frequency dependence of the impedance attributable to the constant A_1 that are not predicted by the theory in the case of specular electron reflection is analyzed. The conditions in which nonresonance Fermi-liquid additions to the impedance and the electrical field at an asymptotically great depth from the metallic surface are examined.

In the fourth chapter the surface impedance with diffuse electron scattering is calculated in a model when only the Fermi-liquid parameter A_2 is nonzero. Such a model allows incorporation of features associated with the possibility for the existence of Fermi-liquid cyclotron waves attributable to the nonzero value of the constant A_2. The frequency dependence of the impedance in the vicinity of cyclotron-wave and cyclotron resonances is investigated and the impedance across the entire range of frequencies is calculated numerically. The possibility for determining the value and sign of the constant A_2 from the experiment is analyzed.

In conclusion we provide the primary results and conclusions of the study and note promising directions for further development of the theory of surface impedance of the electron liquid of metals.

Chapter 1

CYCLOTRON WAVES AND AN INVESTIGATION OF THE FERMI-LIQUID PROPERTIES OF METALS

1.1. Analysis of previous studies and formulation of the research problem

Significant progress was achieved in the theory of degenerate electron liquid (the theory of uncharged Fermi-liquid was developed by L.D. Landau in 1957 [2]) in the years since the publication of the-pioneering studies [1] (1958) and prior to the first experimental confirmations (1966). A number of phenomena for which Fermi-liquid electron interaction either completely determines the properties of the metal or introduces significant differences compared to the gas model were predicted as early as Silin's study [3]. Such phenomena include zero-point sound, the propagation of electromagnetic waves whose frequency is much less than the plasma frequency, and cyclotron and spin waves whose critical frequencies differ from the Larmour frequency and the spin precession frequency, respectively.

In this section in order to determine the position of this study among preceding works we will briefly discuss the fundamental research devoted to the investigation of Fermi-liquid interaction influence on the features of electromagnetic and spin wave propagation in metals, focusing primarily on the problem of the theoretical and experimental investigation of the surface impedance in conditions where cyclotron waves may exist.

Platzman and Wolff [4] were the first to investigate theoretically the possibility for manifestation of Fermi-liquid correlation effects in the surface impedance of a half-bounded metal in a magnetic field perpendicular to the surface with specular electron reflection. They used numerical techniques to analyze the dependence of the impedance singularity near the Doppler-shifted cyclotron resonance on the Fermi-liquid parameters, although the vicinity of cyclotron resonance was not thoroughly investigated.

The discovery of Fermi-liquid spin waves in sodium and potassium by Schults and Dunifer [5] provided a powerful impetus for the development of generalizations and applications of the theory of degenerate electron liquid. The theory of spin waves in Fermi-liquid [6] was developed in study [7] for analysis of experiments on electromagnetic wave propagation through a metallic slab which made it possible based on comparison to experiment [5] to determine the first two constants characterizing the spin-dependent part of Fermi-liquid interaction.

The experiments by Walsh and Platzman [8, 9] to observe electromagnetic oscillations near the cyclotron resonance frequencies in thin alkali metal films prompted extensive investigations of oscillations and waves in electron liquid independent of the spin degrees of freedom. Studies [8, 10] investigate the plasma wave spectrum in potassium in a magnetic field applied along the sample surface and used this as the basis for estimating values of the zero (A_0) and first (A_1) constants of the spin-dependent component of Fermi-liquid interaction. However study [11] pointed out the fact that such a method is improper for determination of A_0 and A_1, since in the experimental conditions of the anomalous skin-effect [8] the dispersion relation does not contain A_0 and A_1 [12].

Subsequently the spectrum of cyclotron waves was investigated in detail theoretically by application of perturbation theory in the parameter $kR \ll 1$ (k is the wave number, R is the cyclotron radius)

[13] that made it possible to find the eigenfrequencies and the k-squared corrections to the spectrum without limiting the number of nonzero Fermi-liquid parameter and by using the technique from studies [14, 15] suitable for long wavelength comparable to the cyclotron radius although accounting for only certain parameters such as A_2 and A_3. Study [16] has carried out a detailed investigation of the spectra of cyclotron waves propagating in a degenerate Fermi-liquid both perpendicular to and along an external constant d.c. magnetic field. Study [17] has found the spectrum of cyclotron waves with an arbitrary angle of inclination of the magnetic field.

The drafting of theoretical dispersion curves of cyclotron waves propagating along the magnetic field for different values of the parameters A_1 and A_3 across the entire range of frequencies up to the termination point of the spectrum [18] has been the focus of special attention from the viewpoint of verifying the possibility for determining the parameters of Fermi-liquid interaction. Thus, study [19] determined the deviation in the frequency of cyclotron waves propagating perpendicular to the magnetic field from the Larmour frequency in the longwave limit based on a comparison of theoretical and experimental data and estimated the values of the constants A_2 and A_3 in a number of simple metals. The essence of the technique used in studies [10, 19] to determine the parameters of Fermi-liquid interaction involves formulating and extrapolating to $k = 0$ the dispersion curve for waves in a thin slab. It was assumed that the maximum and minimum of the surface impedance corresponded to the integer and half-integer numbers of the wavelengths accumulated in the slab. Subsequently this empirical technique was criticized [20-22] in view of the significant inaccuracy in determining the parameters A_l manifest when accounting for the finite frequency of electron collisions, so the values of A_l obtained by this technique cannot be considered reliable.

Another possible method for determining the values of A_l involves observing the resonance singularity of the surface impedance of a sufficiently thick slab in which a Fermi-liquid cyclotron wave may propagate in the direction of the magnetic field. Such a possibility was first investigated in an experiment by Baraff, Grimes, and Platzman [23]. The physical aspect of this effect is rather transparent and appears as the resonance enhancement of absorption in the frequency range where the cyclotron wave transports electromagnetic wave energy into the metal bulk. The qualitative theory of this phenomenon that allows prediction of the position, magnitude, and lineform of the resulting singularity is insufficiently developed and therefore attempts to interpret experimental facts based on existing available theoretical predictions has encountered contradiction.

The position of the peak in the surface resistance in study [23] corresponded to the critical frequency of the Fermi-liquid cyclotron wave if a negative value is taken for the parameter characterizing interelectron interaction. This value did not contradict that obtained by an independent method: extrapolation of the dispersion curve [10, 19]. Estimates of the relative magnitude of the peak in accordance with the theory developed in the case of specular reflection [4] has provided a value that is four orders of magnitude smaller than in experiment [23]. In order to explain the anomalous frequency dependence of the impedance [23] Baraff [24] proposed a model of a rough metal surface in which Fermi-liquid interaction is ignored. A certain enhancement of the fields in the metal and a corresponding increase in the peak is achieved artificially by assuming independence of the electron collision frequency from the distance to the surface. On the one hand with diffuse electron scattering calculation of the impedance in the electron liquid model [25] carried out by means of the approximate variational method proposed in study [26] provides a higher value of the peak than with specular reflection. However the position and contour of the line found in study [25] particularly with the value of the parameter A_2 from studies [10, 19] differ significantly from experimental results [23]. Attempts to achieve agreement between theory and experiment [27] by combining the Fermi-liquid model with the assumption of surface roughness have not been satisfactory, due to the unsuitability of employing a simple superposition of these factors. Moreover, the accuracy of the solution using the variational method depends on selection of the sample functions and, as a rule, is not high.

We will focus on one experimental method that makes it possible in principle to determine the Fermi-liquid interaction parameters: the cyclotron phase resonance technique [32]. This method is based on the resonance propagation of an electromagnetic wave through a slab with a thickness exceeding the skin depth. The theory of electromagnetic wave propagation through a metal slab is rather complex even ignoring the Fermi-liquid interaction, particularly with diffuse electron surface scattering [33-36]. Hence empirical concepts are used in experimental analysis. Study [23] discovered a difference in the magnitude of a signal that has passed through a potassium slab between the resonance frequency and the cyclotron frequency. Further assuming this difference to be proportional to the constant A_1, the authors of study [32] found for it a value an order of magnitude lower than that established by the rule of sums [37] (see also [38]).

However we know that in the anomalous skin-effect regime the constants A_0 and A_1 drop out of the dispersion equation for the cyclotron waves, resulting in the vanishing of the current and the corresponding density perturbation [12]. This makes it necessary to approach such assumptions as used in study [32] critically. Therefore the assumption that the peak in the signal that has passed through the slab corresponds to a conduction peak and that its position is dependent on the magnitude of the constant A_1 was criticized in studies [39, 40]. It turned out that the signal peak corresponds to the cyclotron resonance point and is position is, therefore, independent of A_1. However, an analysis of the signal line contour in study [39] was based on the derived expressions for the electrical field in the slab with diffuse electron scattering and proceeded only from qualitative considerations, and the line contour was not calculated. The calculations were carried out based on the approximate variational method in study [40].

Thus, the problem of determining the magnitude of the parameters A_l [41] is still far from resolution, since the proposed methods for their determination by experiment (such as determination of A_1 based on observation of coupled acoustic and cyclotron waves [42]) are not reliable. As discussed above the value of the parameter A_2 cannot be considered firmly established, due to the insufficient accuracy in the semiempirical method of its determination from an extrapolation of the dispersion curve and in connection with the absence of a comprehensive theory of the line shape of the surface impedance singularity in a thick slab for the case of diffuse scattering.

In this respect the method of calculating the surface impedance of an electron liquid [28] is promising; this is based on its expansion into an asymptotic series [29, 30] in inverse powers of the parameter of the skin-effect anomaly [31] that is free of the inaccuracies of the variational method discussed above. Here the important aspect is the problem of obtaining an asymptotic expansion of the metallic impedance in the Fermi-liquid model with diffuse electron scattering and investigating the frequency dependence of the terms of this expansion near cyclotron resonance, including the range of existence of cyclotron waves.

In this regard our approach involves formulation of a theory of surface impedance of the metal in the electron liquid model in a magnetic field perpendicular to the surface when the possibility for the appearance of the A_2 constant in the high-frequency properties of the metal is most favorable. Here we devote primary attention to the case where the electrons are diffusely scattered on the surface. With the identical geometry of the problem and an identical boundary condition for electron scattering it is interesting to investigate the influence of the A_1 constant both on the surface impedance and the field that has passed through the slab.

We note that methodologically in calculating the impedance it is most advisable to generalize the asymptotic expansion method that utilizes the skin-effect anomaly parameter to the case of electron liquid. The final goal in calculating the impedance involves finding and investigating the resonance singularity of the terms of this expansion in the vicinity of the singularities that are the cyclotron and cyclotron-wave resonance points.

1.2. Approximations and initial equations from degenerate electron liquid theory

Initially we will discuss the general assumptions in which the kinetic equation is derived in an electron liquid model, since it is used as the basis for deriving the material equation. This equation is necessary for calculating the surface impedance where calculation of the high-frequency singularities of the impedance makes it possible to investigate the Fermi-liquid properties of the metal.

Underlying the theory of degenerate Fermi-liquid which is, in some sense, the continuation of research on the kinetic theory of Fermi-gas with weak interaction [43] is the assumption that even with strong interparticle interaction the energy attributable to a single quasi particle depends on the state-distribution of the particles as is the case in the Hartree-Fock approximation, and the energy of the entire system is not an additive accumulation of the energy of the separate quasi particles. The relation between the change in the quasi particle distribution $\delta n(\mathbf{p}, \mathbf{r})$ and the change in quasi particle energy corresponding to it may be written as a momentum- and coordinate-nonlocal integral operator:

$$\delta\varepsilon(\mathbf{p}, \mathbf{r}, t) = \mathrm{Sp} \int \frac{d\mathbf{p}'\, d\mathbf{r}'}{(2\pi\hbar)^3} \hat{F}(\mathbf{p}, \mathbf{p}', \mathbf{r}, \mathbf{r}')\, \delta n(\mathbf{p}', \mathbf{r}', t). \tag{1.1}$$

Here the spin indices are dropped and we assume that this relation is local in time. This corresponds to conditions where the characteristic variation in particle distribution is much greater than $\hbar/\varepsilon_F$, where ε_F is the Fermi energy (for example, $\varepsilon_F/\hbar \gg \omega$, where ω is the electromagnetic wave frequency). One additional general assumption used in the theory of degenerate electron liquid is the assumption of a delta dependence on the coordinates of the part of the function $\hat{F}$ corresponding to the near correlation $f(\mathbf{p}, \mathbf{p}')\delta(\mathbf{r} - \mathbf{r}')$. The other part of the function $\hat{F}$ is nonlocal and corresponds to the long-range self-consistent Coulomb potential. This assumption is justified when the characteristic length over which the particle distribution varies in space significantly exceeds the correlation radius, which for the case of metals coincides with the average distance between electrons (for example, when $k \ll k_F$, where k is the wave number of the electromagnetic wave, $k_F = \frac{1}{\hbar}\sqrt{2m\varepsilon_F} = (3\pi^2 n)^{1/3}, p = \hbar\; k_F$ is the Fermi momentum).

In the majority of cases we may ignore the small spin-orbital interaction, then we may isolate the spin-independent part in the function $f(\mathbf{p}, \mathbf{p}')$:

$$f(\mathbf{p}, \mathbf{p}') = \hat{1}\varphi(\mathbf{p}, \mathbf{p}') + 4(\hat{s}\cdot\hat{s}')\psi(\mathbf{p}, \mathbf{p}'). \tag{1.2}$$

Finally, limiting our examination to isotropic metals, due to the dependence of $\varphi(\mathbf{p}, \mathbf{p}')$ and $\psi(\mathbf{p}, \mathbf{p}')$ on a single parameter - the angle θ between the vectors $\mathbf{p}$ and $\mathbf{p}'$ lying on the Fermi sphere, we obtain the following representation ($m = p/v$ is the effective mass, v is the electron Fermi velocity):

$$\frac{p^2}{\pi^2\hbar^3 v}\varphi(\mathbf{p}, \mathbf{p}') = \sum_{l=0}^{\infty}(2l+1)A_l P_l(\cos\theta). \tag{1.3}$$

Here we have expanded φ into a series in terms of Legendre polynomials. An analogous formula for φ contains the coefficients B_l. The set of parameters A_l and B_l completely determine the function $f(\mathbf{p}, \mathbf{p}')$

and, consequently, F in formula (1.1). The microscopic calculation of the function F related to the zero angle scattering amplitude of the quasi particles according to study [44] is correct only with smallness of the parameter $e^2m/\hbar^2k_F$ of Born's order or perturbation theory in the scattering potential which is greater than unity for metals. However different expansion coefficients may appear in (1.3) in certain experimental conditions and hence in principle in comparing theory to experiment we may determine a set of parameters A_l, B_l and thereby obtain information on the correlation function $f(\mathbf{p}, \mathbf{p}')$.

We will now proceed to a discussion of the kinetic equation for a degenerate electron liquid. Here we limit our examination to the case where interelectron interaction may be described by the function $\varphi(\mathbf{p}, \mathbf{p}')$ and we do not account for spin paramagnetism effects, taking $\psi(\mathbf{p}, \mathbf{p}') = 0$. This may be justified, for example, if the characteristic resonance frequencies related to the spin and orbital degrees of freedom are sufficiently different.

We will consider minor deviations of the electron distribution function $f(\mathbf{p}, \mathbf{r}, t)$ which is the spur in the spin variables of the complete density matrix from the equilibrium Fermi function f_0. Following study [6] we write the equation for such deviations δf and the electrical field $\mathbf{E}$ (we label the induction of the constant d.c. external field as $\mathbf{B}$):

$$\frac{\partial}{\partial t}\delta f + \nu\delta f + \mathbf{v}\frac{\partial}{\partial \mathbf{r}}\left(\delta f - \frac{\partial f_0}{\partial \varepsilon}\delta\varepsilon_1\right) + \frac{e}{c}[\mathbf{v}\times\mathbf{B}]\frac{\partial}{\partial \mathbf{p}}\left(\delta f - \frac{\partial f_0}{\partial \varepsilon}\delta\varepsilon_1\right) + e\mathbf{E}\mathbf{v}\frac{\partial f_0}{\partial \varepsilon} = 0, \tag{1.4}$$

where

$$\delta\varepsilon_1 = \int d\mathbf{p}'\varphi\,(\mathbf{p}, \mathbf{p}')\,\delta f\,(\mathbf{p}'). \tag{1.5}$$

Here in describing electron collisions we use a momentum independent relaxation time approximation $1/\nu$.

We will formulate other particular assumptions that will be used henceforth in solving equation (1.4) relating to the dimensions of the sample and the direction of the external magnetic field. Specifically we will take the dimensions of the sample to be much greater than the free path length of the electrons v/ν so that we may assume the metal is half-bounded occupying an area of say $z > 0$. We will take the magnetic field $\mathbf{B}$ to be perpendicular to the surface of the metal $z = 0$. With this field and sample geometry in the $z = 0$ plane isotropy is conserved by symmetry.

With these assumptions kinetic equation (1.4) for a perturbation stimulated by a monochromatic wave of frequency ω propagating along the magnetic field takes the form (see study [6])

$$(-i\omega + \nu)\,\delta f + e\,(E_x v_x + E_y v_y)\frac{\partial f_0}{\partial \varepsilon} + \left(v_z\frac{\partial}{\partial_z} - \Omega\frac{\partial}{\partial \varphi}\right)\left(\delta f + \frac{\partial f_0}{\partial \varepsilon}\delta\varepsilon_1\right) = 0, \tag{1.6}$$

where $\Omega = eB/mc$ is the Larmour frequency of the electron; φ is the azimuthal angle in the momentum space with the polar axis lying on z. We will search the solution of equation (1.6) as

$$\delta f(\mathbf{p}, \mathbf{r}, t) = e\frac{\partial f_0}{\partial \varepsilon}\Psi(\vartheta, \varphi), \tag{1.7}$$

where

$$\frac{\partial f_0}{\partial \varepsilon} = -\frac{1}{v}\frac{2}{(2\pi\hbar)^3}\delta(p - p_F). \tag{1.8}$$

Subject to (1.3) formula (1.5) for $\delta\varepsilon_1$ takes the form

$$\delta\varepsilon_1(\mathbf{p}) = -\frac{e}{4\pi}\int_0^{2\pi} d\varphi' \int_0^{\pi} \sin\vartheta' \, d\vartheta' \Psi(\vartheta', \varphi') \sum_{l=0}^{\infty} (2l+1) A_l P_l(\cos\theta). \tag{1.9}$$

Using the addition theorem for Legendre polynomials [45] ($P_l^m(\cos\vartheta)$ — are associated Legendre polynomials

$$P_l(\cos\theta) = P_l(\cos\vartheta) P_l(\cos\vartheta') + \\ + 2\sum_{m=1}^{\infty} \frac{(l-m)!}{(l+m)!} P_l^m(\cos\vartheta) P_l^m(\cos\vartheta)\cos m(\varphi - \varphi') \tag{1.10}$$

and converting to the circular components in (1.6) we may easily determine that the substitution

$$\Psi(\vartheta, \varphi) = \frac{v}{2}(\Phi_- e^{i\varphi} + \Phi_+ e^{-i\varphi}) \tag{1.11}$$

determines the dependence of the Ψ solution on the azimuthal angle φ. For the functions $\Phi_\pm(\vartheta, z)$ from (1.6) and (1.9) we have the equation

$$\left[v\cos\vartheta\frac{\partial}{\partial z} - i(\omega \mp \Omega) + \nu\right]\left\{\Phi_\pm(\vartheta, z) + \sum_{l=1}^{\infty}\frac{2l+1}{l(l+1)} A_l P_l^1(\cos\vartheta)\Phi_\pm^{1,l}(z)\right\} = \\ = -E_\pm \sin\vartheta + (-i\omega + \nu)\sum_{l=1}^{\infty}\frac{2l+1}{l(l+1)} A_l P_l^1(\cos\vartheta)\Phi_\pm^{1,l}(z), \tag{1.12}$$

where $E_\pm = E_x \pm iE_y$ are the circular field components and

$$\Phi_\pm^{1,l}(z) = \frac{1}{2}\int_0^{\pi} d\vartheta \sin\vartheta P_l^1(\cos\vartheta)\Phi_\pm(\vartheta, z). \tag{1.13}$$

We note an important circumstance relating to equation (1.13). It follows from the form of solution (1.11) that there is no mode with the azimuthal number $m = 0$ and two independent modes $m = \pm 1$ are excited (the sign corresponds to right and left polarizations of the circularly-polarized wave propagating along the magnetic field). As a result the Fermi-liquid interaction constant A_0 is dropped from the dispersion equation.

We will proceed to a solution of equation (1.12) for the variable z and a formulation of the boundary conditions for $z = 0$. We introduce the conventions:

$$\mathcal{F}(\vartheta, z) = \Phi_{\pm}(\vartheta, z) + \sum_{l=1}^{\infty} \frac{2l+1}{l(l+1)} A_l P_l^1(\cos\vartheta)\, \Phi_{\pm}^{1,l}(z), \tag{1.14}$$

$$\mathcal{E}(\vartheta, z) = -E_{\pm}(z)\sin\vartheta + (-i\omega + \nu) \sum_{l=1}^{\infty} \frac{2l+1}{l(l+1)} A_l P_l^1(\cos\vartheta)\, \Phi_{\pm}^{1,l}(z). \tag{1.15}$$

The general solution of equation (1.12) is written as

$$\mathcal{F}(\vartheta, z) = \frac{1}{v\cos\vartheta} \int_{C(\vartheta)}^{z} dz' \exp\left\{-\frac{z-z'}{v\cos\vartheta}[-i(\omega \mp \Omega) + \nu]\right\} \mathcal{E}(\vartheta, z'). \tag{1.16}$$

The $C(\vartheta)$ constant in (1.16) will be selected from the boundary conditions on the metal surface $z = 0$ and $z = \infty$ in the metal interior. In the latter case it is natural to require $\mathcal{F}(\vartheta, z \to \infty) \to 0$, which corresponds to damping of the perturbation in the metal interior.

In order to describe electron scattering on the surface of the metal $z = 0$ we will use a simple variant of possible boundary conditions for the electron distribution function, and specifically we will take the portion ρ as specularly reflected electrons and the remaining $(1 - \rho)$ to be diffusely scattered (see study [51]). In the case of specular electron scattering on the surface $z = 0$ only the z-component of the momentum reverses and hence the relation

$$\mathcal{F}(v_z, z = 0) = \mathcal{F}(-v_z, z = 0). \tag{1.17}$$

is valid. For electrons diffusely scattered at the boundary $z = 0$ the nonequilibrium addition to the distribution function associated with the electrical field of the wave vanishes and therefore the equality

$$\mathcal{F}(v_z > 0, z = 0) = 0. \tag{1.18}$$

is satisfied. We set $C(\vartheta) = 0$ for electrons traveling from the interior of the metal to the surface ($v_z < 0$, $\pi/2 < \vartheta < \pi$), then we obtain

$$\mathcal{F}_{-}(\vartheta, z) = \frac{1}{v\cos\vartheta} \int_{\infty}^{z} dz' \exp\left\{-\frac{z-z'}{v\cos\vartheta}[-i(\omega \mp \Omega) + \nu]\right\} \mathcal{E}(\vartheta, z') - \tag{1.19}$$

The distribution function of electrons traveling from the surface $v_z > 0$, $0 < \vartheta < \pi/2$), contains the contributions of electrons both specularly and diffusely reflected

$$\mathcal{F}_{+}(\vartheta, z) = \frac{1}{v\cos\vartheta} \int_{0}^{z} dz' \exp\left\{-\frac{z-z'}{v\cos\vartheta}[-i(\omega \mp \Omega) + \nu]\right\} \mathcal{E}(\vartheta, z') +$$

$$+ \frac{\rho}{v\cos\vartheta} \int_{0}^{\infty} dz' \exp\left\{-\frac{z+z'}{v\cos\vartheta}[-i(\omega \mp \Omega) + \nu]\right\} \mathcal{E}^{-}(\vartheta, z'), \tag{1.20}$$

where

$$\mathcal{E}^{-}(\vartheta, z) = -\sin\vartheta E_{\pm} + (-i\omega + \nu) \sum_{l=1}^{\infty} \frac{2l+1}{l(l+1)} A_l P_l^1(\cos\vartheta)\, \Phi_{\pm}^{1,l}(z)(-1)^{l+1}. \tag{1.21}$$

The first term in (1.20) when $z = 0$ vanishes in accordance with the boundary conditions (1.18), and the second term satisfying (1.17) describes the contribution of specularly-reflected electrons.

We will multiply the right and left halves of equation (1.12) by $1/2 \sin\vartheta P_n^1(\cos\vartheta)$ and will integrate with respect to ϑ within the limits from 0 to π. Using formulae (1.19) and (1.20) we obtain the following system of integral equations for the moments of the distribution function $\Phi_{\pm}^{1,l}(z)$

$$(1 + A_n)\Phi_{\pm}^{1,n}(z) - \sum_{l=1}^{\infty} \frac{2l+1}{l(l+1)} A_l \left\{ \int_0^{\infty} dz' Q_{nl}(|z-z'|) \times \right.$$
$$\times \left(\frac{z-z'}{|z-z'|}\right)^{h+l} \Phi_{\pm}^{1,l}(z') + \rho \left. \int_0^{\infty} dz' Q_{nl}(|z+z'|)\Phi_{\pm}^{1,l}(z')(-1)^{l+1} \right\} =$$
$$= \frac{i}{\omega + i\nu} \left\{ \int_0^{\infty} dz' Q_{n1}(|z-z'|) \left(\frac{z-z'}{|z-z'|}\right)^{n+1} E_{\pm}(z') + \rho \int_0^{\infty} dz' Q_{n1}(|z+z'|) E(z') \right\}, \tag{1.22}$$

where

$$Q_{nl}(z) = \frac{1}{2} \int_0^{\pi/2} d\vartheta \sin\vartheta P_n^1(\cos\vartheta) P_l^1(\cos\vartheta) \times$$
$$\times \frac{-i\omega + \nu}{v\cos\vartheta} \exp\left\{ -\frac{z}{v\cos\vartheta} [-i(\omega \mp \Omega) + \nu] \right\}. \tag{1.23}$$

Equation system (1.22) together with Maxwell's equations are used as the basis for investigating the electromagnetic properties of an electron liquid excluding spin paramagnetism when a circularly polarized electromagnetic wave impacts the half-bounded metal in a magnetic field perpendicular to the surface.

1.3. Transverse conductivity of an electron liquid

We will consider the possibility of obtaining a material equation from equation system (1.22) to describe electromagnetic waves propagating along the magnetic field. Due to the transverse nature of the waves we must calculate the transverse (circular) component of the electric current. The latter is expressed through the moment of the distribution function $\Phi_{\pm}^{1,1}(z)$ by a simple formula following from (1.9) and (1.13):

$$j_{\pm}(z) = j_x \pm i j_y = e \int d\mathbf{p} v_{\pm} \left(\delta f - \frac{\partial f_0}{\partial \varepsilon} \delta\varepsilon_1 \right) = \frac{3}{8\pi} \omega_{\mathrm{Le}}^2 (1 + A_1) \Phi_{\pm}^{1,1}(z), \tag{1.24}$$

where

$$\omega_{\mathrm{Le}}^2 = \frac{4e^2 v p^2}{3\pi^2 \hbar^3} = \frac{4\pi n e^2}{m}, \tag{1.25}$$

(ω_{Le} is the Langmuir frequency of the free electron gas). Consequently in order to obtain the material equation it is sufficient to express $\Phi_{\pm}^{1,1}$ through the electrical field $E_{\pm}$.

This is most easily done in the case of specular electron boundary reflection. We note that specular electron surface reflection may be represented as the simultaneous passage through the point $z = 0$ of two electrons with z-components of the momentum of opposite sign and identical magnitude. The only difference from the homogenous zero-boundary medium model is in the existence of surface current in the $z = 0$ plane generated by the external electromagnetic wave.

In order to solve system (1.22) in the case of specular reflection it is convenient to evenly continue all functions of the coordinate to the z negative semiaxis:

$$E_{\pm}(-z) = E_{\pm}(z),\ \Phi_{\pm}^{1,n}(-z) = (-1)^{n+1}\Phi_{\pm}^{1,n}(z). \tag{1.26}$$

Taking $\rho = 1$ in (1.22) and incorporating (1.26) we obtain a system of integral equations with different kernels:

$$(1 + A_n)\Phi_{\pm}^{1,n}(z) - \sum_{l=1}^{\infty}\frac{2l+1}{l(l+1)}A_l\int_{-\infty}^{\infty}dz'C_{nl}(|z-z'|)\left(\frac{z-z'}{|z-z'|}\right)^{n+l}\Phi_{\pm}^{1,l}(z') =$$
$$= \frac{i}{\omega + i\nu}\int_{-\infty}^{\infty}dz'Q_{n1}(|z-z'|)\left(\frac{z-z'}{|z-z'|}\right)^{n+1}E_{\pm}(z'). \tag{1.27}$$

We will provide a solution of system (1.27) by the Fourier transform

$$E_{\pm}(k) = \int_{-\infty}^{\infty}dze^{-ikz}E_{\pm}(z) = 2\int_{0}^{\infty}dz\cos kzE_{\pm}(z), \tag{1.28}$$

$$\Phi_{\pm}^{1,l}(k) = \int_{-\infty}^{\infty}dze^{-ikz}\Phi_{\pm}^{1,l}(z) \tag{1.29}$$

and will arrive at a system of algebraic equations for the Fourier-transforms

$$(1 + A_n)\Phi_{\pm}^{1,n}(k) - \sum_{l=1}^{\infty}\frac{2l+1}{l(l+1)}A_lq_{nl}(k)\Phi_{\pm}^{1,l}(k) = \frac{i}{\omega + i\nu}q_{n1}(k)E_{\pm}(k), \tag{1.30}$$

where

$$q_{nl}(k) = \int_{-\infty}^{\infty}dze^{-ikz}\left(\frac{z}{|z|}\right)^{n+l}Q_{nl}(z). \tag{1.31}$$

The transverse conductivity $\sigma_{\pm}^{tr}$ relating the Fourier-transforms of the circular current and electrical field components

$$j_{\pm}(\omega, k) = \sigma_{\pm}^{tr}(\omega, k)E_{\pm}(k) \tag{1.32}$$

may be expressed in accordance with (1.24) through the solution of equation system (1.30)

$$\Phi_{\pm}^{1,1}(\omega, k) = \frac{i}{\omega + i\nu}\frac{M(\omega, k)}{D(\omega, k)}E_{\pm}(k), \tag{1.33}$$

where

$$D(\omega, k)=\begin{vmatrix} 1+A_1-\frac{3}{2}A_1q_{11}(k), & -\frac{5}{6}A_2q_{12}(k),\ldots \\ -\frac{3}{2}A_1q_{21}(k), & 1+A_2-\frac{5}{6}A_2q_{22}(k),\ldots \\ \ldots & \ldots \end{vmatrix}; \tag{1.34}$$

$$M(\omega, k)=\begin{vmatrix} q_{11}(k), & -\frac{5}{6}A_2q_{12}(k),\ldots \\ q_{21}(k), & 1+A_2-\frac{5}{6}A_2q_{22}(k),\ldots \\ \ldots & \ldots \end{vmatrix}, \tag{1.35}$$

where $q_{nl}(k)$ are familiar functions and the formula for calculating these functions in accordance with (1.23), (1.31) may be written as

$$q_{nl}(k)=\frac{\omega+i\nu}{2}\int_{-1}^{1}dy\frac{P_n^1(y)P_l^1(y)}{\omega\mp\Omega+i\nu-ikvy}. \tag{1.36}$$

Introducing the conventions

$$x=\frac{kv}{\omega\mp\Omega+i\nu}, \quad \eta=\frac{\omega+i\nu}{\omega\mp\Omega+i\nu}, \tag{1.37}$$

for the first three functions we have

$$q_{11}(x)=\eta q(x),\ q_{12}(x)=x^{-1}[3q_{11}(x)-2\eta],\ q_{22}(x)=3x^{-2}q_{12}(x),$$

where

$$q(x)=\frac{1}{2}x^{-3}\left[2x-(1-x^2)\ln\frac{1+x}{1-x}\right]. \tag{1.38}$$

Henceforth we will assume that only the first two parameters A_1 and A_2 are nonzero. There are no fundamental difficulties for incorporating the higher parameters as well such as A_3, although we will take these to be insignificant since they describe the finer structure of the electron correlation function. Then for the transverse conductivity of electron liquid we obtain from (1.34), (1.35) (compare to study [6]):

$$\sigma_{\pm}^{tr}(\omega, k)=\frac{3}{16\pi}\frac{i\omega_{Le}^2}{\omega\mp\Omega+i\nu}\varkappa(x), \tag{1.39}$$

where

$$\varkappa(x)=2\frac{q(x)-\frac{5\eta A_2}{3(1+A_2)}x^{-2}[3q(x)-2]}{1-\frac{3\eta A_1}{2(1+A_1)}q(x)-\frac{5\eta A_2}{2(1+A_2)}\left(1-\frac{\eta A_1}{1+A_1}\right)x^{-2}[3q(x)-2]}. \tag{1.40}$$

The function $\varkappa(x)$ determining the transverse conductivity of the electron liquid of metal defined in accordance with (1.39) will be used henceforth in calculating the spectrum of cyclotron waves and the surface impedance. We note that in the Fermi-gas model ($A_1 = A_2 = 0$) the transverse conductivity is proportional to the function $q(x)$.

1.4. The existence conditions and spectrum of Fermi-liquid cyclotron waves

The dispersion equation for electromagnetic waves propagating in a metal along the magnetic field (the wave vector k is parallel to the z axis) ignoring bias current takes the form [6]

$$k^2 - \frac{4\pi i\omega}{c^2}\sigma^{tr}(\omega, k) = 0. \tag{1.41}$$

With real values of the frequency ω the solutions of equation (1.41) lying near the real axis of the complex variable k (since the imaginary part of the root determined by the collisional damping $i\nu$ is small compared to the real part), corresponds to the weakly-damped electromagnetic excitations. In this case when the transverse conductivity is determined by the function (1.39) helicons [46], zero-point acoustic waves [47] and cyclotron waves [16] may appear in certain conditions.

Calculation of the frequency dependence of the impedance assumes knowledge of the complex roots and the analytic singularities of equation (1.41). Specifically, as in the case of the Fermi gas model, the primary contribution to the surface impedance comes from the poles determining the skin electromagnetic field component that is damped in the metal interior.

This section is devoted to a discussion of the singularities of the Fermi-liquid cyclotron wave spectrum in the region where they are weakly damped. In explaining the consequences deriving from equation (1.41) we will write this equation through the complex variable x (1.37) as

$$x^2 - \xi\varkappa(x) = 0. \tag{1.42}$$

Here the parameter

$$\xi = -\frac{3}{4}\frac{\omega_{Le}^2\omega}{(\omega \mp \Omega + i\nu)^3}\frac{v^2}{c^2} \tag{1.43}$$

has a large modulus value in the frequency range where the skin-effect is anomalous. In the absence of the magnetic field ($\Omega = 0$) the parameter ξ corresponds with the anomalous parameter of the skin-effect,[1] introduced in study [31].

At sufficiently low frequencies such that $\omega \ll \Omega\left(\frac{\Omega c}{\omega_{Le} v}\right)^2$, in the vicinity of the normal skin-effect the roots of equation (1.42) correspond to small values of x when $\varkappa(x \to 0) \to \frac{4}{3}\left(1 - \frac{\eta A_1}{1 + A_1}\right)^{-1}$ and provide a dispersion relation for the helicons [48] ($\nu \ll \omega \ll \Omega$):

---

[1] In the regime $\omega \ll \nu$ the anomalous parameter $\xi = i3/2(l/\delta)^2$, where $l = v/\nu$ is the free path length, and δ is the classical depth of the skin-layer $\delta = c/\sqrt{2\pi\omega\sigma_0}$, $\sigma_0 = \omega_{Le}^2/4\pi\nu$.

$$\omega = k^2 \frac{\Omega c^2}{\omega_{Le}^2}. \tag{1.44}$$

On the other hand we will be interested in the range of high frequencies ω lying near the cyclotron frequency Ω when the skin-effect is anomalous and the parameter ξ has a large modulus value. It follows from (1.42) that the inequality

$$|\xi| \sim \left| \frac{r x^2}{\varkappa(x)} \right| \gg 1. \tag{1.45}$$

must be satisfied. Due to the limits on the quantity $D(\omega, k)$ (1.34) with term accuracy ~$1/\xi$ the roots of equation (1.42) are determined by the vanishing condition of the transverse conductivity $\varkappa(x)$ (1.40):

$$q(x) - \frac{5}{3} \frac{\eta A_2}{1 + A_2} x^{-2} [3q(x) - 2] = 0. \tag{1.46}$$

The solution of equation (1.46) providing near-real values of the wave vector provides a dispersion relation for cyclotron waves whose damping is small and is attributable to collision effects ($\nu \to 0$). We note that the absence of the constant A_1 in the dispersion equation for cyclotron waves (1.46) is due to the zero value of the current [12] analogous to how a zero value of the density perturbation resulting from the transverse position of the waves caused the constant A_0 to be absent.

We will proceed to an investigation based on (1.46) of the region of existence of the branch of weakly-damped cyclotron waves attributable to the Fermi-liquid constant A_2 taking $A_l = 0$, $l \geqslant 3$. The nonzero value of the other constants such as A_3 produces new branches of the cyclotron waves [18].

We will investigate the function $q(x)$ with real values of the argument. It follows from (1.38) that $q(x)$ is real when $|x| < 1$ and

$$\operatorname{Im} q(x) = -\frac{i\pi}{2|x|} \frac{x^2 - 1}{x^2} \theta(|x| - 1). \tag{1.47}$$

The real part

$$\operatorname{Re} q(x) = \frac{1}{2} x^{-3} \left[2x - (1 - x^2) \ln \left| \frac{1 + x}{1 - x} \right| \right] \tag{1.48}$$

is positive everywhere. Graphs of the real and imaginary parts of $q(x)$ are given in Fig. 1.

Two important facts follow from formulae (1.47) and (1.48). When $A_2 = 0$, i.e., in the absence of Fermi-liquid interaction, there is no root of equation (1.46) lying near the real axis, since the equation $q(x) = 0$ has no real solutions. Consequently the existence of the weakly-damped cyclotron waves is attributable to the Fermi-liquid interaction (the nonzero value of the constant A_2). Moreover when $|x| > 1$ the imaginary part of $q(x)$ becomes nonzero; this part is not related to the collisions and this produces additional Landau damping [49] in the solution of equation (1.46). Hence weakly-damped solutions may exist when $0 < x < 1$, or in a frequency range satisfying the inequality

$$3/5 < \eta A_2/(1 + A_2) < 1. \tag{1.49}$$

Since no general formula exists for the solution of equation (1.46) we will henceforth consider the limiting cases of short and long wavelengths, when analytic solutions may be obtained together with corresponding approximate expressions for the cyclotron wave spectrum.

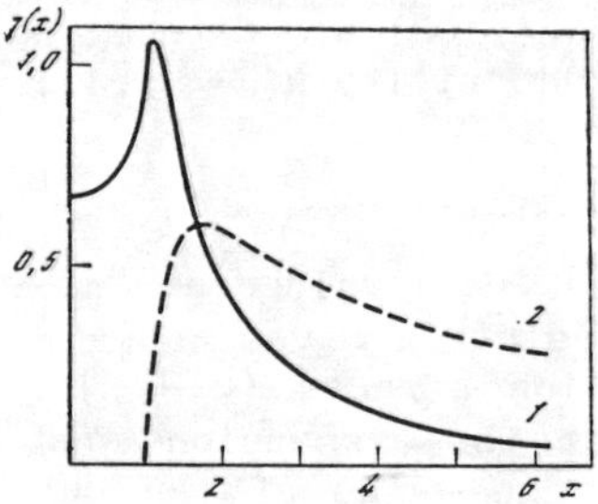

The real (1) and imaginary (2) parts of the function $q(x)$ expressing the transverse conductivity of the electron gas according to (1.38)

The spectrum in the longwave limit. We will first consider the longwave region in which the modulus value of the wave vector may be considered small to satisfy the condition $|x| \ll 1$. Then, after using the expansion

$$q(x) = \frac{2}{3} + \frac{2}{3\cdot 5}x^2 + \frac{2}{5\cdot 7}x^4 + \frac{2}{7\cdot 9}x^6 + \dots, \tag{1.50}$$

we obtain from equation (1.46)

$$x^2 = 5\left(1 - \frac{\eta A_2}{1 + A_2}\right) \Big/ \left[\frac{15\eta A_2}{7(1 + A_2)} - 1\right], \tag{1.51}$$

or, converting to the variables ω and k (1.37),

$$k^2 v^2 = \frac{35(\omega \mp \Omega + i\nu)^2 [\omega \mp \Omega(1 + A_2) + i\nu]}{15(\omega + i\nu)A_2 - 7(1 + A_2)(\omega \mp \Omega + i\nu)}. \tag{1.52}$$

An analysis of equation (1.52) provides the following expression for the frequency of the weakly-damped branch of the cyclotron wave that is the critical frequency when $k = 0$ [3]:

$$\omega_0 = \pm\Omega(1 + A_2), \tag{1.53}$$

corresponding to the equality $\eta A_2/(1 + A_2) = 1$ in conditions of smallness of the collisional damping. Minor k-squared corrections to the frequency (1.53) were first obtained in study [3] and may easily be found from equation (1.52)

$$\omega = \omega_0 + \frac{8}{35}\frac{k^2 v^2}{\Omega}\frac{1 + A_2}{A_2} - i\nu, \tag{1.54}$$

where satisfaction of the inequality (1.55) is assumed. The dispersion curve of (1.54) is shown in Fig. 2.

$$\nu \ll kv \ll |A_2\Omega|. \tag{1.55}$$

The spectrum in the shortwave limit (termination point). We will now consider the other limiting case: shortwaves where the values of the wave vector are sufficiently large to satisfy the inequality $|x| > 1$ (1.47) which results in additional damping in dispersion equation (1.46) that is not related to collisions. As is well known [49] this damping is attributable to Cerenkov resonance absorption of the electromagnetic wave energy by electrons in transit with the wave in the phase $(\omega \mp \Omega)/v = \pm k$. The curves $\omega = \pm\Omega \pm kv$ corresponding to the condition $|x| = 1$ in the limit $v \to 0$ on the ω, k plane is the edge of the collisionless damping region. The frequency ω_k and the wave number k_m corresponding to the point of intersection between the dispersion curve and the edge of the damping region (the termination point of the spectrum) may be found from the equalities:

$$\frac{\eta A_2}{1+A_2} = \frac{3}{5}, \quad \omega_k = \pm\Omega \frac{1+A_2}{1-\frac{3}{2}A_2}. \quad (1.56)$$

The corresponding quantity k_m which is the largest wave number at which collisionless damping begins is equal to [18]

$$k_m = \frac{\omega_k \mp \Omega}{v} = \left|\frac{\Omega}{v}\frac{5A_2}{3-2A_2}\right|. \quad (1.57)$$

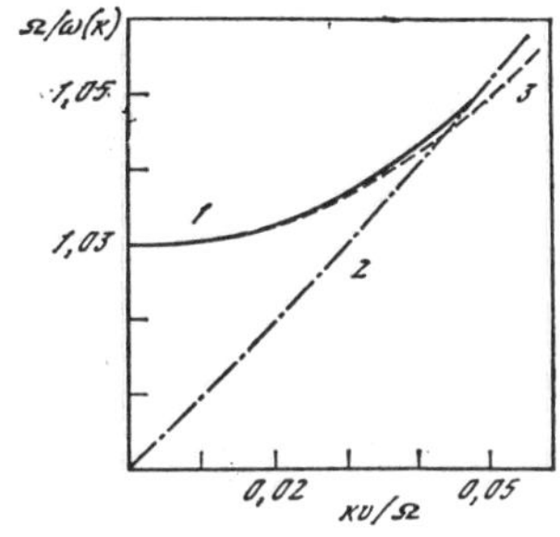

Fig. 2. The dispersion curve of $\Omega/\omega(k)$ for the cyclotron wave branch corresponding to the parameter A_2, for $A_2 = -0.03$, $\nu = 0$
1 - The exact solution of dispersion equation (1.46); 2 - the edge of the collisionless damping region; approximate quadratic relation (1.54)

Formula (1.57) illustrates the fact that collisionless damping will occur in an electron gas ($A_2 = 0$) with any values of the wave number ($k_m = 0$) which makes it impossible for cyclotron waves propagating along the magnetic field to exist. On the other hand, in an electron liquid a "transparency window" arises attributable to interelectron interaction, and such waves may exist within this window.

Since the singularities in the Fermi-liquid wave spectrum are manifest in the frequency dependence of the surface impedance, we will consider the nature of the behavior of the dispersion curve near the termination point of the spectrum. Taking $|x - 1| \ll 1$, we use the expansion

$$\mathrm{Re}\, q(x) = 1-(1-x)\ln\frac{2e^{-2}}{|1-x|} + (1-x)^2\left(1-\frac{5}{2}\ln\frac{2e^{-1}}{|1-x|}\right)+\ldots \quad (1.58)$$

The graphs of the various approximations of the function $q(x)$ are given in Fig. 3.

Setting $x = 1 - \Delta x$, $0 < \Delta x \ll 1$ and using expansion (1.58) we obtain from equation (1.46)

$$\omega = \omega_k + 2k_m v\left(1+\frac{k_m v}{\Omega}\right)\Delta x\left[\ln\frac{2e^{-3}}{|\Delta x|} + \frac{1}{2}\Delta x\left(1-5\ln\frac{2e^{-1}}{|\Delta x|}\right)\right]. \quad (1.59)$$

Since the quantity Δx is independent of both ω and k, equation (1.59) implicitly determines the $\omega(k)$ relation. This equation was first obtained in study [50] ignoring Δx-squared terms.

It follows from (1.59) that when $\Delta x \to 0$ the frequency ω approaches ω_k more slowly than Δx and since, by definition,

$$\Delta x = \frac{\Delta\omega - v\Delta k}{\Delta\omega + k_m v}, \tag{1.60}$$

where

$$\Delta\omega = \omega - \omega_k, \ \Delta k = k - k_m, \tag{1.61}$$

the dispersion curve is tangential to the edge of the damping region. Ignoring the Δx-squared terms we may obtain from (1.59) the explicit relation $\omega(k)$ near the termination point:

$$\Delta\omega = v\Delta k\left[1 - \left(2\left(1 + \frac{k_m v}{\Omega}\right)\ln\frac{2e^{-3}k_m}{|\Delta k|}\right)^{-1}\right]. \tag{1.62}$$

The dispersion curve obtained from (1.62) is given in Fig. 4. Fig. 3 reflects the results from the numerical solution of equation (1.46) for $\nu = 0$ and was used for deriving a dispersion curve across the entire range of existence of the Fermi-liquid cyclotron waves. The curve confirms the smooth $\omega(k)$ relation near the termination point as indicated by (1.62).

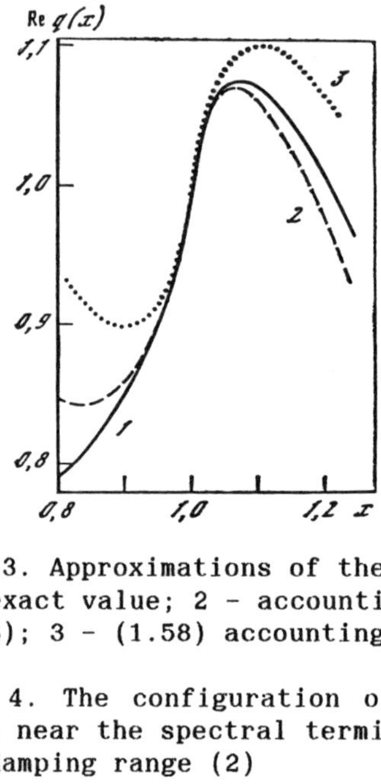

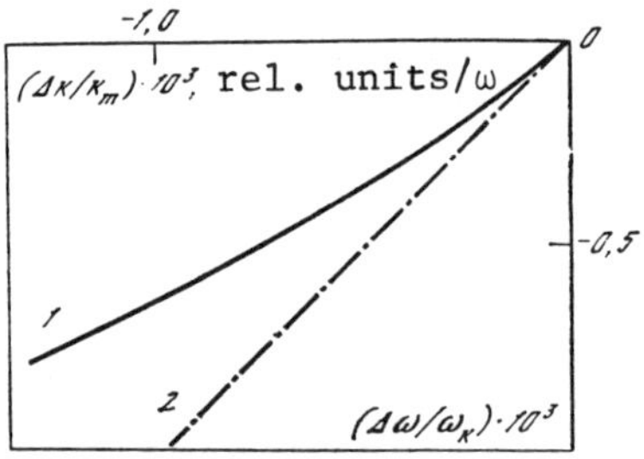

Fig. 3. Approximations of the real part of $q(x)$ near $x = 1$.
1 - exact value; 2 - accounting for the first three terms of expansion (1.58); 3 - (1.58) accounting for two terms

Fig. 4. The configuration of the dispersion curve of the cyclotron waves near the spectral termination point $\Delta\omega = \Delta k = 0$ (1); the edge of the damping range (2)

Thus, the dispersion curve for the weakly-damped branch of Fermi-liquid cyclotron waves corresponding to the constant A_2 is obtained and investigated. The dispersion curve is approximated with good accuracy by the parabola of (1.54). The range of existence of such waves lies in the vicinity of cyclotron frequency (1.49) and is sufficiently narrow if the parameter A_2 is small, which exists, for example, in alkali metals.

We should note that in this section we have not considered the behavior of the complex solutions of dispersion equation (1.46) in the forbidden region, which will be done in the subsequent chapter in calculating the surface impedance.

Conclusions

1. The initial equations for calculating the impedance of an electron liquid may be rather simple and at the same time strictly formulated for two types of boundary conditions on the electron distribution function: specular reflection and diffuse scattering.

2. The constant A_2 is manifest in the dispersion equation for the weakly-damped branch of cyclotron waves propagating along an external magnetic field and attributable to the Fermi-liquid interaction of electrons.

3. The unique nature of the frequency dependence of the impedance may be attributed to cyclotron waves when inequality (1.55) is satisfied, which makes possible experimental determination of A_2.

Chapter 2

CALCULATION OF THE SURFACE IMPEDANCE OF A METAL INCORPORATING THE FERMI-LIQUID CONSTANTS A_1 AND A_2 IN THE CASE OF SPECULAR ELECTRON BOUNDARY REFLECTION

All electromagnetic properties of a half-bounded metal impacted by an external electromagnetic wave (and, specifically, the experimentally-observed quantity of the absorbed power) may be expressed through the surface impedance relating the field strength of the electrical and magnetic fields at the boundary. The impedance is determined in turn by both the volumetrical characteristics (the nonlocal conductivity) of the medium and by the representation of the boundary conditions for electron scattering on the surface.

In this surface we find the impedance of a half-bounded metal in the Fermi-liquid model in the simplest case of electron specular scattering. We provide a general expression for the impedance through the transverse conductivity function in a magnetic field perpendicular to the surface and develop a method for asymptotic expansion of the impedance in the electron liquid model in terms of the inverse powers of the anomaly parameter of the skin-effect and, finally, compare calculation results to experimental data from existing experimental research aimed at measuring the frequency dependence of the impedance in alkali metals.

2.1. Expression of the surface impedance through the transverse conductivity of electron liquid

The concept of surface impedance may be introduced most easily for a half-bounded metal impacted normally by a plane electromagnetic wave. The problem involved determining the amplitude of the reflected wave ($|E_r| = |H_r|$) if the amplitude of the incident wave ($|E_i| = |H_i|$) is given. The following relations are valid for the amplitudes due to the continuity of the electric and magnetic fields at the boundary $z = 0$:

$$E_x(0) = E_i - E_r, \ H_y(0) = H_i + H_r. \tag{2.1}$$

Consequently the complex reflection coefficient r which is the quotient of the amplitudes of the reflected and incident monochromatic waves ($r = E_r/E_i$) is expressed through the ratio of the electric and magnetic field strengths in the metal [51]

$$r = \frac{1 - \frac{c}{4\pi} Z(\omega)}{1 + \frac{c}{4\pi} Z(\omega)}. \tag{2.2}$$

In (2.2) $Z(\omega)$ is the surface impedance

$$Z(\omega) = \frac{4\pi}{c} \frac{E_x(0)}{H_y(0)}. \tag{2.3}$$

The quantity r determines both the portion of reflected energy $|r|^2$ and the portion of energy absorbed in the metal

$$A = 1 - |r|^2. \tag{2.4}$$

The latter is valid for rather thick (half-bounded) samples. In samples of finite thickness (compared to the depth of the skin-layer) the quantity observed in the experiment is also the amplitude of the passed wave [52].

$B = H$ due to neglect of paramagnetism effects. Hence in accordance with Maxwell's equations

$$H_y = \frac{c}{i\omega} \frac{\partial E_x}{\partial z} \tag{2.5}$$

the ratio of the electric and magnetic field strengths in the metal is equal to δ/λ within an order of magnitude, where δ is the characteristic scale of the spatial variation in the field (the skin-layer depth), $\lambda = c/\omega$ is the wavelength in a vacuum and is a small quantity. In this regime the power absorbed by the sample is expressed in accordance with (2.2), (2.4) through the real part of the surface impedance

$$A = \frac{c}{\pi}\,\mathrm{Re}\,Z(\omega). \tag{2.6}$$

We note that with a magnetic field present the surface impedance in the general case is a tensor. However in a magnetic field perpendicular to the surface, as noted above, isotropism is conserved in the plane of the boundary, and the value of the surface impedance $Z_{\pm}$ remains scalar for the circular components.

Equation (2.5) makes it possible to express impedance (2.3) in a form that allows generalization to the circular components of the electrical field as well

$$Z_{\pm}(\omega) = \frac{4\pi i\omega}{c^2}\,\frac{E_{\pm}(0)}{E'_{\pm}(0)}, \tag{2.7}$$

where $E'_{\pm}(0) = \frac{\partial}{\partial z} E_{\pm}(z)|_{z=0}$. Thus, the problem of calculating the impedance involves finding the ratio of the electrical field strength and its derivative at the boundary of the metal.

We may eliminate the magnetic field strength by means of equation (2.5) from Maxwell's equations, which, ignoring the bias current and accounting for the transverse position of the wave (div **E** = 0) we have the equation for the electrical field

$$\frac{\partial^2}{\partial z^2} E_{\pm}(z) = -\frac{4\pi i\omega}{c^2}\, j_{\pm}(z). \tag{2.8}$$

Material equation (1.32) relating the current to the electrical field was obtained in section 1.3 for the case of specular electron reflection at the boundary. In this case as is the case for a zero-boundary medium we may introduce the transverse conductivity function of the electron liquid $\sigma_{\pm}^{tr}(\omega, k)$. Accounting for the jump of the derivative at the boundary $E'(0)$ attributed to the even continuation of field (1.26) we obtain for the Fourier transforms from equation (2.8)

$$k^2 E_{\pm}(k) + 2E'_{\pm}(0) = \frac{4\pi i\omega}{c^2}\,\sigma_{\pm}^{tr}(\omega, k)\, E_{\pm}(k). \tag{2.9}$$

Applying the inverse Fourier transform and dropping the signs ±, we obtain the following formula for the electrical field in the metal

$$E(z) = -\frac{1}{\pi} E'(0) \int_{-\infty}^{\infty} dk \frac{e^{ikz}}{k^2 - \frac{4\pi i\omega}{c^2}\sigma^{tr}(\omega, k)}. \tag{2.10}$$

From here consistent with formula (2.7) the surface impedance is expressed through the transverse conductivity [4]

$$Z(\omega) = \frac{4\omega}{ic^2} \int_{-\infty}^{\infty} \frac{dk}{k^2 - \frac{4\pi i\omega}{c^2}\sigma^{tr}(\omega, k)}. \tag{2.11}$$

Expression (2.11) is the initial expression for calculating the surface impedance with specular reflection, since we know the transverse conductivity of the electron liquid determined by (1.39). We note that the difference between this expression and the corresponding gas model may be found only in the change in the transverse conductivity function σ^{tr}.

2.2. Asymptotic expansion of the surface impedance in inverse powers of the skin-effect anomaly parameter

The expression for the surface impedance through the transverse conductivity of the electron liquid with specular electron boundary reflection given above is identical in form to the corresponding expression in the gas model. Accounting for interelectron interaction changes the transverse conductivity of the metal in accordance with (1.29) and produces new Fermi-liquid singularities in σ^{tr}.

In this section we generalize the asymptotic expansion of the impedance in terms of the anomaly parameter of the skin-effect ξ developed by Dingle [29, 53] for a noninteracting electron model. The need to develop such a technique which utilizes a Mellin transform in ξ is due to the impossibility of obtaining a regular power expansion (when $|\xi| \ll 1$ or $|\xi| \gg 1$) whose terms, with the exception of the first few, are divergent [53]. Since Fermi-liquid interaction is clearly manifest at the higher order terms of the expansion of the impedance when $|\xi| \gg 1$, this method is quite useful for obtaining an asymptotic expansion of the surface impedance.

Using the quantity $y = ix$ as the integration variable, where x is determined by (1.37), we may write the formula for calculating the impedance (2.11), following study [29], as

$$Z(\omega) = \frac{4\pi v}{c^2}\frac{\omega}{\omega - \Omega + i\nu} I_0, \tag{2.12}$$

$$I_0 = \frac{2}{\pi}\int_0^{\infty[\nu + i(\omega-\Omega)]} \frac{dy}{y^2 + \xi\varkappa(y)} \tag{2.13}$$

and where in accordance with (1.43)

$$\xi = \xi_0\left(\frac{\omega}{\omega - \Omega + iv}\right)^3, \quad \xi_0 = -\frac{3}{4}\left(\frac{\omega_{Le}}{\omega}\right)^2\left(\frac{v}{c}\right)^2. \tag{2.14}$$

Equation (2.13) uses the previous conventions of (1.40) for the new function of the variable y:

$$\varkappa(y) = 2\frac{q(y) + Y_2\frac{1}{y^2}\left[\frac{3}{2}q(y) - 1\right]}{1 - Y_1 q(y) + Y_2\left(\frac{3}{2} - Y_1\right)\frac{1}{y^2}\left[\frac{3}{2}q(y) - 1\right]}, \tag{2.15}$$

$$Y_1 = \frac{3}{2}\frac{A_1}{1 + A_1}\frac{\omega + i\nu}{\omega - \Omega + i\nu}, \qquad Y_2 = \frac{10}{3}\frac{A_2}{1 + A_2}\frac{\omega + i\nu}{\omega - \Omega + i\nu}, \tag{2.16}$$

$$q(y) = -\frac{1}{y^2} + \frac{1+y^2}{y^3}\operatorname{arctg} y. \qquad (2.17)$$

The function $q(y)$ is real and positive for real y (Fig. 5) and has branching points on the imaginary axis when $y = \pm i$. In order to evaluate integral (2.13) following study [29] we must shift the integration contour to the real axis y. In the gas model ($Y_1 = Y_2 = 0$), when $\varkappa(y) = 2q(y)$ in the absence of a magnetic field in this contour mixing process there is no intersection of the singularities of the integrand [31]. In the electron liquid model due to the appearance of new singularities attributable to Fermi-liquid interaction, the possibility for intersection of singularities with shifting of the contour in (2.13) requires further examination. We will analyze the position of the singularities in the collisionless limit $\nu \to 0$, when the parameters ξ, Y_1 and Y_2 are real.

The pole of the integrand functions as the roots of the equation (compare to (1.42))

$$y^2 + \xi\varkappa(y) = 0. \qquad (2.18)$$

With large absolute values of the parameter ξ, i.e., in the range of the anomalous skin effect of interest to us, the roots are divided in accordance with two possibilities: 1) the vanishing of the function $\varkappa(y)$ (cyclotron poles y_{φ}) and 2) the limit $y \to \infty$ in the case $\varkappa(y) \neq 0$ (the pole of the skin-effect y_{ξ}).

We will first consider the poles of the skin-effect. In the limit $|y| \gg 1$ we use the expansion ($y > 0$) for the function $\varkappa(y)$:

$$\varkappa(y) = \frac{\pi}{y}\left(1 + C_1\frac{1}{y} + C_2\frac{1}{y^2} + \ldots\right), \qquad (2.19)$$

where

$$C_1 = \frac{4}{\pi}\left(1 - \frac{\pi^2}{8}Y_1 + \frac{1}{2}Y_2\right), \qquad (2.20)$$

$$C_2 = 1 + 3Y_2\left(1 - \frac{2}{3}Y_1\right) - 2Y_1\left(2 + \frac{\pi^2}{8}Y_1\right), \qquad (2.21)$$

and using the expansion

$$q(y) = \frac{\pi}{2y}\left(1 - \frac{4}{\pi}\frac{1}{y} + \frac{1}{y^2} - \frac{4}{3\pi}\frac{1}{y^3} + \ldots\right), \quad y \gg 1. \qquad (2.22)$$

It follows from formula (2.19) that the solution of equation (2.18) in this limit may be obtained by iterations using the solution of the equation

$$y_0^3 = -\pi\xi, \qquad (2.23)$$

as the zeroeth solution, i.e.,

$$y_0 = (-\pi\xi)^{1/3} = (-\pi\xi_0)^{1/3}\frac{\omega}{\omega - \Omega + i\nu}\exp\frac{2\pi n}{3}i, \quad n = 0, \pm 1. \qquad (2.24)$$

Subject to the first two iterations we obtain from (2.13) and (2.19)

$$y_{\xi}=y_0+\frac{C_1}{3}+\frac{1}{3y_0}\left(C_2-\frac{2}{3}C_1^2\right). \tag{2.25}$$

Thus the position of the skin-effect poles with $|\xi|^{-1/3}$ accuracy is determined by formula (2.24).

We will now consider the cyclotron-wave poles whose appearance is attributable to Fermi-liquid interaction and the vanishing of the function $\varkappa(y)$ (2.15). The equation corresponding to this condition

$$\varphi(y)-Y_2=0, \tag{2.26}$$

where

$$\varphi(y)=\frac{y^2 q(y)}{1-\frac{3}{2}q(y)}, \tag{2.27}$$

may be analyzed in the longwave ($|y| \ll 1$) and the shortwave ($|y| \gg 1$) regions. First let $|y| \ll 1$, then using the expansion

$$q(y)=\frac{2}{3}-\frac{2}{15}y^2+\frac{2}{35}y^4-\ldots \tag{2.28}$$

and limiting the examination to squared terms in the solution of (2.26), we obtain

$$y_{\varphi}=\pm\sqrt{\frac{35}{8}\left(\frac{3Y_2}{10}-1\right)}=\pm\sqrt{\frac{35}{8}\,\frac{\Omega(1+A_2)-\omega-i\nu}{(1+A_2)(\omega-\Omega+i\nu)}}. \tag{2.29}$$

It follows from formula (2.29) that in the range of existence of cyclotron waves determined by the inequality ($\nu \to 0$)

$$[\omega-\Omega(1+A_2)]A_2>0, \tag{2.30}$$

we have the imaginary root $y_{\varphi}^{(1)}$ (corresponding to the real values of the wave vector of the cyclotron wave):

$$y_{\varphi}^{(1)}=\pm\sqrt{\frac{35}{8}\left|\frac{\omega-\Omega(1+A_2)}{(\omega-\Omega)(1+A_2)}\right|}\left\{i-\frac{\nu}{2[\omega-\Omega(1+A_2)]}\right\}. \tag{2.31}$$

Here collisional damping is assumed to be small:

$$\nu \ll |\omega-\Omega(1+A_2)| \ll |\omega-\Omega|. \tag{2.32}$$

Formula (2.31) corresponds to the longwave limit in the cyclotron wave spectrum (1.51). The behavior of this branch of the root near the termination point is examined in section 1.4.

In the forbidden region when an inequality inverse to (2.30) is satisfied the root of equation (2.26) is real and in the longwave limit is equal to:

$$y_{\varphi}^{(2)}=\pm\sqrt{\frac{35}{8}\left|\frac{\omega-\Omega(1+A_2)}{(\omega-\Omega)(1+A_2)}\right|}\left\{1+\frac{i\nu}{2[\omega-\Omega(1+A_2)]}\right\}. \tag{2.33}$$

In the shortwave range for this branch of the root using expansion (2.22) we obtain ($Y_2 \gg 1$)

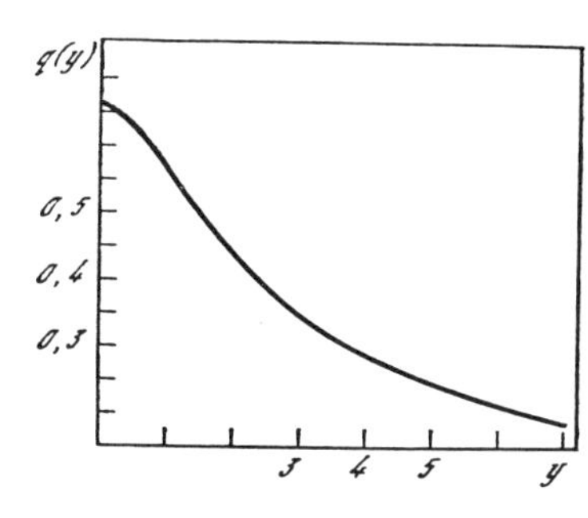

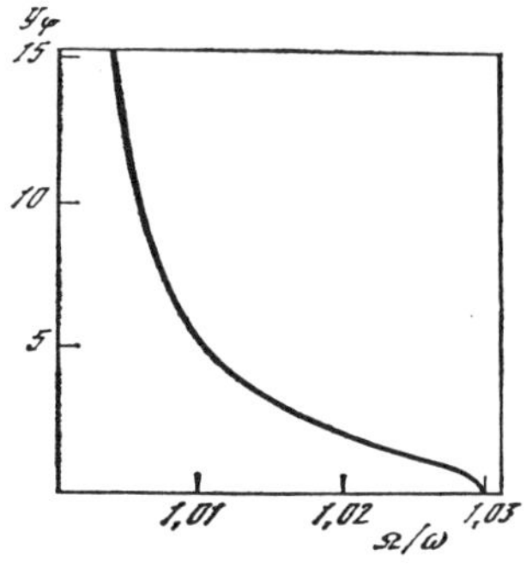

Fig. 5. Graph of function (2.17) representing the transverse conductivity (1.38) on the imaginary axis $y = ix$

Fig. 6. Solution of equation (2.26) in the forbidden range

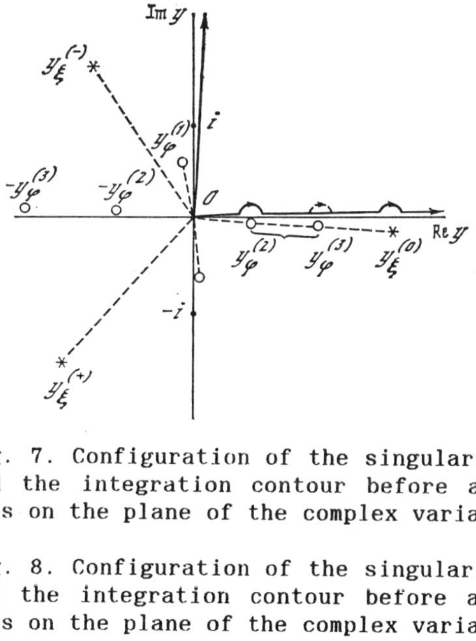

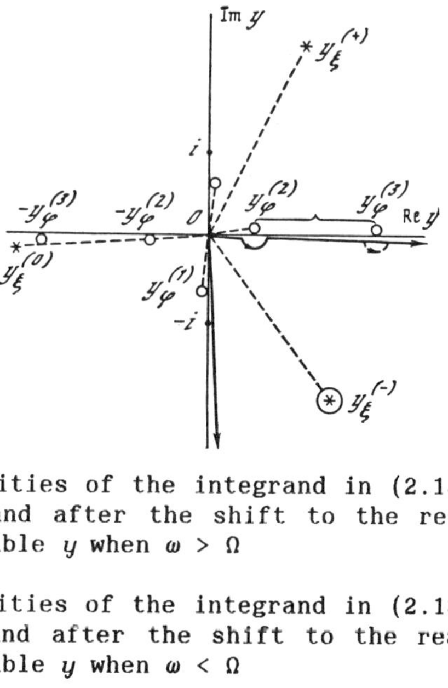

Fig. 7. Configuration of the singularities of the integrand in (2.13) and the integration contour before and after the shift to the real axis on the plane of the complex variable y when $\omega > \Omega$

Fig. 8. Configuration of the singularities of the integrand in (2.13) and the integration contour before and after the shift to the real axis on the plane of the complex variable y when $\omega < \Omega$

$$y_\varphi^{(3)} = \pm \frac{2}{\pi} Y_2 = \pm \frac{20}{3\pi} \frac{A_2}{1+A_2} \frac{\omega}{\omega-\Omega}\left(1-\frac{i\nu}{\omega-\Omega}\right). \tag{2.34}$$

The result of the numerical solution of equation (2.26) in the forbidden region when $\nu = 0$ is shown in Fig. 6. Formulae (2.33), (2.34) correspond to the longwave and shortwave asymptotes of the solution. Thus, the position of the poles of the integrand in (2.13) is determined by formulae (2.24), (2.31), (2.33), and (2.34) and is shown in Fig. 7, 8.

The integration contour should be shifted to the real positive y semiaxis in (2.13) in a manner that avoids intersecting the branch points $y = \pm i$, i.e., in a clockwise direction when $\omega > \Omega$ and in a counterclockwise direction when $\omega < \Omega$. Here, as indicated by Fig. 8, when $\omega < \Omega$ the integral I_0 is equal to the integral in the sense of the principal Cauchy value on the real axis, and the contribution from the residue at the skin-effect pole $y_\xi^{(-)}$ (when $n = -1$) and the contribution from the semiresidue at the cyclotron-wave pole y_φ (on the real axis). In the case $\omega > \Omega$ the contributions from the singularities are determined by the semiresidues on the real axis from the skin-effect pole with opposite sign $y_\xi^{(0)}$ and from the cyclotron-wave pole y_φ. Consequently, when $\omega < \Omega$

$$I_0 = \frac{2}{\pi}\int_0^\infty \frac{dy}{y^2 + \xi\varkappa(y)} + \frac{2i}{\left|y_\xi^{(-)} + \frac{1}{2}\xi\varkappa'(y_\xi^{(-)})\right|} + \frac{i}{|\xi|}\int_0^\infty dy y^2 \delta[\varphi(y) - Y_2]. \tag{2.35}$$

Expansion of the part of the integral that is determined by the skin-effect pole in powers of the anomaly parameter may be obtained by means of formulae (2.19) and (2.25):

$$\begin{aligned}&\frac{4}{3\sqrt{3}}(-\pi\xi_0)^{-1/3}\frac{\omega-\Omega}{\omega}\left(\frac{3}{2} - i\frac{\sqrt{3}}{2}\right) - \frac{32}{9\pi\sqrt{3}}(-\pi\xi_0)^{-2/3}\left(\frac{\omega-\Omega}{\omega}\right)^2 \times \\ &\times \left(\frac{3}{2} + i\frac{\sqrt{3}}{2}\right)\left(1 - \frac{\pi^2}{8}Y_1 + \frac{Y_2}{2}\right) - \frac{4i}{3\pi\xi_0}\left(\frac{\omega-\Omega}{\omega}\right)^3 \times \\ &\times \left[\frac{16}{\pi^2} - 1 + \left(\frac{16}{\pi^2} - 3\right)Y_2 + \frac{4}{\pi^2}Y_2^2\right].\end{aligned} \tag{2.36}$$

Consequently, in the case $\omega > \Omega$ after shifting of the contour

$$I_0 = \frac{2}{\pi}\int_0^\infty \frac{dy}{y^2 + \xi\varkappa(y)} - \frac{i}{|\xi|}\int_0^\infty dy y^2 \delta[\varphi(y) - Y_2] - \frac{1}{\left|y_\xi^{(0)} + \frac{1}{2}\xi\varkappa'\right|}. \tag{2.37}$$

The asymptotic expansion of the contribution from the semiresidue at the skin-effect pole on the real axis is equal to

$$\begin{aligned}&-\frac{2i}{3}\left(\frac{\omega-\Omega}{\omega}\right)(-\pi\xi_0)^{-1/3} - \frac{16i}{9\pi}\left(\frac{\omega-\Omega}{\omega}\right)^2(-\pi\xi_0)^{-2/3}\left[1 - \frac{\pi^2}{8}Y_1 + \frac{Y_2}{2}\right] + \\ &+ \frac{2i}{3\pi\xi_0}\left(\frac{\omega-\Omega}{\omega}\right)^3\left[\frac{16}{\pi^2} - 1 + \left(\frac{16}{\pi^2} - 3\right)Y_2 + \frac{4}{\pi^2}Y_2^2\right].\end{aligned} \tag{2.38}$$

We will now evaluate the proper part of the integrals in formulae (2.35), (2.37). We first note that when $\omega < \Omega$ the parameter $\xi > 0$, while when $\omega > \Omega$, $\xi < 0$. Moreover, if $y < y_\varphi$, then $\varkappa(y) < 0$, i.e., $\varkappa(y) = -|\varkappa(y)|$ and $\varkappa(y) = |\varkappa(y)|$ when $y > y_\varphi$.

First let $\omega > \Omega$, then for the integral $I_0^{(-)}$ we have by definition of the proper part

$$I_0^{(-)} = \frac{2}{\pi}\lim_{\varepsilon_1\to 0,\ \varepsilon_2\to 0}\left\{\int_0^{y_\varphi^{(2)}-\varepsilon_1} \frac{dy}{y^2 + |\xi||\varkappa(y)|} + \int_{y_\varphi^{(2)}+\varepsilon_1}^{1} \frac{dy}{y^2 - |\xi||\varkappa(y)|} + \right.$$

$$+\int_{1}^{y_{\xi}^{(0)}-\varepsilon_2}\frac{dy}{y^2-|\xi||\varkappa(y)|}+\int_{y_{\xi}^{(0)}+\varepsilon_2}^{\infty}\frac{dy}{y^2-|\xi||\varkappa(y)|}\Bigg\}. \quad (2.39)$$

It is assumed in writing (2.39) that $Y_2 > 10/3$, then $|y_{\varphi}^{(2)}| \ll 1$, if $y_2 \gg 1$, when $|y_{\varphi}^{(3)}| \gg 1$, the limits of integration will change correspondingly.

We will consider the integral $I_0^{(-)}$ (2.39) as a function of the parameter $|\xi|$ and will carry out the Mellin transform

$$M_0^{(-)}(z)=\int_0^{\infty}|\xi|^{z-1}I_0^{(-)}d|\xi|. \quad (2.40)$$

The integral in $|\xi|$ in (2.40) may easily be taken since

$$\int_0^{\infty}d|\xi|\frac{|\xi|^{z-1}}{y^2\pm|\xi||\varkappa|}=\pi y^{2z-2}\varkappa^{-z}\begin{cases}\operatorname{cosec}\pi z,\\ \operatorname{ctg}\pi z.\end{cases} \quad (2.41)$$

By means of an inverse Mellin transform

$$I_0^{(-)}(\xi)=\frac{1}{2\pi i}\int_{C-i\infty}^{C+i\infty}dz\,|\xi|^{-z}M_0^{(-)}(z) \quad (2.42)$$

we may obtain the expression for $I_0^{(-)}$ as

$$I_0^{(-)}(\xi)=\lim_{\varepsilon_1\to 0}\frac{1}{\pi i}\int_{C-i\infty}^{C+i\infty}dz\,|\xi|^{-z}\Bigg\{\int_0^{y_{\varphi}^{(2)}-\varepsilon_1}dy\,y^{2z-2}\frac{|\varkappa|^{-z}}{\sin\pi z}+\int_{y_{\varphi}^{(2)}+\varepsilon_1}^{\infty}dy\,y^{2z-1}\varkappa^{-z}\operatorname{ctg}\pi z\Bigg\}. \quad (2.43)$$

In order to take the y integral we will use a power expansion for the function

$$\left[\frac{3}{4}\varkappa(y)\right]^{-z}\sum_{\mu=0}^{\infty}Q_{\mu}(z)\,y^{\mu},\ |y|\ll 1, \quad (2.44)$$

$$\left[\frac{y}{\pi}\varkappa(y)\right]^{-z}-\sum_{\nu=0}^{\infty}R_{\nu}(z)\,y^{-\nu},\ y>1. \quad (2.45)$$

As a result we arrive at the formula

$$I_0^{(-)}(\xi)=\frac{1}{\pi i}\int_{C-i\infty}^{C+i\infty}dz\left[\left(\frac{3}{4|\xi|}\right)^{z}\sum_{\mu=0}^{\infty}\frac{Q_{\mu}(z)}{\mu+2z-1}\left\{\frac{(-1)^z}{\sin\pi z}(y_{\varphi}^{(2)}-\varepsilon_1)^{\mu+2z-1}+\right.\right.$$
$$\left.\left.+\operatorname{ctg}\pi z\,[1-(y_{\varphi}^{(2)}+\varepsilon_1)^{\mu+2z-1}]\right\}+(\pi|\xi|)^{-z}\sum_{\nu=0}^{\infty}\frac{R_{\nu}(z)}{-3z+\nu+1}\operatorname{ctg}\pi z\right]. \quad (2.46)$$

Formula (2.46) makes it possible to obtain an asymptotic expansion of $I_0^{(-)}(\xi)$ in diminishing powers of $\xi(|\xi| \gg 1)$ whose terms are the contributions from the poles to the right of the integration line

Re $z = C$. The first component in (2.46) has the poles (with integer values $z = m$) from $(-1)^z/\sin \pi z$ and ctg πz minus $1/\pi$, so that the higher order term ($m = 1$) is

$$-\frac{3}{2|\xi|}\frac{1}{\pi}\sum_{\mu=0}\frac{Q_\mu(1)}{\mu+1}[(y_\varphi^{(2)}-\varepsilon_1)^{\mu+1}-(y_\varphi^{(2)}+\varepsilon_1)^{\mu+1}+1]. \quad (2.47)$$

The sum represented in (2.47) is transformed to the principal value of the integral in accordance with (2.44)

$$\sum_\mu Q_\mu(1)\Big[\int_0^{y_\varphi^{(2)}-\varepsilon_1} dyy^\mu + \int_{y_\varphi^{(2)}+\varepsilon_1}^{1} dyy^\mu\Big] \to \frac{4}{3}\int_0^1 \frac{dy}{\varkappa(y)}. \quad (2.48)$$

The second component in (2.46) has poles when $z = \frac{1+\nu}{3}$ so that with a noninteger value z this pole with residue $-1/3$ makes the contribution

$$\frac{2}{3}(\pi|\xi|)^{-\frac{1+\nu}{3}} R_\nu\Big(\frac{1+\nu}{3}\Big)\mathrm{ctg}\Big(\frac{1+\nu}{3}\pi\Big). \quad (2.49)$$

When $z = m$ ($m = 1, 2, \ldots$) a second order pole arises. Here we will write out the higher order term $m = 1$:

$$\frac{2}{3\pi}\Big\{\frac{\partial}{\partial z}[(\pi|\xi|)^{-z}R_2(z)]\Big\}_{z=1} = \frac{2}{3\pi}(\pi|\xi|)^{-1}[R_2'(1) - R_2(1)\ln(\pi|\xi|)]. \quad (2.50)$$

Finally, when $z = m$ the contribution from the other values of ν in (2.46) is

$$-\frac{2}{\pi}(\pi|\xi|)^{-m}\sum_{\nu\neq 3m-1}\frac{R_\nu(m)}{-3m+\nu+1}. \quad (2.51)$$

In accordance with definition (2.45) the sum in (2.51) may be transformed to the form (form $m = 1$)

$$\sum_{\nu\neq 2}\frac{R_\nu(1)}{\nu-2} = -\frac{1}{2}R_0(1) - R_1(1) + \sum_{\nu=3}R_\nu(1)\int_1^\infty dyy^{-\nu+1} =$$
$$= -\frac{1}{2}R_0(1) - R_1(1) - \int_1^\infty dy\Big[\frac{\pi}{x(y)} - yR_0(1) - R_1(1) - \frac{1}{y}R_2(1)\Big]. \quad (2.52)$$

Now using definition (2.15) as well as the values of the coefficients from (2.45) (knowledge of the form of Q_μ as indicated by (2.48) is not required):

$$R_0(z) = 1, \quad R_1(z) = \frac{4}{\pi}z\Big(1 - \frac{\pi^2}{8}Y_1 + \frac{1}{2}Y_2\Big),$$
$$R_2(z) = \frac{8z(z+1)}{\pi}\Big(1 - \frac{\pi^2}{8}Y_1 + \frac{1}{2}Y_2\Big)^2 -$$
$$-z\Big[1 + 3Y_2\Big(1 - \frac{2}{3}Y_1\Big) - 2Y_1\Big(2 - \frac{\pi^2}{8}Y_1\Big)\Big], \ldots \quad (2.53)$$

we obtain the desired asymptotic representation of the proper part of the integral (with $|\xi|^{-1}$ accuracy):

$$
\begin{aligned}
I_0^{(-)} = {} & \frac{2}{3\sqrt{3}}(\pi|\xi|)^{-1/3} - \frac{16}{9\pi\sqrt{3}}(\pi|\xi|)^{-2/3}\left(1+\frac{\pi^2}{8}Y_1+\frac{Y_2}{2}\right) - \\
& -(\pi|\xi|)^{-1}\left\{\frac{2}{3\pi}\ln(\pi|\xi|)\left[\frac{16}{\pi^2}-1+\left(\frac{16}{\pi^2}-3\right)Y_2+\frac{4}{\pi^2}Y_2^2\right] - \right. \\
& -\frac{2}{3\pi}\left[\frac{12}{\pi}\left(\frac{2}{\pi}+1\right)+\frac{1}{2}-2Y_1\left(1+\frac{3\pi}{4}-\frac{\pi^2}{16}Y_1\right) - \right. \\
& \left. -3Y_2\left(1-\frac{8}{\pi^2}-\frac{2}{\pi}-\frac{2}{\pi^2}Y_2\right)-2Y_1Y_2\right]-\frac{1}{\pi}+\frac{8}{\pi^2}\left(1+\frac{1}{2}Y_2 - \right. \\
& \left. -\frac{\pi^2}{4}Y_1\right)+2\int_0^1\frac{dy}{\varkappa(y)}+\frac{2}{\pi}\int_1^\infty dy\left[\frac{\pi}{\gamma(y)}-y-\frac{4}{\pi}\left(1-\frac{\pi^2}{8}Y_1+\frac{1}{2}Y_2\right) - \right. \\
& \left.\left. -\frac{1}{y}\left(\frac{16}{\pi^2}-1-3Y_2\left(1+\frac{16}{3\pi^2}\right)+\frac{4}{\pi^2}Y_2^2\right)\right]\right\}. \qquad (2.54)
\end{aligned}
$$

Formula (2.54) is also valid for $|Y_2| \gg 1$, when $|y_\varphi^{(3)}| \gg 1$, and here, as we may easily determine, the integral in y is the integral in the sense of the principal value. We note that the frequency range below cyclotron resonance ($\omega < \Omega$) may be examined analogously (see Fig. 8) based on a formula similar to (2.39) for the integral $I_0^{(+)}$. The calculation result is different from that given in (2.54) in the sign in front of $(\pi|\xi|)^{-1}$ and the factor 2 in front of the first and second components. In accordance with (2.35) and (2.37) adding to the principal value integrals $I_0^{(-)}$ or $I_0^{(+)}$ the components determined by the residues of (2.36) and (2.37), respectively, we obtain an expression for the quantity I_0 with an arbitrary sign of $\omega - \Omega$. We will write the result in a form that allows us to identify the corrections attributable to the Fermi-liquid interaction:

$$
\begin{aligned}
I_0 = {} & \frac{4}{3\sqrt{3}}\frac{\omega-\Omega}{\omega|\pi\xi_0|^{1/3}}\left(\frac{1}{2}-i\frac{\sqrt{3}}{3}\right)-\frac{32}{9\pi\sqrt{3}}\frac{(\omega-\Omega)^2}{\omega^2|\pi\xi_0|^{2/3}}\left(\frac{1}{2}+i\frac{\sqrt{3}}{2}\right)\times \\
& \times\left(1-\frac{\pi^2}{8}Y_1+\frac{1}{2}Y_2\right)-\frac{(\omega-\Omega)^3}{\omega^3|\pi\xi_0|}\left\{\frac{2}{3\pi}\left[\frac{16}{\pi^2}-1+\left(\frac{16}{\pi^2}-3\right)Y_2 + \right.\right. \\
& \left. +\frac{4}{\pi^2}Y_2^2\right]\left[\ln|\pi\xi_0|+3\ln\frac{\omega}{|\omega-\Omega|}+i\pi\left(1-3\theta(\Omega-\omega)\right)\right] - \\
& -\frac{8}{\pi^2}\left(\frac{2}{\pi}+1\right)-\frac{1}{3\pi}+F_0+\frac{4}{3\pi}Y_1+\frac{2}{\pi}\left(1-\frac{2}{\pi}\left(1+\frac{4}{\pi}\right)\right)Y_2 - \\
& \left. -\frac{\pi}{12}Y_1^2-\frac{4}{\pi^3}Y_2^2+\frac{2}{3\pi}Y_1Y_2+Y_2F\right\}, \qquad (2.55)
\end{aligned}
$$

where

$$
F_0=\int_0^1\frac{dy}{q(y)}+\int_1^\infty dy\left[\frac{1}{q(y)}-\frac{2y}{\pi}-\frac{8}{\pi^2}\right];
$$

$$
\begin{aligned}
F = {} & I_1+I_3+(I_2+I_4)Y_2+Y_2^2\left\{\int_0^\infty\frac{dy\,y^2}{\varphi^3(y)[\varphi(y)-Y_2]} - \right. \\
& \left. -\frac{i\pi}{Y_2^3}\int_0^\infty dy\,y^2\delta[\varphi(y)-Y_2]\operatorname{sgn}(\omega-\Omega)\right\}, \qquad (2.56)
\end{aligned}
$$

$$
I_1=\int_0^1 dy\frac{y^2}{\varphi(y)}, \qquad I_2=\int_0^1 dy\frac{y^2}{\varphi^3(y)}; \qquad (2.57)
$$

$$I_3 = \int_1^\infty dy \left\{ \frac{y^2}{\varphi^2(y)} - \frac{4}{\pi^2} - \frac{6}{\pi y}\left(\frac{16}{3\pi^2} - 1\right) \right\},$$

$$I_4 = \int_1^\infty dy \left\{ \frac{y^2}{\varphi^3(y)} - \frac{8}{\pi y} \right\}. \tag{2.58}$$

Finally, evaluation of integral (2.56) yields the result

$$F_0 - \frac{8}{\pi^2}\left(\frac{2}{\pi} + 1\right) - \frac{1}{3\pi} = 0{,}085, \tag{2.59}$$

which makes it possible to write a general asymptotic expression for the impedance when $\nu = 0$ [54]:

$$Z(\omega) = \frac{4\pi v}{c^2}\left\{ \frac{2}{3\sqrt{3}} \frac{1 - i\sqrt{3}}{|\pi\xi_0|^{1/3}} - \frac{16}{9\pi\sqrt{3}} \frac{1 + i\sqrt{3}}{|\pi\xi_0|^{2/3}} \times \right.$$
$$\times \left[\frac{\omega - \Omega}{\omega} - \frac{3\pi^2}{16} \frac{A_1}{1 + A_1} + \frac{5}{3} \frac{A_2}{1 + A_2} \right] - \frac{1}{\pi|\xi_0|} \left[\left[\frac{2}{3\pi}\left(\frac{16}{\pi^2} - 1\right) \times \right.\right.$$
$$\times \left(\frac{\omega - \Omega}{\omega}\right)^2 + \frac{20}{3\pi}\left(\frac{16}{3\pi^2} - 1\right) \frac{A_2}{1 + A_2} \frac{\omega - \Omega}{\omega} +$$
$$+ \frac{800}{27\pi^3}\left(\frac{A_2}{1 + A_2}\right)^2 \left[\ln|\pi\xi_0| + 3\ln\frac{\omega}{|\omega - \Omega|} + i\pi - \right.$$
$$- 3i\pi\theta(\Omega - \omega) + 0{,}085\left(\frac{\omega - \Omega}{\omega}\right)^2 + \left(\frac{\omega - \Omega}{\omega}\right)\left[\frac{2}{\pi} \frac{A_1}{1 + A_1} + \right.$$
$$+ \frac{20}{3\pi}\left(1 - \frac{2}{\pi}\left(1 + \frac{4}{\pi}\right)\right) \frac{A_2}{1 + A_2} \left.\right] - \frac{3\pi}{16}\left(\frac{A_1}{1 + A_1}\right)^2 -$$
$$- \frac{400}{9\pi^3}\left(\frac{A_2}{1 + A_2}\right)^2 + \frac{10}{3\pi} \frac{A_1 A_2}{(1 + A_1)(1 + A_2)} +$$
$$\left. \left. \left. + \frac{10}{3} \frac{A_2}{1 + A_2} \frac{\omega - \Omega}{\omega} F\left(\frac{10}{3} \frac{A_2}{1 + A_2} \frac{\omega}{\omega - \Omega}\right) \right]\right]\right\}. \tag{2.60}$$

2.3. Analysis of the frequency dependence of the impedance in the vicinity of the critical frequency of cyclotron waves

Formula (2.60) which represents the first three terms of the asymptotic expansion of the surface impedance near the anomalous skin-effect with specular electron reflection makes it possible to investigate the frequency dependence determined by the function $F\left(\frac{10}{3} \frac{A_2}{1 + A_2} \frac{\omega}{\omega - \Omega}\right)$. This function (2.56) must be tabulated numerically, although in certain limiting cases analytic expressions may be obtained for it.

First we will consider the limiting process to the case of an electron gas when A1 = 0, A_2 = 0. Accounting for the fact that $I_1 + I_3 = 0.754$ and $I_2 + I_4 = 0.442$; in this limit we will obtain from (2.57)

$$F = 0{,}754 - 0{,}442 Y_2 + Y_2^2 \int_0^\infty dy \frac{y^2}{\varphi^4(y)} + O(Y_2^3). \tag{2.61}$$

Consistent with formula (2.61) we have the following expression for the impedance in the gas limit

$$Z_{03}(\omega)=\frac{4\pi v}{c^2}\Big\{0{,}770\,|\pi\xi_0|^{-1/3}\left(\frac{1}{2}-i\frac{\sqrt{3}}{2}\right)-0{,}653\,|\pi\xi_0|^{-2/3}\times$$
$$\times\left(\frac{1}{2}+i\frac{\sqrt{3}}{2}\right)\frac{\omega-\Omega}{\omega}+\frac{1}{\pi\xi_0}\Big(0{,}132\left[\ln|\pi\xi_0|+3\ln\frac{\omega}{|\omega-\Omega|}+i\pi-\right.$$
$$\left.-i3\pi\theta(\Omega-\omega)+0{,}085\right)\left(\frac{\omega-\Omega}{\omega}\right)^2+\ldots\Big\}. \quad (2.62)$$

Expression (2.62) corresponds to the result in study [29] obtained by Dingle excluding the magnetic field (compare to study [55] as well). Consistent with (2.26) the surface impedance in the gas model contains resonance components of the type $(\omega - \Omega)^2 \ln(\omega - \Omega)$ in the expansion term $\sim\xi_0^{-1}$. The influence of the magnetic field is evident in the resonance vanishing of terms of the order $\xi_0^{-2/3}$ and higher when $\omega \to \Omega$.

We will consider how the Fermi-liquid interaction influences the frequency dependence of the impedance near cyclotron resonance $\omega = \Omega$. In sufficient proximity of cyclotron resonance, when $|Y_1| \gg 1$ and $|Y_2| \gg 1$, expression (2.57) for the function may be represented as

$$F=F'+iF'',\qquad F'=F_1'+F_2',\qquad F_1'=I_1+I_2Y_2+Y_2^2\Phi_1,$$
$$F_2'=I_3+I_4Y_2+Y_2^2\Phi_2,$$
$$\Phi_1=\int_0^1 dy\Phi(y),\qquad \Phi_2=\int_1^\infty dy\Phi(y),$$
$$\Phi(y)=\frac{y^2}{\varphi^3(y)\,[\varphi(y)-Y_2]}. \quad (2.63)$$

Expanding Φ_1 in terms of $1/Y_2$ we have

$$\Phi_1=-\frac{1}{Y_2}\left[I_2+\frac{1}{Y_2}I_1+\frac{1}{Y_2^2}\int_0^1 dy\frac{y^2}{\varphi(y)}+O\left(\frac{1}{Y_2^3}\right)\right], \quad (2.64)$$

so $F_1'\sim\frac{1}{Y_2}\int_0^1 dy\frac{y^2}{\varphi(y)}\sim\mathrm{const}/Y_2$. In order to expand the function Φ_2 in terms of $1/Y_2{}^0$ we will apply a Mellin transform to Y, where $Y_2 = Y\,\mathrm{sgn}\,Y_2$:

$$\Psi(z)=\int_0^\infty dY\,Y^{z-1}\Phi_2(Y)=\frac{\pi(\cos\pi z)^{\theta(Y_2)}}{\sin\pi z}\int_1^\infty dy\,y^2\varphi^{z-4}. \quad (2.65)$$

Introducing the convention $C_\nu(z)$ for the coefficients of the expansion

$$\varphi^z=\left(\frac{\pi y}{2}\right)^z\sum_{\nu=0}^\infty C_\nu(z)\,y^{-\nu},\quad y>1, \quad (2.66)$$

we obtain

$$\Psi(z)=-\frac{\pi(\cos\pi z)^{\theta(Y_2)}}{\sin\pi z}\left(\frac{\pi}{2}\right)^{z-4}\sum_{\nu=0}^\infty\frac{C_\nu(z-4)}{z-\nu-1}, \quad (2.67)$$

from which, by means of an inverse Mellin transform,

$$\Phi_2 = \frac{1}{2\pi i}\int_{C-i\infty}^{C+i\infty} dz Y^{-z}\Psi(z) = -\frac{1}{2\pi i}\frac{16}{\pi^4}\int_{C-i\infty}^{C+i\infty} dz \frac{\pi(\cos \pi z)^{\theta(Y_2)}}{\sin \pi z} \times$$

$$\times \left(\frac{\pi}{2Y}\right)^z \sum_{\nu=0}^{\infty} \frac{C_\nu (z-4)}{z-\nu-1}. \tag{2.68}$$

In order to obtain the expansion of Φ_2 with Y^{-3} accuracy we must incorporate the poles to the right of the line Re $z = C$, where $0 < C < \mu + 1$. The closest simple pole $z = 1$, $\nu \neq 0$ makes the contribution

$$\frac{16}{\pi^4}\frac{\pi}{2Y_2}\sum_{\nu\neq 0}\frac{C_\nu(-3)}{-\nu} = -\frac{1}{Y_2} I_4. \tag{2.69}$$

The contribution of the double pole $z = 1$, $\nu = 0$ (we account for $C_0 = 1$) is:

$$\frac{16}{\pi^4}\frac{d}{dz}\left\{\pi(z-1)\frac{(\cos \pi z)^{\theta(Y_2)}}{\sin \pi z}\left(\frac{\pi}{2Y}\right)^z\right\}_{z=1} = \frac{8}{\pi^3 Y_2}\ln\frac{\pi}{2Y}. \tag{2.70}$$

The simple pole $z = 2$, $\nu \neq 1$ yields

$$\frac{16}{\pi^4}\frac{\pi^2}{4Y^2}\sum_{\nu\neq 1}\frac{C_\nu(-2)}{1-\nu} = \frac{4}{\pi^2 Y_2^2} - \frac{1}{Y_2^2} I_3, \tag{2.71}$$

where we account for the fact that $C_1(z) = \frac{3\pi}{4} z(1 - 16/3\pi^2)$. Then we account for the double pole $Z = 2$, $\nu = 1$

$$\frac{16}{\pi^4}\frac{d}{dz}\left\{\pi(z-2)\frac{(\cos \pi z)^{\theta(Y_2)}}{\sin \pi z}\left(\frac{2Y}{\pi}\right)^{-z} C_1(z-4)\right\}_{z=2} =$$

$$= \frac{1}{Y_2^2}\left\{\frac{3}{\pi} - \frac{16}{\pi^3} - \frac{6}{\pi}\left(1 - \frac{16}{3\pi^2}\right)\ln\frac{\pi}{2Y}\right\}. \tag{2.72}$$

Thus, with $\sim 1/Y$ accuracy we obtain

$$F' = \frac{8Y_2}{\pi^3}\ln\frac{\pi}{2|Y_2|} + \frac{3}{\pi} + \frac{4}{\pi^2} - \frac{16}{\pi^3} - \frac{6}{\pi}\left(1 - \frac{16}{3\pi^2}\right)\ln\frac{\pi}{2|Y_2|} + O(1/Y). \tag{2.73}$$

The asymptotic formula for the imaginary part F'' is determined by the contribution of the cyclotron-wave pole $y_\varphi^{(3)}$:

$$iF'' = -\operatorname{sgn}(\omega - \Omega)\frac{i\pi}{Y_2^2}\int_0^\infty dy y^2 \delta[\varphi(y) - Y_2] =$$

$$= -\pi \operatorname{sgn}(\omega - \Omega)\left.\frac{y^2\theta(Y_2)}{Y_2\varphi'(y)}\right|_{y=y_\varphi^{(3)}} \tag{2.74}$$

and may be obtained easily using an expression that supplements (2.34) with higher order terms in $1/y$:

$$y_\varphi^{(3)} = \frac{2Y_2}{\pi} - \frac{3\pi}{4}\left(1 - \frac{16}{3\pi^2}\right) - \frac{\pi}{2Y_2}\left[\frac{9}{16}\pi^2\left(1 - \frac{16}{3\pi^2}\right) - 2\right] + O(1/Y). \tag{2.75}$$

As a result

$$F''(Y_2\to\infty) = -\theta(Y_2)\,\mathrm{sgn}\,(\omega-\Omega)\left[\frac{8}{\pi^2}Y_2+\frac{32}{\pi^2}-6+\frac{1}{Y_2}\left(\frac{32}{\pi^2}-2\right)\right]+$$
$$+O\left(\frac{1}{Y_2^2}\right). \qquad (2.76)$$

Formula (2.73) and (2.76) makes it possible to write, based on (2.60), in the limit $\omega \to \Delta$ the following expression for the impedance [54]:

$$Z(\omega)=\frac{4\pi v}{c^2}\left\{\frac{2}{3\sqrt{3}}\frac{1-i\sqrt{3}}{|\pi\xi_0|^{1/3}}+\frac{16}{9\pi\sqrt{3}}\frac{1+i\sqrt{3}}{|\pi\xi_0|^{2/3}}\times\right.$$
$$\times\left[\frac{3\pi^2}{16}\frac{A_1}{1+A_1}-\frac{5}{3}\frac{A_2}{1+A_2}\right]+\frac{1}{\pi\xi_0}\left[\frac{800}{27\pi^3}\left(\frac{A_2}{1+A_2}\right)^2\times\right.$$
$$\times\left\{\ln\left|\pi\xi_0\left(\frac{3\pi(1+A_2)}{20A_2}\right)^3\right|+i\pi\right\}-\frac{3\pi}{16}\left(\frac{A_1}{1+A_1}\right)^2+$$
$$\left.\left.+\frac{10}{3\pi}\frac{A_1}{1+A_1}\frac{A_2}{1+A_2}-\frac{400}{9\pi^3}\left(\frac{A_2}{1+A_2}\right)^2\right]\right\}+\delta Z_1(\omega), \qquad (2.77)$$

where the singular part of the impedance in the order of ξ_0^{-1} may be identified; this part may contain resonance terms when $\omega = \Omega$

$$\delta Z_1(\omega)=-\frac{4\pi i v}{c^2|\pi\xi_0|}\left[\theta(\Omega-\omega)+\mathrm{sgn}\,(\omega-\Omega)\,\theta\left(A_2(\omega-\Omega)\right)\right]\times$$
$$\times\left\{\frac{800}{9\pi^2}\left(\frac{A_2}{1+A_2}\right)^2+\left(\frac{320}{3\pi^2}-20\right)\frac{A_2}{1+A_2}\frac{\omega-\Omega}{\omega}+\left(\frac{32}{\pi^2}-2\right)\left(\frac{\omega-\Omega}{\omega}\right)^2\right\}. \qquad (2.78)$$

Comparing the singular part of (2.78) to the resonance part in the electron gas model (2.62) we may conclude that interelectron interaction smooths out the resonance singularity when $\omega = \Omega$, since the function $\theta(\Omega - \omega) + \mathrm{sgn}\,(\omega - \Omega)\,\theta\,(A_2(\omega - \Omega)) = \theta(A_2)$ is independent of frequency.

We will now proceed to investigate the asymptotics of the function F and the surface impedance in the vicinity of the critical frequency of the Fermi-liquid cyclotron waves (cyclotron-wave resonance). In the Fermi-liquid model utilized here this frequency is $\omega = \Omega(1 + A_2)$, since the parameter Y_2 is close to 10/3. Let $Y_2 = 10/3 + \delta$, $|\delta| \ll 1$, then

$$\Phi_1+\Phi_2=\int_0^\infty\frac{dy\,y^2}{\varphi^3(y)\left[\varphi(y)-\frac{10}{3}-\delta\right]}=\Phi_0+\Delta\Phi, \qquad (2.79)$$

$$\Phi_0=\int_0^\infty\frac{dy\,y^2}{\varphi(\varphi-{}^{10}/_3)}=0{,}099,$$

$$\Delta\Phi=\delta\int_0^\infty\frac{dy\,y^2}{\varphi^3(\varphi-{}^{10}/_3)(\varphi-{}^{10}/_3-\delta)}=\left(\frac{3}{10}\right)^4\frac{35}{8}\frac{\sqrt{21}}{4}\frac{1}{\sqrt{|\delta|}}\frac{\pi}{2}\theta(-\delta), \qquad (2.80)$$

where we have used the expansion for the function (2.27) with small y as well as the formula

$$\frac{2}{\pi}\int_0^\infty\frac{dy}{y^2+\mathrm{sgn}\,\alpha}=\theta(\alpha). \qquad (2.81)$$

Thus, incorporating $I_1+I_3+\frac{10}{3}(I_2+I_4)+\frac{100}{9}\Phi_0=0{,}38$, in this limit we obtain

$$F'=0{,}38-\frac{63\pi}{640}\sqrt{\frac{35}{2}}\sqrt{1-\frac{3Y_2}{10}}\,\theta\left(\frac{10}{3}-Y_2\right). \tag{2.82}$$

The asymptotics of the imaginary part of the function F may be found by the formula

$$F''(Y_2)=-\operatorname{sgn}(\omega-\Omega)\frac{\pi}{Y_2}\frac{y^2}{\varphi'(y)}\bigg|_{y=y_\varphi^{(2)}}=\frac{63\pi}{640}\sqrt{\frac{35}{2}}\sqrt{\frac{3Y_2}{10}-1}\,\theta\left(Y_2-\frac{10}{3}\right) \tag{2.83}$$

By substituting (2.82) and (2.83) into (2.60) we obtain the following expression for the singular part of the impedance [54]:

$$\delta Z_2(\omega)=\frac{4\pi v}{c^2|\pi\xi_0|}\left(\frac{A_2}{1+A_2}\right)^2\frac{21\pi\sqrt{35}}{64\sqrt{2}}\sqrt{\left|\frac{\omega-(1+A_2)\Omega}{A_2(1+A_2)\Omega}\right|}\times$$
$$\times\left\{\operatorname{sgn}(\omega-\Omega)\,i\theta\left[\frac{\Omega(1+A_2)-\omega}{A_2(1+A_2)\Omega}\right]+\theta\left[\frac{\omega-\Omega(1+A_2)}{A_2(1+A_2)\Omega}\right]\right\}. \tag{2.84}$$

It follows from formula (2.84) that interelectron interaction produces an impedance singularity in the vicinity of cyclotron-wave resonance, while the root component in (2.84) has a real value in the frequency range for which cyclotron waves (2.30) may exist, and an imaginary value in the forbidden range.

Thus, with specular electron reflection by the metal surface the frequency dependence of the surface impedance of the electron liquid of a metal is described by smooth function (2.78) in the proximity of cyclotron resonance and contains the principal singularity at the cyclotron-wave resonance point (2.84).

In this section we will not consider details related to the generalization of the impedance calculation method to the case of the final values of the electron collision frequency, although we note that in order for singularity (2.84) to appear, we must require consistent with (2.55) satisfaction of the inequality

$$\left|\frac{\omega}{\nu}A_2\right|\gg 1, \tag{2.85}$$

characterizing the relative significance of the interelectron interaction and collisional damping effects.

2.4. Comparison of impedance calculation results for the case of electron specular reflection to experimental data

The experimental investigation of the frequency dependence of the impedance of a metal when Fermi-liquid cyclotron waves may exist is a complex problem in view of the difficulties of maintaining the necessary experimental conditions (strong magnetic fields, superpure samples, high-sensitivity spectrometric equipment), and to date only one attempt has been made [23]. In this section we will discuss the experimental results of Baraff, Grimes and Platzman [23] to investigate the frequency dependence of the surface resistance (the real part of the surface impedance) of potassium at 1.2 K.

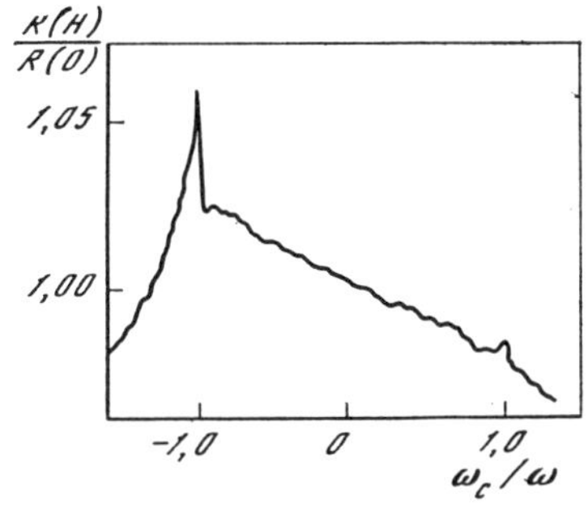

Fig. 9. The real part of the surface impedance $R(H)$ = Re $Z(\omega)$ of potassium plotted as a function of the external magnetic field strength based on data from study [23]
The negative sign of the electron charge is accounted for in determining the quantity $\omega_c = -|e|B/mc$, so the cyclotron resonance condition is $\omega = -\omega_c$

First we will focus on the experimental conditions. The measurements were carried out on samples 0.1-0.5 cm thick using a spectrometer operating at ω = 24 GHz employing a circularly-polarized wave. The magnetic field **B** perpendicular to the surface was tuned over a broad range, including the cyclotron resonance range $\omega = \Omega$. The effective mass $m = 1.21\ m_e$ (which corresponds to $\Omega/B = 1.45\cdot10^7$ Oe^{-1} s^{-1}) is determined from independent experiments by observing shortwave cyclotron waves [19]. The relative residual conductivity of the samples was $\sigma_{1,2}/\sigma_{300} = (5\text{-}13)\cdot10^3$, which corresponded to a very high potassium purity, since the electron collision frequency was $\nu = (1.6\text{-}4)\cdot$ $\cdot10^9$ s^{-1} with an electron density of $n = 1.4\cdot10^{27}$ cm^{-3} ($\omega_{Le} = 6.07\cdot$ $\cdot10^{15}$ s^{-1}). The free path length $\lambda = v/\nu$ determined by the Fermi velocity $v = 7.1\cdot10^7$ cm$\cdot$s^{-1} is equal to $3\cdot10^{-2}$ in this case, which is greater than the depth of the skin-layer ($\delta = c/\omega_{Le} \sim 10^{-5}$ cm), since the skin-effect is an anomalous effect (the parameter $|\pi\xi_0|$ = 21,600). On the other hand, the dimensions of the sample far exceed the free path length, so the second boundary may be ignored and the sample may be considered to be half-bounded.

In these experimental conditions a peak (see Fig. 9) with a relative magnitude of $\Delta R(B)/R(0) = 2\cdot10^{-2}$ is detected in the frequency dependence of the surface resistance. The position of the peak when $\omega_c/\omega = -1.025 \pm 0.005$ corresponds to the cyclotron-wave resonance point, i.e., the critical frequency of the Fermi-liquid cyclotron wave $\omega = \Omega(1 + A_2)$, if for the parameter A_2 we employ the value -0.025 ± 0.005 coinciding with the measured value from independent experiments extrapolating the dispersion curve to $k = 0$ for cyclotron waves propagating perpendicular to the magnetic field [19].

The observation of a relatively large and sharp peak in the vicinity of the critical frequency of the Fermi-liquid cyclotron wave indicates that the electron correlation effects are large compared to the collisional damping effects responsible for broadening of the peak (i.e,. the parameter $|A_2\Omega/\nu|$ has a large modulus value (2.85).

In the experimental conditions of study [23] this parameter was approximately 3-10). In this case the surface impedance has a resonance singularity at the frequency $\Omega(1 + A_2)$ and not at the cyclotron resonance frequency. This fact corresponds to the impedance calculation results given in section 2.3. An estimate of the relative contribution of the impedance singularity represented by formula (2.84) for parameters characteristic of experiment [23] yields the following result:

$$\frac{\delta \operatorname{Re} Z}{\operatorname{Re} Z} \sim 5\,|\pi\xi_0|^{-2/3}\left(\frac{A_2}{1+A_2}\right)^2\left|\frac{\nu}{A_2\Omega}\right|^{1/2} \sim 10^{-6}. \tag{2.86}$$

This quantity is four orders of magnitude smaller than the value of the experimentally-obtained peak. The terms in the impedance of $\sim|\pi\xi_0|^{-2/3}$ (see formula (2.60)) attributable to Fermi-liquid interaction, although they are three orders of magnitude larger than (2.86) are not resonance terms.

Thus, we may conclude from a comparison of the calculation results for the impedance of the electron liquid of a metal for specular electron surface reflection to experimental data on a potassium slab in a magnetic field perpendicular to the surface that there is agreement between theory and experiment with respect to the position of the singularity in the vicinity of cyclotron-wave resonance. At the same time theory in the case of specular reflection yields a singularity value that is four orders of magnitude smaller than that discovered by experiment. This fact makes it necessary to abandon specular electron reflection to explain the value of the peak obtained in experiment.

Conclusions

1. With specular electron reflection the surface impedance of a metal in the Fermi-liquid model incorporating the parameters A_1 and A_2 may be represented as an asymptotic expansion in terms of the skin-effect anomalous parameter ξ_0.

2. Incorporation of interelectron interaction eliminates the logarithmic singularity of the impedance in the vicinity of cyclotron waves in the Fermi-liquid of the metal.

3. An experimental investigation of the frequency dependence of the real part of the impedance characterizing absorption of the electromagnetic wave in a thick slab reveals a peak with cyclotron-wave resonance, which makes it possible to obtain information on the sign and magnitude of the parameter A_2. However such a possibility requires further investigation, since the relative magnitude of the peak discovered experimentally and found theoretically in the case of specular reflection differs by an order of magnitude.

Chapter 3

THE SURFACE IMPEDANCE OF A METAL IN A MODEL WITH A SINGULAR NONZERO PARAMETER A_1 WITH DIFFUSE ELECTRON BOUNDARY SCATTERING

In calculating the surface impedance in the case of specular electron boundary reflection in the preceding section, a lack of correlation between the relative resonance singularity at the cyclotron-wave resonance point and its experimental values was discovered. In the electron gas model the nature of electron scattering at the surface does not change the higher order component in the expansion of the impedance in the anomalous skin-effect, although the frequency dependence of subsequent terms of the asymptotic expansion change significantly [55]. Hence we may hope that calculating the impedance in the electron gas model for the case of diffuse scattering will change the value of the resonance singularity attributable to Fermi-liquid interaction appearing in the higher terms of the expansion. When writing equations for the field in the case of diffuse electron scattering we obtain integral equations with difference kernels in semi-infinite limits; the Wiener-Hopf method is used for their solution [57]. The application of this method to solving the equation in the gas model yields more complex expressions for the impedance than those obtained with specular electron reflection. Implementation of the Wiener-Hopf method in an electron liquid with nonzero parameters A_1 and A_2 when it is necessary to consider a system of two integral equations is even more cumbersome. In this regard in the subsequent sections we will carry out a calculation of the impedance of the electron liquid accounting for the constants A_1 and A_2 separately, which allows us to reduce the problem to solving a single equation in each case.

In this chapter we consider a model with a nonzero parameter A_1 which is of interest in connection with the possible for identifying impedance singularities in the case of diffuse scattering that may be related to this constant. We demonstrated in the preceding chapter that there are no such resonance singularities in the case of specular reflection.

3.1. Solution of the Wiener-Hopf problem for an electric field in the electron fluid of a metal

An electric field in electron liquid obeys Maxwell's equation (2.1) that may be written, incorporating the displacement current, as

$$\frac{\partial^2}{\partial z^2} E(z) + \frac{\omega^2}{c^2} E(z) = -\frac{4\pi i\omega}{c^2} j(z). \tag{3.1}$$

Expressing the current density $j(z)$ through the function Φ in accordance with (1.24) and taking $\omega = 0$, $l \geqslant 2$ in equation system (1.22) with diffuse electron scattering (ρ = 0), we arrive at the following integrodifferential equation for the field $E(z)$

$$\frac{\partial^2}{\partial z^2}E+\frac{\omega^2}{c^2}E-\frac{3}{2}\frac{A_1}{1+A_1}\int_0^\infty dz'Q_{11}(z-z')\left[\frac{\partial}{\partial z'^2}E(z')+\frac{\omega^2}{c^2}E(z')\right]=$$
$$=\frac{1}{C}\int_0^\infty dz'Q_{11}(z-z')E(z'), \tag{3.2}$$

where, in accordance with (1.23),

$$Q_{11}(z)=\frac{a}{2}\int_1^\infty dss^{-1}(1-s^{-2})\exp(-sb|z|),$$
$$a=-i(\omega+i\nu)v^{-1},\; b=[-i(\omega-\Omega)+\nu]v^{-1}; \tag{3.3}$$
$$C=\frac{2c^2(\omega+i\nu)}{3\omega_{Le}^2\omega}. \tag{3.4}$$

In order to solve equation (3.2) which is a Wiener-Hopf equation consistent with study [57] we must continue the electric field region to the negative semiaxis z, after setting $E(z)$ = 0 where z < 0. Then the following equation is valid on the entire axis $-\infty < z < \infty$:

$$\frac{\partial^2}{\partial z^2}E(z)+\frac{\omega^2}{c^2}E(z)-\frac{3}{2}\frac{A_1}{1+A_1}\int_{-\infty}^{\infty}dz'Q_{11}(z-z')\left[\frac{\partial}{\partial z'^2}E(z')+\frac{\omega^2}{c^2}E(z')\right]-$$
$$-\frac{1}{C}\int_{-\infty}^{\infty}dz'Q_{11}(z-z')E(z')=\mathcal{F}(z), \tag{3.5}$$

$$\mathcal{F}(z)=-\frac{3}{2}\frac{A_1}{1+A_1}\int_0^\infty dz'Q_{11}(z-z')\left[\frac{\partial}{\partial z'^2}E(z')+\frac{\omega^2}{c^2}E(z')\right]-$$
$$-\frac{1}{C}\int_0^\infty dz'Q_{11}(z-z')E(z')\quad \text{при}\quad z<0,$$
$$\mathcal{F}(z)=0\quad \text{при}\quad z\geqslant 0. \tag{3.6}$$

Using the double Fourier transform

$$E(k)=\int_{-\infty}^{\infty}dze^{-ikz}E(z)=\int_0^\infty dze^{-ikz}E(z), \tag{3.7}$$

$$f(k)=\int_{-\infty}^{\infty}dze^{-ikz}\mathcal{F}(z)=\int_{-\infty}^{0}dze^{-ikz}\mathcal{F}(z), \tag{3.8}$$

$$q_{11}(k)=\int_{-\infty}^{\infty}dze^{-ikz}Q_{11}(z). \tag{3.9}$$

Then we have from equation (3.5)

$$C\left[1-\frac{3}{2}\frac{A_1}{1+A_1}q_{11}(k)\right]\left[E'(0)+ik(0)+\left(k^2-\frac{\omega^2}{c^2}\right)E(k)\right]+$$
$$+q_{11}(k)E(k)+Cf(k)=0, \tag{3.10}$$

where $E(0)$ and $E'(0)$ are the values of the field and the field derivative when z = 0. It is convenient to rewrite equation (3.10) in a form that will allow direct application of the factorization method underlying the Wiener-Hopf technique

$$\left[k^2-\frac{\omega^2}{c^2}-\frac{3}{2}\frac{A_1}{1+A_1}\left(k^2-\frac{\omega^2}{c^2}-\tilde{k}_0^2\right)q_{11}(k)\right]\left[E(k)+\frac{E'(0)+ikE(0)}{k^2-\frac{\omega^2}{c^2}-\tilde{k}_0^2}\right]-$$

$$-\frac{\tilde{k}_0^2}{k^2-\frac{\omega^2}{c^2}-\tilde{k}_0^2}[E'(0)+ikE(0)]+f(k)=0, \tag{3.11}$$

where we isolate the pole

$$k_0^2\equiv(k_1+ik_2)^2\equiv\tilde{k}_0^2+\left(\frac{\omega}{c}\right)^2=\frac{2(1+A_1)}{3A_1C}+\left(\frac{\omega}{c}\right)^2. \tag{3.12}$$

In carrying out the factorization of equation (3.11) we will use the analytic properties of the functions entering into this equation. Since the electric field diminishes with growth of z, it follows from (3.5) that the function $E(k)$ is regular for Im $k < 0$. Then it follows from (3.3) that $Q_{11}(z) < \exp(-\alpha|z|)$, where $\alpha \sim \nu/\upsilon$ and as a result $q_{11}(k)$ is regular at the pole $-\alpha <$ Im $k < \alpha$. We note that $q_{11}(k)$ has branch points $k = \pm ib = \pm(\omega - \Omega + i\nu)/\upsilon$. It follows from definition (3.4) and the properties of $E(z)$ and $Q_{11}(z)$ noted above that the function $f(k)$ is regular for Im $k > -\alpha$. Finally we may state that the equation

$$\left(k^2-\frac{\omega^2}{c^2}\right)\left[1-\frac{3}{2}\frac{A_1}{1+A_1}q_{11}(k)\right]+\frac{1}{C}q_{11}(k)=0 \tag{3.13}$$

has no roots lying on the real k axis. This corresponds to the fact that there are no intrinsic oscillations whose wave vector is real in an electron liquid with a single nonzero constant A_1. We introduce the function

$$\tau(k)=\left\{\left(k^2-\frac{\omega^2}{c^2}\right)\left[1-\frac{3}{2}\frac{A_1}{1+A_1}q_{11}(k)\right]+\frac{1}{C}q_{11}(k)\right\}\Big/(k^2+\delta^2). \tag{3.14}$$

It is obvious that $\tau(k)$ has no roots and is regular at the pole $-\alpha <$ Im $k < \alpha$. Using the Cauchy formula we may represent $\tau(k)$ as

$$\tau(k)=\frac{\tau_+(k)}{\tau_-(k)}, \tag{3.15}$$

where $(0 < \beta < \alpha)$:

$$\tau_\pm(k)=\exp\left\{\frac{1}{2\pi i}\int\limits_{\mp i\beta-\infty}^{\mp i\beta+\infty}\frac{dk'}{k'-k}\ln\tau(k')\right\}, \tag{3.16}$$

where the function τ_+ is regular and has no zeros in the range Im $k > -\beta$, while τ_- is likewise in the range Im $k < \beta$. Formula (3.15) makes it possible to write (3.11) as

$$\frac{k-i\delta}{\tau_-(k)}E(k)+\frac{E'(0)+ikE(0)}{k^2-k_0^2}\left[\frac{k-i\delta}{\tau_-(k)}-\frac{k^2-\left(\frac{\omega}{c}\right)^2}{\tau_+(k)(k+i\delta)}\right]+$$

$$+\frac{E'(0)+ikE(0)}{\tau_+(k)(k+i\delta)}+\frac{f(k)}{C\tau_+(k)(k+i\delta)}=0. \tag{3.17}$$

We will determine the functions

$$h^{(1)}(k)=\left[\frac{k-i\delta}{\tau_-(k)}-\frac{k^2-\left(\frac{\omega}{c}\right)^2}{\tau_+(k)(k+i\delta)}\right](k^2+\delta^2)^{-1}, \tag{3.18}$$

$$h^{(2)}(k)=kh^{(1)}(k), \tag{3.19}$$

regular at the pole $-\gamma < \mathrm{Im}\, k < \gamma$ where $\gamma < \min\{\beta, k_2\}$. We will decompose these into components that are analogous in terms of their analytic properties to the functions $\tau_{\pm}(k)$:

$$h^{(i)}(k) = h_+^{(i)}(k) - h_-^{(i)}(k),\ i = 1, 2. \tag{3.20}$$

Accounting for the fact that the functions

$$h_+^{(i)}(k)=\frac{1}{2\pi i}\int_{-i\gamma-\infty}^{-i\gamma+\infty}\frac{dk'}{k'-k}h^{(i)}(k') \tag{3.21}$$

are regular for $\mathrm{Im}\, k > -\gamma$, while the functions

$$h_-^{(i)}(k)=\frac{1}{2\pi i}\int_{i\gamma-\infty}^{i\gamma+\infty}\frac{dk'}{k'-k}h^{(i)}(k') \tag{3.22}$$

are regular when $\mathrm{Im}\, k < \gamma$, we may transpose to the right all components in equation (3.17) that are regular in the upper half-plane; the terms that are regular in the lower half-plane will be left in the left half:

$$\frac{k-i\delta}{\tau_-(k)}E(k)-h_-^{(1)}(k)E'(0)-h_-^{(2)}(k)iE(0)=h_+^{(1)}(k)E'(0)-$$
$$-h_+^{(2)}(k)iE(0)-\frac{f(k)+C[E'(0)+ikE(0)]}{C\tau_+(k)(k+i\delta)}. \tag{3.23}$$

Bearing in mind that with large values of $|k|$ as indicated by (3.7) the Fourier component of the electric field has the asymptotics

$$E(k)=-\frac{i}{k}E(0)-\frac{1}{k^2}E'(0)+0\left(\frac{1}{k^3}\right), \tag{3.24}$$

We find that the left half of equation (3.23) when $|k| \to \infty$ is $-iE(0)$. The left half of (3.23) is regular in the lower half-plane $\mathrm{Im}\, k < 0$, while the right half is regular in the upper half-plane and at the pole of width γ below the real axis, $\mathrm{Im}\, k > -\gamma$. Since the regularity regions of the left and right halves of (3.2) overlap, we thereby determine the function that is regular across the entire plane of the complex variable k equal to the constant at infinity. Consequently, both halves of equation (3.23) are identically equal to the constant $-iE(0)$; specifically,

$$\frac{k-i\delta}{\tau_-(k)}E(k)-h_-^{(1)}(k)E'(0)-h_-^{(2)}(k)iE(0)\doteq-iE(0). \tag{3.25}$$

Since the unknown function $f(k)$ does not enter into equation (3.25), it provides the solution for the Fourier-transform of the field:

$$E(k)=\frac{\tau_-(k)}{k-i\delta}\left[iE(0)+h_-^{(1)}(k)E'(0)+h_-^{(2)}(k)iE(0)\right]. \qquad (3.26)$$

The formula of the inverse Fourier-transform

$$E(z)=\frac{1}{2\pi}\int_{-\infty}^{\infty} dk e^{ikz}E(k) \qquad (3.27)$$

and expression (3.26) make it possible to find the electric field distribution in the half-bounded sample. The surface impedance expressed by (2.7) through the ratio of the field strength and its derivative when $z = 0$ may be obtained from (3.26) by means of the relation:

$$E'(0)=-\lim_{k\to\infty} k\left[kE(k)+iE(0)\right], \qquad (3.28)$$

following from (3.24). Isolating terms of identical order of smallness, we obtain

$$\frac{iE(0)}{E'(0)}=-\left(1+\lim_{k\to\infty}\frac{k^2h_-^{(1)}(k)\tau_-(k)}{k-i\delta}\right)\left\{\lim_{k\to\infty}\left[k\left(1-\frac{k\tau_-(k)}{k-i\delta}\right)+\frac{k^2h_-^2(k)\tau_-(k)}{k-i\delta}\right]\right\}^{-1}. \qquad (3.29)$$

We will evaluate the integrals $h_-^{(i)}(k)$ after taking the residues at the poles $k = \pm k_0$ of the integrands in (3.18) and (3.19). Since we may satisfy inequality $\alpha < \delta$ Im $k_0 \equiv k_2$ by proper selection of a sufficiently small value of δ, then taking $\gamma >$ Im k, we obtain

$$h_-^{(1)}(k)=\frac{k-i\delta}{k^2-k_0^2}\frac{1}{\tau_-(k)}+\frac{k_0+i\delta}{2k_0(k-k_0)}\frac{1}{\tau_-(-k_0)}+\frac{\tilde{k}_0^2}{2k_0(k-k_0)(k_0+i\delta)\tau_+(k_0)}, \qquad (3.30)$$

$$h_-^{(2)}(k)=1-\frac{k(k-i\delta)}{(k^2-k_0^2)\tau_-(k)}-\frac{k_0+i\delta}{2(k+k_0)\tau_-(-k_0)}+\frac{\tilde{k}_0^2}{2(k_0+i\delta)(k-k_0)\tau_+(k_0)}. \qquad (3.31)$$

Finally, using the following representation for the functions $\tau_\pm(k)$

$$\tau_+(k)=\frac{k}{k+i\delta}\tilde{\tau}_+(k), \qquad (3.32)$$

$$\tau_-(k)=\frac{k-i\delta}{k}\tilde{\tau}_-(k), \qquad (3.33)$$

$$\tilde{\tau}_\pm=\exp\left\{\frac{1}{2\pi i}\int_{-\infty}^{\infty}\frac{dk'}{k'-k\pm i\delta}\ln\left[\left(1-\frac{\omega^2}{ck'^2}\right)\times\right.\right.$$
$$\left.\left.\times\left(1-\frac{3}{2}\frac{A_1}{1+A_1}q_{11}(k')\right)+\frac{1}{Ck'^2}q_{11}(k')\right]\right\}, \qquad (3.34)$$

we find, based on (3.29), a general formula for the surface impedance

$$Z(\omega)=\frac{4\pi i\omega}{c^2}\frac{E(0)}{E'(0)}=\frac{4\pi\omega}{c^2k_0}\left\{1-\frac{2}{1-\tilde{\tau}_+^2(k_0)\left[1-\left(\frac{\omega}{ck_0}\right)^2\right]^{-1}}\right\}, \qquad (3.35)$$

where we take advantage of the fact that $\tau_-(-k) = \tau_+^{-1}(k)$. We note that in formula (3.35) the limiting process of the electron gas A_2 corresponds to $|k_0| \to \infty$, when in accordance with (3.34),

$$\tau_{\pm}(\pm k_0) = 1 \pm \frac{i}{k_0\lambda}, \tag{3.36}$$

where

$$\lambda^{-1} = \frac{1}{2\pi} \int_{-\infty}^{\infty} dk \ln\left[1 - \frac{\omega^2}{c^2k^2} - \frac{q_{11}(k)}{Ck^2}\right]. \tag{3.37}$$

It is easy to determine that this limit yields the result of regular electron gas theory [29], i.e., $E(0)/E'(0) = -\lambda$.

We will briefly discuss the expression for the electric field in the metal interior. It follows from formula (3.27) that for values of $z = L$ greater than the depth of the skin layer $\lambda_0 = c\omega_{Le}$, $(L \gg \lambda_0)$ the primary contribution to the field comes from the Fourier component $k = 0$ [32]

$$T \equiv E(L) \sim \frac{1}{2\pi i L} \lim_{k\to\infty} E(k). \tag{3.38}$$

Eliminating the field derivative $E'(0)$ from formula (3.26) for $E(k)$ by means of relation (3.35), we arrive at the expression:

$$E(k) = -\frac{iE(0)}{k^2 - k_0^2}\left\{k - k_0 \frac{1 - \left(\frac{\omega}{ck_0}\right)^2 - \tilde{\tau}_+^2(k_0)}{1 - \left(\frac{\omega}{ck_0}\right)^2 + \tilde{\tau}_+^2(k_0)} - \frac{\tau(k)}{k} \frac{2k_0\tilde{\tau}_+(k_0)}{1 + \tilde{\tau}_+^2(k_0)\left[1 - \left(\frac{\omega}{ck_0}\right)^2\right]^{-1}}\right\}. \tag{3.39}$$

In order to calculate $E(k)$ we must determine the limit

$$\lim_{k\to 0} \frac{\tau_-(k)}{k} = \lim_{k\to 0} \frac{1}{k} \exp\left\{\frac{1}{2\pi i} \int_{-\infty}^{\infty} \frac{dk}{k} L(k) - \frac{1}{2} L(k)\right\}, \tag{3.40}$$

where

$$L(k) = \ln\left[\left(1 - \frac{\omega^2}{c^2k^2}\right)\left(1 - \frac{3}{2} \frac{A_1}{1 + A_1} q_{11}(k)\right) + \frac{1}{Ck^2} q_{11}(k)\right]. \tag{3.41}$$

Since the integral in the sense of the principal value in (3.40) is equal to zero due to the even value of $L(k)$, we obtain

$$\lim_{k\to 0} \frac{\tau_-(k)}{k} = \left[\frac{2a}{3bC} - \frac{\omega^2}{c^2}\left(1 - \frac{a}{b} \frac{A_1}{1 + A_1}\right)\right]^{-1/2} \tag{3.42}$$

and as a result of study [59] we have

$$T = -\frac{iE(0)}{\pi L}\left\{\left[\frac{2a}{3bC} - \frac{\omega^2}{c^2}\left(1 - \frac{a}{b} \frac{A_1}{1 + A_1}\right)\right]^{-1/2} \times \right.$$
$$\left. \times \frac{2\tilde{\tau}_+(k_0)}{1 + \tilde{\tau}_+^2(k_0)\left[1 - \left(\frac{\omega}{ck_0}\right)^2\right]^{-1}} + \frac{1 - \left(\frac{\omega}{ck_0}\right)^2 - \tilde{\tau}_+^2(k_0)}{k_0\left[1 - \left(\frac{\omega}{ck_0}\right)^2 + \tilde{\tau}_+(k_0)\right]}\right\}. \tag{3.43}$$

Thus, formulae (3.35) and (3.43) expresses the surface impedance in an electric field at a depth significantly exceeding the skin-layer depth through the function $\tilde{\tau}_+(k_0)$ for which we will obtain the asymptotic expansion below in the anomalous skin effect regime.

3.2. Asymptotic calculation of the impedance and the electric field passed through a slab

In this section we will show that the integral $\tilde{\tau}_+(k_0)$ (3.32) and, consequently, the surface impedance in the RF range when we may ignore the displacement current ($\omega^2 \ll c^2k^2$) and take the skin-effect to be extremely anomalous ($|\xi_0| \gg 1$), may be represented as an asymptotic series. From this expansion we may obtain an explicit dependence of the surface impedance on the magnetic field.

Employing the dimensionless variable $t = -k/b$ and $k_0 = k_0 = -bt_0$, where

$$t_0^2 = X = 2\xi_0 \frac{\omega^3}{(\omega - \Omega + i\nu)^3 Y},$$

$$Y = \frac{3}{2}\frac{A_1}{1+A_1}\frac{\omega + i\nu}{\omega - \Omega + i\nu}, \tag{3.44}$$

we may write the following formula:

$$\tau_+(-b\sqrt{X}) = \exp\left\{\frac{1}{2\pi i}\int_{-\infty/b}^{\infty/b} \frac{dt}{t - \sqrt{X} + \frac{i\delta}{b}} \ln\left[1 - Y\left(1 - \frac{X}{t^2}\right)\tilde{q}_{11}(t)\right]\right. . \tag{3.45}$$

An examination of the integrand in (3.45) where $\tilde{q}_{11}(t) = b/a \; q_{11}(t)$ shows that the integration contour may be rotated and coincide with the real axis. As a result we have

$$\tau_+(-b\sqrt{X}) = \exp\{i\sqrt{X}\, I(X, Y)\}, \tag{3.46}$$

where

$$I(X, Y) = \frac{1}{\pi}\int_0^\infty \frac{dt}{X - t^2}\ln\left[1 - Y\left(1 - \frac{X}{t^2}\right)\tilde{q}_{11}(t)\right]. \tag{3.47}$$

For asymptotic evaluation of integral (3.47) we will use the Mellin transform (compare to study [53])

$$M(z, Y) = \int_0^\infty dX X^{z-1} I(X, Y) = \frac{1}{\pi}\int_0^\infty d\psi \frac{\psi^{z-1}}{\psi - 1}\int_0^\infty dt t^{2z-1}\ln[1 - (1 - \psi) Y \tilde{q}_{11}(t)], \tag{3.48}$$

where $X = \psi t^2$. Carrying out integration by parts in (3.48) and noting that by virtue of $\tilde{q}_{11}(0) = 2/3$ and $\tilde{q}_{11}(t \to \infty) = \pi/2t$, the component related to the substitution of the integration limits vanishes if

$$1/2 < \operatorname{Re} z < 1, \tag{3.49}$$

we obtain

$$M(z, Y) = -\frac{1}{2z-1}\int_0^\infty dt\, t^{2z-1}\frac{\tilde{q}'_{11}(t)}{\tilde{q}_{11}(t)}\frac{1}{\pi}\int_0^\infty \frac{d\psi\, \psi^{z-1}}{\psi - 1 + \frac{1}{Y\tilde{q}_{11}(t)}}. \tag{3.50}$$

Then using the relation

$$\int_0^\infty \frac{dx\, x^{s-1}}{x+\alpha} = \alpha^{s-1}\frac{\pi}{\sin \pi s},\ |\arg \alpha| < \pi,\ 0 < s < 1 \tag{3.51}$$

and representing (3.50) as

$$M(z, Y) = -\frac{1}{(2z-1)\sin \pi z}\int_0^\infty dt\, t^{2z-1}\frac{\tilde{q}'_{11}(t)}{\tilde{q}_{11}(t)}\left[\frac{1}{Y\tilde{q}_{11}(t)} - 1\right]^{z-1}. \tag{3.52}$$

It will be convenient below to use the expansions

$$\frac{\tilde{q}'_{11}(t)}{\tilde{q}_{11}(t)}\left[\frac{1}{Y\tilde{q}_{11}(t)} - 1\right]^{z-1} = \begin{cases} \sum\limits_{\mu=0}^{\infty} a_\mu(z-1)\, t^{\mu+1}, & |t| < 1; \\ \sum\limits_{\nu=0}^{\infty} b_\nu(z-1)\, t^{-\nu+z-2}, & t > 1. \end{cases} \tag{3.53}$$

These formulae make it possible to write integral (3.52) in the form of series

$$M(z, Y) = -\frac{1}{(2z-1)\sin \pi z}\left\{\sum_{\mu=0}^{\infty}\frac{a_\mu(z-1)}{2z+\mu+1} + \sum_{\nu=0}^{\infty}\frac{b_\nu(z-1)}{\nu+2-3z}\right\}. \tag{3.54}$$

It is assumed in deriving (3.54) that z satisfies the inequality

$$1/2 < \operatorname{Re} z < 2/3. \tag{3.55}$$

Expansion of $I(X, Y)$ in terms of diminishing powers of X when $|X| \gg 1$ is given by the inverse Mellin transform

$$I(X, Y) = \frac{1}{2\pi i}\int_{C-i\infty}^{C+i\infty} dz\, X^{-z} M(z, Y),\ 1/2 < C < 2/3, \tag{3.56}$$

if the integration contour is shifted to the right and if the poles of (3.54) lying to the left of the line Re $z = C$ are incorporated.

The first series in braces in (3.54) has no poles for $z > 0$ and hence the poles arise from $(\sin \pi z)^{-1}$ at the points $z - m$ ($m = 1, 2, \ldots$). Specifically, the contribution from the pole $z = 1$ to integral (3.56) is equal to

$$-\frac{1}{\pi X}\sum_{\mu=0}^{\infty}\frac{a_\mu(0)}{\mu+3} = -\frac{1}{\pi X}\int_0^1 dt\, \frac{\tilde{q}'_{11}(t)}{\tilde{q}_{11}(t)} = \frac{0{,}0314}{X}. \tag{3.57}$$

The second series in (3.54) has poles for $z > 0$. Hence the contributions to integral (3.56) are of three different types: the contributions of the simple poles yielding fractional powers of X arising when $z = (2 + \nu)/3$, the contributions from the double pole $z = n$ when $\nu = 3n - 2$ and, finally, the contributions from the simple poles $(\sin \pi z)^{-1}$ when $z = n$, $\nu \neq 3n - 2$. Therefore we have from the fractional poles:

$$-\frac{2}{\sqrt{3}X^{1/3}} b_0\left(-\frac{1}{3}\right) = \frac{2}{\sqrt{3}X}\left(\frac{\pi XY}{2}\right)^{1/3} \text{ when } \nu = 0, \tag{3.58}$$

$$\frac{2}{5\sqrt{3}X^{4/3}} b_2\left(\frac{1}{3}\right) = -\frac{2}{3\sqrt{3}X}\left(\frac{2}{\pi XY}\right)^{1/3}\left[1 - \frac{32}{\pi^2} + \frac{2Y}{3} - \frac{\pi^2 Y^2}{60}\right] \quad \text{when } \nu = 2 \tag{3.59}$$

The contribution of the double pole $z = 1$ when $\nu = 1$ yields

$$\frac{1}{3\pi}\frac{d}{dz}\left\{X^{-z}\frac{1}{2z-1}b_1(z-1)\right\}_{z=1} = \frac{1}{3\pi X}\left[-b_1(0)\ln X + b_1'(0) - 2b_1(0)\right] =$$

$$= -\frac{1}{3\pi X}\left\{\frac{4}{\pi}\ln\frac{\pi XY}{2} + \frac{12}{\pi} - \frac{\pi Y}{2}\right\}. \tag{3.60}$$

Finally, the contribution of the simple pole when $z = 1$ takes the form

$$\frac{1}{\pi X}\sum_{\nu\neq 1}\frac{b_\nu(0)}{1-\nu} = \frac{1}{\pi X}\left\{b_0(0) - \int_1^\infty dt\left[\frac{\tilde{q}_{11}'(t)}{\tilde{q}_{11}(t)}t - b_0(0) - \frac{b_1(0)}{t}\right]\right\} =$$

$$= \frac{1}{\pi X}\left\{1 + \int_1^\infty dt\left[\frac{\tilde{q}_{11}'(t)}{\tilde{q}_{11}(t)}t + 1 - \frac{4}{\pi t}\right]\right\} = \frac{1}{X}\left(-\frac{1}{\pi} + 0{,}158\right). \tag{3.61}$$

We used the following values of the expansion coefficients of (3.59) in formulae (3.58) and (3.61)

$$b_0(0) = -1,\quad b_0\left(-\frac{1}{3}\right) = -\left(\frac{\pi Y}{2}\right)^{1/3},\quad b_1'(0) = -\frac{4}{\pi} + \frac{\pi Y}{2} -$$

$$-\frac{4}{\pi}\ln\frac{\pi Y}{2},\quad b_1(0) = \frac{4}{\pi},$$

$$b_2\left(\frac{1}{3}\right) = \frac{5}{3}\left(\frac{2}{\pi Y}\right)^{1/3}\left[-1 + \frac{32}{3\pi^2} - \frac{2Y}{3} + \frac{\pi^2 Y^2}{60}\right]. \tag{3.62}$$

Collecting all these contributions and incorporating the fact that $\xi = 1/2\ XY$ with $X^{-4/3}$ accuracy we obtain the asymptotic expansion:

$$I(X, Y) = \frac{1}{X}\left\{\frac{2}{\sqrt{3}}(\pi\xi)^{1/3} - \frac{4}{3\pi^2}\ln\pi\xi - 0{,}534 + \frac{Y}{6} + \right.$$

$$\left. + \frac{2}{3\sqrt{3}}(\pi\xi)^{-1/3}\left(\frac{32}{3\pi^2} - 1 - \frac{2Y}{3} + \frac{\pi^2 Y^2}{60}\right)\right\}, \tag{3.63}$$

We will estimate the order of the corrections that arise in the impedance and the field when accounting for the displacement current. In this case in place of (3.47) we have

$$I(\tilde{X}, Y) = \frac{1}{\pi}\int_0^\infty \frac{dt}{\tilde{X} - t^2}\ln\left[1 - \frac{B}{t^2} - Y\left(1 - \frac{\tilde{X}}{t^2}\right)\tilde{q}_{11}(t)\right], \tag{3.64}$$

where $\tilde{X} = (k_0/b)^2 = B + X$, $B = (\omega/cb)^2$. Consequently we may incorporate the displacement current by substituting X with $\tilde{X}$ in (3.47) and by adding the component $-Bt^{-2}$ under the logarithm sign. This last addition results in terms of $\sim B\xi^{-2}$. The contribution of the pole $t^2_{\sim} = \tilde{X}$, (when $B = 0$ there is no such contribution) yields a correction $\sim \tilde{X}^{-1/2}$ ln $(1 - B\tilde{X}^{-1}) \sim B\xi^{-3/2}$. Thus, the asymptotic expansion of the integral I with $\sim U^{-4/3}$ term accuracy does not contain the contribution of the displacement current.

Bearing in mind result (3.63) we obtain the asymptotic expansion of surface impedance (3.35). For this we use the relation

$$\frac{\tilde{\tau}_+^2\,(k_0)}{1-\left(\frac{\omega}{ck_0}\right)^2} = \exp\left\{2i\left[\tilde{X}^{1/2} I(\tilde{X},\,Y) - \frac{1}{2i}\ln\left(1-\frac{B}{\tilde{X}}\right)\right]\right\}. \tag{3.65}$$

Then for the impedance we obtain

$$Z(\omega) = \frac{4\pi\omega}{c^2 ib}\,\tilde{X}^{-1/2}\,\mathrm{ctg}\left\{\tilde{X}^{1/2} I(\tilde{X},\,Y) - \frac{1}{2i}\ln\left(1-\frac{B}{\tilde{X}}\right)\right\}. \tag{3.66}$$

When $|\tilde{X}| \gg 1$ the logarithm in (3.66) yields a correction $\sim B\tilde{X}^{-3/2}$ and then, bearing in mind that

$$X^{1/2} I(X,\,Y) = \pi^{1/3}\xi_0^{-1/6}\left(\frac{A_1}{1+A_1}\right)^{1/2}\left[1+\frac{\sqrt{3}}{2}(\pi\xi)^{-1/3}+\cdots\right], \tag{3.67}$$

with the accuracy of calculation (3.63) we limit our examination to the first three terms in the series expansion of ctg $(\sqrt{X}I)$. Then [59]

$$\begin{aligned} Z(\omega) = \frac{4\pi\omega}{c^2 ib}\{&0{,}8660\,(\pi\xi)^{-1/3} + 0{,}1013\,(\pi\xi)^{-2/3}\ln\pi\xi + (0{,}400 - 0{,}729Y)(\pi\xi)^{-2/3} + \\ &+ 0{,}0119\,(\pi\xi)^{-1}\ln^2(\pi\xi) + (0{,}0935 + 0{,}0416Y)(\pi\xi)^{-1}\ln\pi\xi + \\ &+ (0{,}1625 + 0{,}3562Y - 0{,}2010Y^2)(\pi\xi)^{-1}\}. \end{aligned} \tag{3.68}$$

Formula (3.68) yields the desired asymptotic representation for the impedance of an electron liquid with $A_0 \neq 0$ in the case of electron diffuse scattering.

We will now address the asymptotics of expression (3.43) for which we may write the following formula ignoring the displacement current

$$\begin{aligned} T = \frac{E(0)}{2\pi L}\frac{c}{\omega_{Le}}&\left\{\sqrt{\frac{A_1}{1+A_1}\frac{\omega+i\nu}{\omega}}\,\mathrm{tg}[\sqrt{X}I(X,\,Y)] - \right. \\ &\left. - i\sqrt{\frac{\omega-\Omega+i\nu}{\omega}}\sec[\sqrt{X}I(X,\,Y)]\right\}. \end{aligned} \tag{3.69}$$

The asymptotic expansion of the quantity $I(X, Y)$ given by formula (3.63) makes it possible to determine T in the form of the following expansion:

$$\begin{aligned} T = \frac{E(0)}{2\pi L}\frac{c}{\omega_{Le}}&\left\{\frac{A_1}{1+A_1}\frac{\omega+i\nu}{\sqrt{(\omega-\Omega+i\nu)\,\omega}}\frac{\sqrt{3}}{2}\left[\frac{2\pi^{1/3}}{\sqrt{3}\xi^{1/6}} - \right.\right. \\ &\left. - \frac{4\ln\pi\xi}{3\pi\xi^{1/2}} - \frac{0{,}534 - \frac{1}{6}Y}{\xi^{1/2}} + \cdots\right] - i\sqrt{\frac{\omega-\Omega+i\nu}{\omega+i\nu}}\times \\ &\times\left.\left[1+\frac{1}{2}\frac{A_1}{1+A_1}\frac{\omega+i\nu}{\omega-\Omega+i\nu}\frac{\pi^{2/3}}{\xi^{1/3}}+\cdots\right]\right\}. \end{aligned} \tag{3.70}$$

We note that expression (3.63) makes it possible to also find subsequent terms of expansion (3.70).

3.3. Comparison of results from impedance and field calculations to experiments on cyclotron phase resonance

Study [32] discovered resonance enhancement of a signal passed through a sufficiently thick (greater than the skin-depth) potassium slab. The authors of study [32] have demonstrated that the peak of this signal corresponds to the frequency at which the transverse conductivity $\sigma^{tr}(\omega, k)$ of the electron liquid when $k = 0$ has a resonance singularity. In the model $A_1 \neq 0$, as indicated by (1.40) the transverse conductivity when $k = 0$ is proportional to $\left(1-\frac{A_1}{1+A_1}\frac{\omega+i\nu}{\omega-\Omega+i\nu}\right)^{-1}$, and, consequently, the resonance value of the frequency is $\omega = \Omega(1 + A_1)$.

In order to identify the role of Fermi-liquid interaction in the frequency dependence of the impedance and the field passed through the slab, we will write formulae (3.68) and (3.70) in the form of study [59] which explicitly demonstrates their dependence on the magnetic field $\left(\Omega=\frac{|e|B}{mc}\right)$:

$$\begin{aligned} Z(\omega) = {} & \frac{4\pi v}{c^2}\Big\{0{,}8660\,(\pi\xi_0)^{-1/3} + 0{,}1013\,(\pi\xi_0)^{-2/3}\frac{\omega-\Omega+i\nu}{\omega}\times \\ & \times\left[\ln(\pi\xi_0) - 3\ln\frac{\omega-\Omega+i\nu}{\omega}\right] + (\pi\xi_0)^{-2/3}\times \\ & \times\left[0{,}400\,\frac{\omega-\Omega+i\nu}{\omega} - 1{,}095\frac{A_1}{1+A_1}\frac{\omega+i\nu}{\omega}\right] + 0{,}0119\,(\pi\xi_0)^{-1}\times \\ & \times\left(\frac{\omega-\Omega+i\nu}{\omega}\right)^2\left[\ln(\pi\xi_0) - 3\ln\left(\frac{\omega-\Omega+i\nu}{\omega}\right)\right]^2 + \\ & + \left[0{,}0937\frac{\omega-\Omega+i\nu}{\omega} + 0{,}0662\frac{A_1}{1+A_1}\frac{\omega+i\nu}{\omega}\right]\times \\ & \times(\pi\xi_0)^{-1}\frac{\omega-\Omega+i\nu}{\omega}\left[\ln(\pi\xi_0) - 3\ln\frac{\omega-\Omega+i\nu}{\omega}\right] + \\ & + (\pi\xi_0)^{-1}\left[0{,}1620\left(\frac{\omega-\Omega+i\nu}{\omega}\right)^2 + 0{,}5346\frac{A_1}{1+A_1}\times \right. \\ & \left.\times\frac{(\omega+i\nu)(\omega-\Omega+i\nu)}{\omega^2} - 0{,}4525\left(\frac{A_1}{1+A_1}\right)^2\left(\frac{\omega+i\nu)}{\omega}\right)^2\right]\Big\}, \end{aligned} \qquad (3.71)$$

$$\begin{aligned} T(\omega) = {} & \frac{E(0)}{\pi L}\frac{c}{\omega_{Le}}\Big\{-i\sqrt{\frac{\omega-\Omega+i\nu}{\omega}}\left(1+\frac{A_1}{1+A_1}\frac{\pi^{2/3}}{2\xi_0^{1/3}}\frac{\omega+i\nu}{\omega}\right) + \\ & + \frac{A_1}{1+A_1}\frac{\omega+i\nu}{\omega}\left[\pi^{1/3}\xi_0^{-1/6} - \frac{\sqrt{3}}{2}\xi_0^{-1/2}\frac{\omega-\Omega+i\nu}{\omega}\times \right. \\ & \times\left(\frac{4}{3\pi^2}\ln(\pi\xi_0) - \frac{4}{\pi^2}\ln\frac{\omega-\Omega+i\nu}{\omega} - 0{,}534\right) + \\ & \left. + \frac{\sqrt{3}}{8}\xi_0^{-1/2}\frac{A_1}{1+A_1}\frac{\omega+i\nu}{\omega}\right]\Big\}. \end{aligned} \qquad (3.72)$$

The first term of the expansion of the surface impedance (3.71) in powers of the anomaly parameter $\sim U_0^{-2/3}$ is independent of the magnetic field and coincides with the term obtained in study [29] for the electron gas model (compare to study [55] as well). The higher order terms $\sim U_0^{-2/3}$ and $\sim\xi_0^{-1}$ for $A_1 = 0$ also coincide with the corresponding terms of study [29] and diminish resonantly when $\omega \to \Omega$. A correction $\sim U_0^{-2/3}$ contains the Fermi-liquid interaction effects which is manifest in the appearance of a magnetic field-independent component. Thus, as in the case of the theory utilizing the representation of specular electron scattering (see formula (2.60) for $A_2 = 0$) incorporating the Fermi-liquid constant A_1 does not produce resonance dependencies of the impedance.

The Fermi-liquid effects in the expression for $T(\omega)$ determine the two correction terms of the expansion $\sim\xi_0^{-1/6}$ and $\xi^{-1/3}$ that do not arise in the electron gas model. At the same time as indicated by formula (3.69) the primary component in the electron gas model arises from the expansion of the secant, while the Fermi-liquid correction $\sim\xi_0^{-1/6}$ arises from the expansion of the tangent there is, consequently, some promise when they are comparable in magnitude. In the regime of experiment [32] this possibility may not exist when

$$\left|\frac{A_1}{1+A_1}\right| |\xi_0|^{-1/6} \sim \sqrt{\left|\frac{\omega-\Omega}{\omega}\right|} > \sqrt{\frac{\nu}{\omega}}. \tag{3.73}$$

However the Fermi-liquid corrections in T related to the constant A_1 are not resonance corrections which makes it impossible to interpret the experimentally observed frequency dependence of the signal assuming diffuse electron scattering [32] as resulting from interelectron interaction.

Conclusion

1. In the electron fluid model with a single parameter $A_1 \neq 0$ and diffuse electron boundary scattering the equation for the electric field is solved by means of the Wiener-Hopf method, which made it possible to obtain an analytic expression for the surface impedance.

2. The representation of the impedance as an asymptotic series in terms of the anomaly parameter and the investigation of the frequency dependence of the terms of this series leads us to the conclusion that, as in the case of specular reflection, accounting for the constant A_1 does not produce resonance dependencies, but rather produces small nonresonance corrections.

3. Nonresonance corrections attributable to the constant A_1. arise in the expression for the electric field that has passed through a thick slab, assuming proportionality to $E(k = 0)$; these corrections in certain experimental regimes may be comparable in magnitude to the principal term obtained in the electron gas model.

Chapter 4

INVESTIGATION OF THE RESONANCE SINGULARITIES OF THE SURFACE IMPEDANCE OF A METAL IN A MODEL WITH A NONZERO PARAMETER A_2 WITH DIFFUSE ELECTRON BOUNDARY SCATTERING

In the preceding chapter the theory of the surface impedance of the electron liquid of a metal for the case of diffuse electron scattering was formulated in a model for which only the parameter A_1 was assumed to be nonzero. The necessity for formulating such a theory was due to the appearance of the A_1 parameter in experimentally-observed resonance dependencies of the impedance and of the field passed through the metal slab. In this chapter we will consider an electron liquid model in which only the constant A_2 is nonzero. Such a model, on the one hand, is sufficiently simple for solving a system of two integral equations (unlike a model with two nonzero parameters A_1 and A_2) and on the other hand allows the possibility for the existence of the branch of a Fermi-liquid cyclotron wave propagating along the magnetic field. Thus, the singularities of the impedance in the vicinity of cyclotron and cyclotron-wave resonances may be analyzed for the case of diffuse scattering in this model as well.

4.1. Solution of a system of two Wiener-Hopf equations for the electric field

Initial system of integral equations (1.22) for the moments of the distribution function $\Phi_{\pm}^{1,l}(z) = \Phi_l(z)$ in Maxwell's equation (2.8) takes the form [60]:

$$E''(z) = -\frac{3i\omega}{4c^2}\,\omega_{\mathrm{Le}}^2\Phi_1(z), \tag{4.1}$$

$$\Phi_1(z) = \frac{i}{\omega + i\nu}\int_0^\infty dz'\, Q_{11}(z-z')\,E(z') + \frac{5}{6}A_2\int_0^\infty dz'\,Q_{22}(z-z')\,\Phi_2(z'), \tag{4.2}$$

$$\Phi_2(z)(1 + A_2) = \frac{i}{\omega + i\nu}\int_0^\infty dz'\,Q_{12}(z-z')\,E(z') + \frac{5}{6}A_2\int_0^\infty dz'\,Q_{22}(z-z')\,\Phi_2(z'). \tag{4.3}$$

Here in accordance with (1.23)

$$Q_{nl}(z) = \frac{a}{2}\left(\frac{z}{|z|}\right)^{n+l}\int_0^1 dx\,P_n^1(x)\,P_l^1(x)\,\frac{1}{x}\exp\left(-\frac{|z|\,b}{x}\right), \tag{4.4}$$

where $a = (-i\omega + \nu)\,v^{-1}$, $b = [-i(\omega - \Omega) + \nu]\,v^{-1}$. Eliminating the function Φ_1 by means of (4.1) and incorporating the conventions

$$\Phi(z) = (\nu - i\omega)\,\Phi_2(z),\quad C = \frac{2c^2(\omega + i\nu)}{3\omega_{\mathrm{Le}}^2\omega}, \tag{4.5}$$

we obtain the following system of two integrodifferential Wiener-Hopf equations:

$$CE''(z) - \frac{5}{6} A_2 \int_0^\infty dz'\, Q_{12}(z-z')\,\Phi(z') = \int_0^\infty dz'\, Q_{11}(z-z')\,E(z'), \tag{4.6}$$

$$(1 + A_2)\,\Phi(z) - \frac{5}{6} A_2 \int_0^\infty dz'\, Q_{22}(z-z')\,\Phi(z') = \int_0^\infty dz'\, Q_{12}(z-z')\,E(z'). \tag{4.7}$$

It follows from (4.4) that the kernel of equation system (4.6) and (4.7) has the properties

$$Q'_{22}(z) = -3bQ_{12}(z), \qquad Q'_{12}(z) = 3bQ_{11}(z) + 2a\delta(z). \tag{4.8}$$

Bearing in mind formula (4.8) we may easily obtain the following differential equation following from (4.6) and (4.7):

$$CE''(z) - \frac{2a}{3b} E(z) + \frac{1 + A_2}{3b} \Phi'(z) = 0. \tag{4.9}$$

Substituting the solution of this equation

$$\Phi(z) = -\frac{3bC}{1 + A_2} E'(z) + \frac{2a}{1 + A_2} \int_\infty^z dz'\, E(z') \tag{4.10}$$

into (4.7) and introducing the conventions

$$\psi = \frac{15}{2} \frac{A_2 b^2 C}{1 + A_2}, \qquad k_0^2 = -\frac{5A_2 ab}{(1 + A_2)(1 + \psi)}, \tag{4.11}$$

we arrive at the following integrodifferential equation of the field

$$CE''(z) - C(1+\psi)k_0^2 E(z) - (1+\psi)\int_0^\infty dz'\, E(z') \times$$
$$\times \left\{ Q_{11}(z-z') + \frac{k_0^2}{9b^2} Q_{22}(z-z') \right\} = g(z), \tag{4.12}$$

where

$$g(z) = (\psi/3b)\,Q_{12}(z)\,E(0) - \frac{k_0^2(1+\psi)}{9b^2} Q_{22}(z) \int_0^\infty dz'\, E(z'), \tag{4.13}$$

($E(0)$ and $E'(0)$ are the values of the field and its derivative when $z = 0$).

The integral operator of the left half of equation (4.12) is the difference operator. On the other hand, the operator of the right half corresponds to the degenerate integral equation. Hence in searching the solution of equation (4.12) we may use methods of solving an inhomogeneous Hopf-Wiener equation [58]. We emphasize that such inhomogeneity is solely the result of Fermi-liquid interaction.

Bearing in mind that equation (4.12) is valid when $z \geqslant 0$, we will take $E(z) = 0$ with $z < 0$. For equation (4.12) to be valid on the entire axis $z(-\infty < z < \infty)$, we extend the definition of the function

$$g(z) = -(1+\psi)\int_0^\infty dz' E(z')\left\{Q_{11}(z-z') + \frac{k_0^2}{9b^2} Q_{22}(z-z')\right\} \text{ when } z < 0 \quad (4.14)$$

The solution may then be obtained by means of a Fourier transform

$$E(k) = \int_0^\infty dz\, e^{-ikz} E(z), \quad (4.15)$$

$$F(k) = \frac{1}{C}\int_{-\infty}^0 dz\, e^{-ikz} g(z), \quad (4.16)$$

$$q_{nl}(k) = \int_{-\infty}^\infty dz\, e^{-ikz} Q_{nl}(z). \quad (4.17)$$

Here from (4.12) for the Fourier transforms we obtain

$$-E(k)(k^2+\delta^2)t(k) = E'(0) + ikE(0) + F(k) + \\ + \frac{k_0^2(1+\psi)ikE(0) + D}{(1+\psi)(k^2-k_0^2)}\left\{1 - \frac{\psi k^2}{k_0^2(1+\psi)} + \frac{\psi(k^2+\delta^2)}{k_0^2(1+\psi)} t(k)\right\}, \quad (4.18)$$

where

$$t(k) = \frac{1}{k^2+\delta^2}\left\{k^2 + k_0^2(1+\psi) - \frac{k_0^4(1+\psi)^2}{k^2\psi} - \frac{3k_0^2(1+\psi)^2}{2\psi}\left(1 - \frac{k_0^2}{k^2}\right) q\left(\frac{k}{b}\right)\right\}, \quad (4.19)$$

$$q_{11}(k) = \frac{a}{b} q\left(\frac{k}{b}\right),\ q(x) = -\frac{1}{x^2} + \frac{1+x^2}{x^3}\operatorname{arctg} x, \quad (4.20)$$

$$D = \frac{k_0^2(1+\psi)^2}{\psi}\frac{1}{2\pi i}\int_{-\infty+i\varkappa}^{\infty+i\varkappa} \frac{dk}{k} E(k). \quad (4.21)$$

Since the function $E(k)$ is regular for Im $k < \beta$, integration over k in formula (4.21) is taken on the contour located in the upper half-plane of the complex variable ($\varkappa < \beta$), so

$$\int_0^\infty dz\, E(z) = -\frac{1}{2\pi i}\int_{-\infty+i\varkappa}^{\infty+i\varkappa} dk \frac{E(k)}{k}. \quad (4.22)$$

As determined by (4.20) the function $q_{11}(k)$ is regular for $-\alpha <$ Im $k < \alpha$, where $0 < \alpha < \nu/\upsilon$, i.e., at the pole including the real axis. Consequently, when $0 < \beta < \delta < \alpha$ the function (4.19) will be regular and will not have zeros at the pole $-\beta <$ Im $k < \beta$. They will subsequently vanish by virtue of the smallness of β and δ.

In accordance with the Wiener-Hopf method we will represent the function $t(k)$ as $(0 < \varkappa < \beta < \delta \to +0)$:

$$t(k) = \frac{t_+(k)}{t_-(k)},\ t_\pm(k) = \exp\left\{\frac{1}{2\pi i}\int_{-\infty\mp i\beta}^{\infty\mp i\beta} \frac{dk'}{k'-k}\ln t(k')\right\}, \quad (4.23)$$

where the function $t_+(k)$ is regular for Im $k > -\beta$ while $t^-(k)$ when Im $k < \beta$. We will transpose all terms that are regular in the upper half-plane to the right half of equation (4.18), and will transpose the terms that are regular in the lower half-plane to the left half.

Here by virtue of the overlapping of the regularity regions, the right and left halves of the equation determine the polynomial which, by virtue of the asymptotics of the Fourier-transform of the electric field

$$E(k) = -\frac{i}{k}E(0) - \frac{1}{k^2}E'(0) \qquad \text{when } k \to \infty \tag{4.24}$$

is equal to a constant. As a result to $E(k)$ we obtain

$$E(k) = -\frac{\psi[k_0^2(1+\psi)ikE(0)+D]}{(1+\psi)^2k_0^2(k^2-k_0^2)} -$$
$$-\frac{iE(0)t_-(k)}{(1+\psi)(k-i\delta)}\left[1+\frac{k_0^2}{2(k_0+i\delta)(k-k_0)t_+(k_0)}+\frac{\psi(k_0+i\delta)}{2(k+k_0)t_-(-k_0)}\right] -$$
$$-\frac{Dt_-(k)}{k_0(1+\psi)^2(k-i\delta)}\left[\frac{1}{2(k_0+i\delta)(k-k_0)t_+(k_0)}-\frac{\psi(k_0+i\delta)}{2k_0^2(k+k_0)t_-(-k_0)}\right]. \tag{4.25}$$

Here we have agreed that $\mathrm{Im}\, k_0 > 0$. Substituting this expression into formula (4.21) and using the values of the integrals obtained from the shifting downwards of the integration contours and incorporation of the poles, we obtain the following equation for the quantity D:

$$D = iE(0)k_0^3(1+\psi)\left[1-\frac{k_0}{2(k_0+i\delta)t_+(k_0)}+\frac{\psi(k_0+i\delta)}{2k_0t_-(-k_0)}\right] \times$$
$$\times\left[\frac{k_0}{2(k_0+i\delta)t_+(k_0)}+\frac{\psi(k_0+i\delta)}{2k_0t_-(-k_0)}\right]^{-1}. \tag{4.26}$$

The expression for $E(k)$ (4.25) in conjunction with (4.26) and the formula of the inverse Fourier transform

$$E(z) = \frac{1}{2\pi}\int_{-\infty+i\varkappa}^{\infty+i\varkappa} dk\, e^{ikz}E(k) \tag{4.27}$$

is the solution of equation system (4.6) and (4.7) for the electric field in a metal occupying the region $z > 0$ for diffuse electron scattering at the boundary $z = 0$ [60].

4.2. Expression for the surface impedance and its asymptotic representation in the anomalous skin-effect regime

In order to determine the surface impedance of the metal $Z(\omega) = (4\pi\omega/c^2)\cdot iE(0)/E'(0)$ we will use formula (4.24). Substituting expression (4.25) into this formula and using (4.26), we obtain

$$\frac{E'(0)}{iE(0)} = \frac{1}{1+\psi}\left\{i\xi + k_0\left[2\psi+\frac{1-\psi}{2t_+(k_0)}-\frac{\psi}{2}(1-\psi)t_+(k_0)\right]\left[\frac{1}{2t_+(k_0)}+\right.\right.$$
$$\left.\left.+\frac{\psi}{2}t_+(k_0)\right]^{-1}\right\}. \tag{4.28}$$

Here we have passed to the limit $\delta \to 0$, accounting for $t^-(-k_0) = [t_+(k_0)]^{-1}$ and introduce the conventions

$$i\zeta = \lim_{k\to\infty} k\,[t_+(k) - 1] = \frac{i}{2\pi}\int_{-\infty}^{\infty} dk \ln t(k). \tag{4.29}$$

It is obvious that the surface impedance is determined by two quantities: ζ and $t_+(k_0)$, which are functions of the frequency and the magnetic field. For these functions in the anomalous skin-effect mode when the quantity (2.14) $\xi_0 = -3\omega_{Le}^2 v^2/(4\omega^2 c^2)$ is large in absolute value we obtain the asymptotic expansion.

In formulating the expansions we use the parameters:

$$\xi = \xi_0\left(\frac{\omega}{\omega - \Omega + i\nu}\right)^3, \quad Y_2 = \frac{10}{3}\frac{A_2}{1 + A_2}\frac{\omega + i\nu}{\omega - \Omega + i\nu}, \tag{4.30}$$

that may easily be used on place of

$$\psi = \frac{9Y_2}{8\xi}, \quad \left(\frac{k_0}{b}\right)^2 = -\frac{3}{2}Y_2\left(1 + \frac{9Y_2}{8\xi}\right)^{-1}. \tag{4.31}$$

Expression (4.28), with ~ $1/\xi$ term accuracy of the initial functions, takes the form

$$\frac{iE(0)}{E'(0)} = \left\{i\zeta + k_0\frac{8\xi - 9Y_2 t_+^2(k_0)}{8\xi + 9Y_2 t_+^2(k_0)}\right\}^{-1},$$

where

$$\zeta = \frac{b}{\pi}\int_0^{\infty/b} dx \ln t(x), \tag{4.32}$$

$$\ln t_+(k_0) = \frac{k_0}{b\pi i}\int_0^{\infty/b} \frac{dx \ln t(x)}{x^2 + k_0^2/b^2}, \tag{4.33}$$

$$t(x) = 1 + \frac{2\xi\,[\varphi(x) - Y_2] - \frac{3}{2}Y_2 x^2}{x^2\left[x^2 + \frac{3}{2}\varphi(x)\right]}. \tag{4.34}$$

The determination of the function $\varphi(x)$ is given by formulae (2.27) and (2.17).

We will first obtain an asymptotic representation of the quantity ≃ (4.33). Using integration by parts for ζ, we have

$$\zeta = -\frac{b}{\pi}(I_1 + I_2), \tag{4.35}$$

where

$$I_1 = \int_0^{|b|/b} dx\, x\frac{d}{dx}\ln t(x), \quad I_2 = \int_{|b|/b}^{\infty/b} dx\, x\frac{d}{dx}\ln t(x). \tag{4.36}$$

The integrand in I_1 may be expanded into a series in terms of $1/\xi$ and integrated termwise

$$I_1 = -2\frac{|b|}{b} + \int_0^{|b|/b} dx\, x\frac{d}{dx}\Phi_1(x, Y_2) + O(1/\xi), \tag{4.37}$$

where

$$\Phi_1(x,\ Y_2)=\frac{\varphi(x)-Y_2}{x^2+\frac{3}{2}\varphi(x)}. \tag{4.38}$$

We obtain $I_2 = \xi I(\xi)$ for I_2 with $\sim 1/\xi$ term accuracy,

$$I(\xi)=2\int\limits_{|b|/b}^{\infty/b} dx\, x\frac{d}{dx}\frac{\Phi_1(x,\ Y_2)}{x^2}\left\{\left(\frac{2\xi}{x^2}+\frac{3}{2}\right)\Phi_1(x,\ Y_2)+\frac{x^2}{x^2+\frac{3}{2}\varphi(x)}\right\}^{-1}. \tag{4.39}$$

we will use the Mellin transform for the asymptotic expansion of $I(\xi)$:

$$I(\xi)=\int\limits_{C-i\infty}^{C+i\infty}\frac{dz}{2\pi i}|\xi|^{-z}M(z),\quad M(z)=\int\limits_0^\infty d|\xi||\xi|^{z-1}I(\xi). \tag{4.40}$$

For the quantity $M(z)$ we have

$$M(z)=e^{-i\arg\xi}\int\limits_{|b|/b}^{\infty/b} dx\,\frac{x^3}{\Phi_1}\left(\frac{d}{dx}\frac{\Phi_1}{x^2}\right)\int\limits_0^\infty d|\xi||\xi|^{z-1}\times$$
$$\times\left\{|\xi|+e^{-i\arg\xi}\frac{x^2}{2}\left[\frac{3}{2}+\frac{x^2}{\left(x^2+\frac{3}{2}\varphi\right)\Phi_1}\right]\right\}^{-1}. \tag{4.41}$$

The integral over $|\xi|$ is a tabular integral (see study [45], (3.224)) and is evaluated with satisfaction of the inequality

$$\left|\arg\left\{\frac{x^2}{\xi}\left[\frac{3}{2}+\frac{x^2}{\varphi(x)-Y_2}\right]\right\}\right|<\pi,\ 0<\operatorname{Re}z<1. \tag{4.42}$$

In order to take the integral over x we use the formal expansion

$$\frac{x^2}{\Phi_1}\left(\frac{d}{dx}\frac{\Phi_1}{x^2}\right)\left\{\frac{x^4}{\varphi(x)-Y_2}+\frac{3}{2}x^2\right\}^{z-1}=-\frac{3\pi}{2x^4}\left(\frac{2x^3}{\pi}\right)\sum_{\mu=0}^{\infty}b_\mu(z)\,x^{-\mu}, \tag{4.43}$$

whose substitution into the formula for $M(z)$ yields

$$M(z)=\frac{3\pi^2 e^{-iz\arg\xi}}{\pi^2\sin\pi z}\sum_{\mu=0}^{\infty}\frac{b_\mu(z)}{3z-\mu-2}e^{-i(3z-\mu-2)\arg b}. \tag{4.44}$$

Using this expression to determine $I(\xi)$ by means of the Mellin transform, we will shift the contour of integration with respect to $z(1/2 < \operatorname{Re}z < 1)$ to the right half-plane. The pole $M(z)$ will produce powers of the parameter ξ. The higher order term of the expansion arises from the pole $z = 2/3$, $\mu = 0$:

$$-\pi^2(\pi|\xi|)^{-2/3}\exp\left\{-\frac{2}{3}i\arg\xi\right\}\operatorname{cosec}\frac{2\pi}{3}. \tag{4.45}$$

The simple pole $z = 1$, $\mu \neq 1$ makes the contribution

$$\frac{3}{|\xi|}e^{-i\arg\xi}\sum_{\mu\neq 1}^{\infty}\frac{b_\mu(1)}{\sin\pi z}e^{-i(1-\mu)\arg b}=$$

$$
=\frac{1}{|\xi|}e^{-i\arg\xi}\left\{\int\limits_{|b|/b}^{\infty/b}dx\left[\frac{x^3}{\Phi_1}\frac{d}{dx}\frac{\Phi_1}{x^2}+3+3\frac{b_1(1)}{x}\right]+3\frac{|b|}{b}\right\}. \tag{4.46}
$$

The contribution of the double pole $z = 1$, $\mu = 1$

$$
-\pi^2\frac{d}{dz}\left\{(|\pi\xi|)^{-z}\,\frac{b_1(z)}{\sin\pi z}(z-1)\,e^{-iz\arg\xi-i(3z-3)\arg b}\right\}_{z=1}=
$$

$$
=\frac{1}{|\xi|}e^{-i\arg\xi}\,\{b_1'(1)-b_1(1)\ln(\pi|\xi|)-b_1(1)\,i\arg(\xi b^3)\}. \tag{4.47}
$$

The pole $z = 4.3$, $\mu = 2$ yields $\sim(\pi|\xi|)^{-4/3}$ which we will ignore, together with all higher order powers of the expansion. Accounting for

$$
b_0(z)=1,\ b_1(1)=-\frac{4}{3\pi}\left(1+\frac{Y_2}{2}\right),\ b_1'(1)=\frac{4}{\pi}\left(1+\frac{Y_2}{2}\right) \tag{4.48}
$$

and summing all contributions, we obtain

$$
\begin{aligned}
\zeta/b=&\frac{2}{\sqrt{3}}(\pi|\xi|)^{1/3}e^{i/3\arg\xi}-\frac{4}{\pi^2}\left(1+\frac{Y_2}{2}\right)-\frac{4}{3\pi^2}\left(1+\frac{Y_2}{2}\right)[\ln(\pi|\xi|)+i\arg(\xi b^3)]-\\
&-\frac{1}{\pi}\left\{\frac{|b|}{b}+\int\limits_0^{|b|/b}dx\,x\frac{q'(x)}{q(x)}+\int\limits_{|b|/b}^{\infty/b}dx\left[x\frac{q'(x)}{q(x)}+1-\frac{4}{\pi x}\right]+\right.\\
&\left.+Y_2\left(\int\limits_0^{|b|/b}dx\,x\frac{\varphi'(x)}{\varphi^2(x)}+\int\limits_{|b|/b}^{\infty/b}dx\left[x\frac{\varphi'(x)}{\varphi^2(x)}-\frac{2}{\pi x}\right]\right)+Y_2^2\int\limits_0^{\infty/b}\frac{dx\,x\varphi'(x)}{\varphi^2(x)[\varphi(x)-Y_2]}\right\}.
\end{aligned} \tag{4.49}
$$

We note that by shifting the integration contours with the real axis we may make all integrals entering into this formula, with the exception of the last integral, independent of b, and using the tabulation results:

$$
\begin{aligned}
&\int\limits_0^1 dt\,t\frac{q'(t)}{q(t)}=-0{,}10, &&\int\limits_1^{\infty}dt\left[t\frac{q'(t)}{q(t)}+1-\frac{4}{\pi t}\right]=-0{,}50,\\
&\int\limits_0^1 dt\,t\,\frac{\varphi'(t)}{\varphi^2(t)}=0{,}03, &&\int\limits_1^{\infty}dt\left[t\frac{\varphi'(t)}{\varphi^2(t)}-\frac{2}{\pi t}\right]=-0{,}90,
\end{aligned} \tag{4.50}
$$

arrive at the following expression:

$$
\begin{aligned}
\zeta/b=&\frac{2}{\sqrt{3}}(\pi|\xi|)^{1/3}e^{\frac{i}{3}\arg\xi}-\frac{4}{3\pi^2}\left(1+\frac{Y_2}{2}\right)[\ln(\pi|\xi|)+i\arg\xi+3]-\\
&-\frac{1}{\pi}[0{,}40-0{,}87Y_2+Y_2^2F_0(Y_2)],
\end{aligned} \tag{4.51}
$$

where

$$
F_0=\int\limits_0^{\infty/b}\frac{dx\,x\varphi'(x)}{\varphi^2(x)[\varphi(x)-Y_2]}. \tag{4.52}
$$

The argument of the parameter ξ which may vary by a quantity that is an even multiple of 2π is still undetermined in formula (4.51). It is possible to eliminate this uncertainty of the branch of the logarithm and the root, following [29, 31] in accordance with the limiting process

$$\arg \xi \to -\frac{\pi}{2} \quad \text{when} \quad \frac{\omega - \Omega}{\nu} \to 0. \tag{4.53}$$

We may examine the opposite limiting case $\nu \to 0$ in the same manner as in section 2.2 by shifting the contour in integral (4.33) to the real axis (Fig. 7, 8). We may obtain the result for the quantity ζ analogous to the process of deriving formula (2.55):

$$\zeta/b = \frac{2}{\sqrt{3}}\left(\frac{1}{2} + i\frac{\sqrt{3}}{2}\right)\frac{\omega}{\omega - \Omega}(-\pi\xi_0)^{1/3} - \frac{4}{3\pi^2}\left(1 + \frac{Y_2}{2}\right) \times$$
$$\times \left[\ln(-\pi\xi_0) + 3\ln\frac{\omega}{|\omega - \Omega|} + 3 + i\pi\left(1 - 3\theta(\Omega - \omega)\right)\right] -$$
$$-\frac{1}{\pi}[0{,}40 - 0{,}87Y_2 + Y_2^2 F_0(Y_2)]. \tag{4.54}$$

The function $F_0(Y_2)$ with real values of the argument ($\nu \to 0$) may be written by shifting the integration contour to the real axis as

$$F_0(Y_2) = \int_0^\infty \frac{dx\, x\varphi'(x)}{\varphi^2(x)[\varphi(x) - Y_2]} - \frac{i\pi}{Y_2^2} x_\varphi(Y_2)\operatorname{sgn}(\omega - \Omega), \tag{4.55}$$

where $x_\varphi(Y_2)$ is the root of the equation $\varphi(x) = Y_2$.

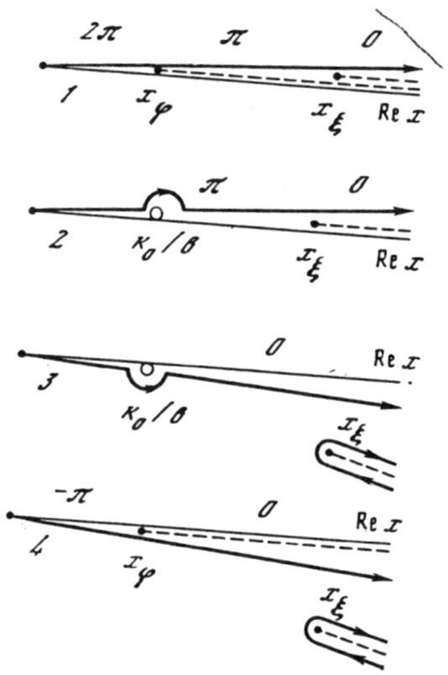

Fig. 10. Argument of the logarithmic function in (4.34) after shifting of the integration contour to the real axis
1 - when $\omega > \Omega$, $A_2 > 0$; 2 when $\omega > \Omega$, $A_2 < 0$; 3 - when $\omega < \Omega$, $A_2 > 0$; 4 when $\omega < \Omega$, $A_2 < 0$. The arrows indicate the integration path, while the dotted lines represent the branch lines

The calculation of the quantity $t_+(k_0)$ in (4.34) is somewhat more complex in the limit $\nu \to 0$. We will rotate the integration contour in integral (4.3) until it coincides with the positive real semiaxis. In this case for $\omega > \Omega$ we will rotate in a clockwise direction, and when $\omega < \Omega$ we will rotate in a counterclockwise direction. Accounting for the inequality $\sqrt{-Y_2}$ in the determination of the root ($\nu \ll |\omega - \Omega|$), we obtain for the pole of the integrand

$$\sqrt{-\frac{3}{2}Y_2} = \pm\frac{k_0}{b} =$$
$$= \pm\sqrt{\left|\frac{3}{2}Y_2\right|}\{i\theta(Y_2) - \theta(-Y_2)\operatorname{sgn}(\omega - \Omega)\}\left[1 - \frac{i\nu}{2(\omega - \Omega)}\right]. \tag{4.56}$$

The other singularities will be the branch points x_ξ and x_φ ($|x_\xi| \gg |x\varphi|$). The following expression (compare to (2.24)) is valid for the quantity x_φ:

$$x_\xi = |\pi\xi_0|^{1/3} \frac{\omega}{\omega-\Omega}\left\{1-\frac{iv}{\omega-\Omega}\right\} e^{\frac{2\pi i}{3}n}, \quad n=0, \pm 1. \tag{4.57}$$

With this shifting of the contour the intersection of the point $x_\xi(n = 1)$ when $\omega < \Omega$ produces an integral over the banks of the branch line equal within an order of magnitude to $\sim|\xi_0|^{-1/3}$. We may ignore such small quantities, and only incorporate the expressions that do not decay with growth of ξ. Then after rotation of the contour we obtain

$$\ln t_+(k_0) = \frac{1}{2}\theta(-Y_2)\ln t\left(\sqrt{\left|\frac{3}{2}Y_2\right|}\right) + \frac{1}{2\pi i}\int_0^\infty dx\left(\frac{1}{x-k_0/b} - \frac{1}{x+k_0/b}\right)\ln t(x). \tag{4.58}$$

Substituting the complex function of the real variable In $t(x)$ in the form $\ln|t(x)| + i \arg t(x)$ and accounting for the fact that (Fig. 10) when $x \to \infty$ $t(x) \to 0$ and

$$\arg t(x) = \pi\{\theta(\omega-\Omega)\theta(|\pi\xi|^{1/3} - x) + \operatorname{sgn}(\omega-\Omega)\theta(Y_2 - \varphi(x))\}, \tag{4.59}$$

we represent formula (4.58) as

$$\ln t_+(k_0) = \frac{1}{2}\theta(-Y_2)\ln\left|\frac{8\xi}{9Y_2}\right| + \frac{i\pi}{2}\theta(\omega-\Omega) +$$
$$+ \frac{k_0}{b}\frac{1}{i\pi}\int_0^\infty \frac{dx}{x^2-(k_0/b)^2}\left\{i\pi\operatorname{sgn}(\omega-\Omega)\theta(Y_2-\varphi(x)) +\right.$$
$$\left. + \ln\left|1-\frac{3Y_2}{2x^2}\left(1-\frac{3}{2}q(x)\right) - \frac{2\xi Y_2}{x^4}\left[1-\left(\frac{3}{2}+\frac{x^2}{Y_2}\right)q(x)\right]\right|\right\}. \tag{4.60}$$

Then, accounting for

$$\frac{1}{\pi}\int_0^\infty \frac{dx}{x^2+\frac{3}{2}Y_2}\ln\left|\frac{2\xi Y_2}{x^4}\right| = \frac{1}{\sqrt{\left|\frac{3}{2}Y_2\right|}}\left\{\frac{1}{2}\theta(Y_2)\ln\left|\frac{8\xi}{9Y_2}\right| - \pi\theta(-Y_2)\right\}, \tag{4.61}$$

we write (4.60) as

$$\ln t_+(k_0) = \frac{1}{2}\ln\left|\frac{8\xi}{9Y_2}\right| + \frac{i\pi}{2}[\theta(\omega-\Omega) - 2\operatorname{sign}(\omega-\Omega)\theta(-Y_2) +$$
$$+ \operatorname{sgn}(\omega-\Omega)\theta(Y_2)] - \frac{ik_0}{b}F_{10}\left(\frac{3}{2}Y_2\right) - \frac{ik_0}{b}I, \tag{4.62}$$

where

$$F_{10}\left(\frac{3}{2}Y_2\right) = \frac{1}{\pi}\int_0^\infty \frac{dx}{x^2+\frac{3}{2}Y_2}\left\{\ln\left|1-\frac{3}{2}q(x)\left(1+\frac{2x^2}{3Y_2}\right)\right| -\right.$$
$$\left. - i\pi\theta(Y_2)\operatorname{sgn}(\omega-\Omega)\theta(x-x_\varphi)\right\}, \tag{4.63}$$

$$I = \frac{1}{\pi}\int_0^\infty \frac{dx}{x^2+\frac{3}{2}Y_2}\ln\left|1+\frac{x^2}{\xi\varkappa(x, Y_2)}\right|. \tag{4.64}$$

The function $\varkappa(x, Y_2)$ is determined in accordance with (2.15) when $A_1 = 0$

$$\varkappa(x,\ Y_2)=2\,\frac{q(x)-\frac{Y_2}{x^2}\left[1-\frac{3}{2}q(x)\right]}{1-\frac{3}{2}\frac{Y_2}{x^2}\left[1-\frac{3}{2}q(x)\right]}. \tag{4.65}$$

Using the asymptotics of the function $\varkappa$ (2.19) we may easily demonstrate that the quantity I does not contain contributions that do not vanish when $|\xi| \gg 1$ and it may be dropped. Indeed,

$$I=\frac{1}{\pi}\int_0^\infty dx\int_0^{1/\xi} d\varepsilon\,\frac{1}{x^2+\frac{3}{2}Y_2}\,\frac{x^2}{\varepsilon x^2+\varkappa}=$$

$$=\frac{1}{\pi}\int_0^{1/\xi} d\varepsilon\left\{I_0(\varepsilon)+I_1(\varepsilon)-\frac{3}{2}Y_2I_2(\varepsilon)+\frac{9}{4}Y_2^2I_3(\varepsilon)\right\}, \tag{4.66}$$

where

$$I_0(\varepsilon)=\int_0^1\frac{dx\,x^2}{\left(x^2+\frac{3}{2}Y_2\right)(x^2\varepsilon+\varkappa)}\sim\varepsilon^0,$$

$$I_1(\varepsilon)=\int_1^\infty\frac{dx}{x^2\varepsilon+\varkappa}\sim\left|\frac{\pi}{\varepsilon}\right|^{1/2},$$

$$I_2(\varepsilon)=\int_1^\infty\frac{dx}{x^2(x^2\varepsilon+\varkappa)}\sim\frac{1}{3\pi}\ln\left|\frac{\pi}{\varepsilon}\right|,$$

$$I_3(\varepsilon)=\int_1^\infty\frac{dx}{x^2\left(x^2+\frac{3}{2}Y_2\right)(x^2\varepsilon+\varkappa)}\sim\varepsilon^0. \tag{4.67}$$

In this case we may write a simple formula for t_+

$$-\frac{9Y_2}{8\xi}\,t_+^2(k_0)=\exp\left[-2i\sqrt{-\frac{3}{2}Y_2}F_{10}\left(\frac{3}{2}Y_2\right)\right], \tag{4.68}$$

that makes it possible subject to (4.54) to obtain the desired asymptotic expansion for the surface impedance in the case of diffuse scattering

$$Z_g(\omega)=\frac{4\pi v}{c^2}\left\{\frac{\sqrt{3}}{2}\left(\frac{1}{2}-i\frac{\sqrt{3}}{2}\right)|\pi\xi_0|^{-1/3}-\frac{3}{4}\left(\frac{1}{2}+i\frac{\sqrt{3}}{2}\right)|\pi\xi_0|^{-2/3}f(\omega,\ \Omega)\right\}, \tag{4.69}$$

where

$$f(\omega,\ \Omega)=\frac{\omega-\Omega}{\omega}\left\{\frac{4}{3\pi^2}\left(1+\frac{Y_2}{2}\right)\Big[\ln|\pi\xi_0|+3\ln\left|\frac{\omega}{\omega-\Omega}\right|+\right.$$
$$\left.+3+i\pi-3i\pi\theta(\Omega-\omega)\Big]+\frac{1}{\pi}\,[0{,}40-0{,}87Y_2+Y_2^2F_0(Y_2)]+\right.$$
$$\left.+\sqrt{-\frac{3}{2}Y_2}\,\mathrm{ctg}\left[\sqrt{-\frac{3}{2}Y_2}F_{10}\left(\frac{3}{2}Y_2\right)\right]\right\}. \tag{4.70}$$

and where $\sqrt{-Y_2}$ is determined by (4.56) while for the functions F_0 and F_{10} definitions (4.55) and (4.63) are utilized.

We note that in order to investigate the frequency dependence of the impedance, in addition to the function F_{10}, in certain situations it is convenient to use

$$\widetilde{F}_{10}\left(\frac{3}{2}Y_2\right)=\frac{1}{\pi}\int\limits_0^\infty \frac{dx}{x^2+\frac{3}{2}Y_2}\left\{\ln\left|\frac{3}{2}q(x)\left(1+\frac{3Y_2}{2x^2}\right)-\frac{3Y_2}{2x^2}\right|+\right.$$

$$\left.+i\pi\,\mathrm{sgn}\,(\omega-\Omega)\,\theta\,(x_\varphi-x)\right\}. \qquad (4.71)$$

Here

$$\sqrt{-\frac{3}{2}Y_2}\left[F_{10}\left(\frac{3}{2}Y_2\right)-\widetilde{F}_{10}\left(\frac{3}{2}Y_2\right)\right]=\frac{\pi}{2}\,\mathrm{sgn}\,(\omega-\Omega)\,\mathrm{sgn}\,Y_2. \qquad (4.72)$$

The analysis of the behavior of the impedance near the cyclotron and cyclotron-wave resonance frequencies based on formula (4.69) will be represented in the next section. We will then give results from a numerical calculation in a frequency range that includes these resonances with finite values of the collision frequency ν.

4.3. Limit formulae for the surface impedance in the vicinity of cyclotron and cyclotron wave resonances

We will first consider the limit of small interelectron interaction. We note that the quantity ζ (4.32) is an analog of the corresponding expression in the gas model [29], at the same time that the cotangent in (4.70) is a fundamentally new component compared to the gas case in the general expression for the impedance of a degenerate electron liquid.

Taking $|Y_2| \ll 1$, we account for the fact that

$$F_0(0)=\int\limits_0^\infty \frac{dx\,x\varphi'(x)}{\varphi^3(x)}=0{,}12, \qquad (4.73)$$

$$\widetilde{F}_{10}(0)=\frac{1}{\pi}\int\limits_0^\infty \frac{dx}{x}\ln\left[\frac{3}{2}q(x)\right]=-\frac{0{,}88}{\pi}=-0{,}28, \qquad (4.74)$$

and as a result (see (4.72))

$$F_{10}\left(\frac{3}{2}Y_2\to 0\right)\to\frac{\pi}{2\sqrt{-\frac{3}{2}Y_2}}-0{,}28. \qquad (4.75)$$

Consequently, in this limit

$$\sqrt{-\frac{3}{2}Y_2}\,\mathrm{ctg}\left(\sqrt{-\frac{3}{2}Y_2}F_{10}\right)\to -0{,}42Y_2. \qquad (4.76)$$

Then, retaining all Y_2 linear terms ($|Y_2| \ll 1$) in (4.70) we obtain

$$f=f_0+\delta f, \qquad (4.77)$$

where

$$f_0 = \frac{4}{3\pi^2}\frac{\omega-\Omega}{\omega}\left[\ln|\pi\xi_0| + 3\ln\left|\frac{\omega}{\omega-\Omega}\right| + 3 + i\pi - 3i\pi\theta(\Omega-\omega)\right] + $$
$$+ 0{,}128\frac{\omega-\Omega}{\omega}, \tag{4.78}$$

$$\delta f = \frac{A_2}{1+A_2}\left\{\frac{20}{9\pi^2}\left[\ln|\pi\xi_0| + 3\ln\left|\frac{\omega}{\omega-\Omega}\right| + 3 + i\pi - 3i\pi\theta(\Omega-\omega)\right] - 2{,}3\right\}. \tag{4.79}$$

Correction (4.79) describes the interelectron correlation effects rather far from cyclotron resonance $\omega = \Omega$. We note that the dependence of the impedance on the magnetic field and the parameter A_2 is contained in the second component of (4.69) $\sim|\pi\xi_0|^{-2/3}$ In the limit $A_2 = 0$ expression (4.69) becomes the familiar result for the impedance of the electron gas [29, 55] (compare to (2.62):

$$Z_{O\pi}(\omega) = \frac{4\pi v}{c^2}\left\{\frac{\sqrt{3}}{2}\left(\frac{1}{2} - i\frac{\sqrt{3}}{2}\right)|\pi\xi_0|^{-1/3} - \frac{3}{4}\left(\frac{1}{2} + i\frac{\sqrt{3}}{2}\right) \times\right.$$
$$\times |\pi\xi_0|^{-2/3}\frac{\omega-\Omega}{\omega}\left(\frac{4}{3\pi^2}\left[\ln|\pi\xi_0| + 3\ln\left|\frac{\omega}{\omega-\Omega}\right| +\right.\right.$$
$$\left.\left.\left. + i\pi - 3i\pi\theta(\Omega-\omega) + 3\right] + 0{,}128\right)\right\}. \tag{4.80}$$

We then examine the vicinity of cyclotron-wave resonance when $\omega \to \Omega(1 + A_2)$ and, consequently, Y_2 is close to the value 10/3. For such values of Y_2 in accordance with (4.55) we may write

$$F_0(Y_2) = F_0\left(\frac{10}{3}\right) + \delta F_0, \tag{4.81}$$

where

$$F_0\left(\frac{10}{3}\right) = \int_0^\infty \frac{dx\, x\varphi'(x)}{\varphi^2(x)\,[\varphi(x) - 10/3]} = 0{,}36. \tag{4.82}$$

According to (4.55)

$$\operatorname{Im}\delta F_0 = -i\pi\frac{9}{100}\operatorname{sgn}(\omega-\Omega)\, x_\varphi, \tag{4.83}$$

where we may use the formula derived from (2.33) for x_φ:

$$x_\varphi = \sqrt{\frac{35}{8}\left(\frac{3Y_2}{10} - 1\right)}\,\theta\left(Y_2 - \frac{10}{3}\right). \tag{4.84}$$

We may obtain the expression for Re δF_0 analogously to (2.79) ($\delta \equiv Y_2 - 10/3$):

$$\operatorname{Re}\delta F_0 = \delta\int_0^\infty \frac{dx\, x\varphi'(x)}{\varphi^2(x)\left[\varphi(x) - \frac{10}{3}\right]\left[\varphi(x) - \frac{10}{3} - \delta\right]} =$$
$$= -\frac{9\pi}{100}\sqrt{\frac{35}{8}}\sqrt{\left|\frac{3Y_2}{10} - 1\right|}\,\theta\left(\frac{10}{3} - Y_2\right). \tag{4.85}$$

As a result and bearing in mind that $Y_2 = \frac{10}{3}\frac{A_2}{1+A_2}\frac{\omega}{\omega-\Omega}$, we obtain

$$F_0\left(\frac{10}{3}\,\frac{A_2}{1+A_2}\,\frac{\omega}{\omega-\Omega}\right)-F_0\left(\frac{10}{3}\right)=-\frac{9\pi\sqrt{35}}{200\sqrt{2}}\sqrt{\left|\frac{\omega-(1+A_2)\,\Omega}{(1+A_2)\,A_2\Omega}\right|}\times$$
$$\times\{\theta\,(A_2\,[\omega-\Omega\,(1+A_2)])+i\,\mathrm{sgn}\,A_2\theta\,(A_2\,[-\omega+\Omega\,(1+A_2)])\}. \qquad (4.86)$$

We now proceed to the calculation of $\tilde{F}_{10}(3Y_2/2)$ for which when Y_2 = 10/3 as a result of tabulation we obtain

$$\tilde{F}_{10}(5)=\frac{1}{\pi}\int_0^\infty\frac{dx}{x^2+5}\ln\left|\frac{3}{2}\,q\,(x)\left(1+\frac{5}{x^2}\right)-\frac{5}{x^2}\right|=-0{,}5. \qquad (4.87)$$

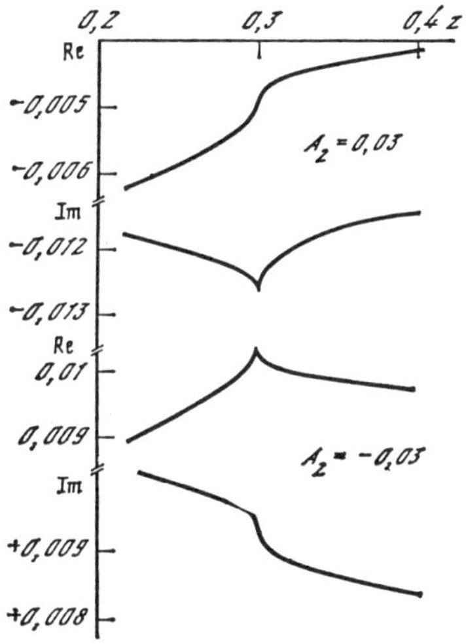

Fig. 11. The real and imaginary parts of the relative addition to the impedance in accordance with (4.69), (4.93) in the vicinity of the cyclotron-wave resonance point z = 0.3 (z = $0.3(1 + A_2)/(1 - \Omega/\omega)$ when ν = 0)

Further, taking

$$\tilde{F}_{10}\left(\frac{3}{2}\,Y_2\right)=\tilde{F}_{10}(5)+\delta\tilde{F}_{10}\left(\frac{3}{2}\,Y_2\right), \qquad (4.88)$$

for the imaginary part of the addition and proceeding from (4.71), we obtain

$$\mathrm{Im}\,\delta\tilde{F}_{10}\left(\frac{3}{2}\,Y_2\right)=\mathrm{sgn}\,(\omega-\Omega)\int_0^{x_\varphi}\frac{dx}{x^2+\frac{3}{2}\,Y_2}=$$
$$=0{,}5\,\sqrt{0{,}7}\,\mathrm{sgn}\,A_2\sqrt{\frac{3Y_2}{10}-1}\,\theta\left(Y_2-\frac{10}{3}\right). \qquad (4.89)$$

Using the following formula for the real part of the addition

$$\mathrm{Re}\,\delta\tilde{F}_{10}\left(\frac{3}{2}\,Y_2\right)=\frac{1}{\pi}\int_0^\infty\frac{dx}{x^2+5}\ln\left|\frac{\varphi\,(x)-Y_2}{\varphi\,(x)-10/3}\right|+$$
$$+\frac{1}{\pi}\int_0^\infty dx\left[\frac{1}{x^2+\frac{3}{2}\,Y_2}-\frac{1}{x^2+5}\right]\ln\left|\frac{3}{2}\,q\,(x)\left(1+\frac{3Y_2}{2x^2}\right)-\frac{3Y_2}{2x^2}\right|, \qquad (4.90)$$

we obtain

$$\mathrm{Re}\,\delta\tilde{F}_{10}\left(\frac{3}{2}\,Y_2\right)=\frac{1}{2}\sqrt{0{,}7}\sqrt{1-\frac{3Y_2}{10}}\,\theta\left(1-\frac{3Y_2}{10}\right). \qquad (4.91)$$

As a result we have

$$\tilde{F}_{10}\left(\frac{5A_2}{1+A_2}\,\frac{\omega}{\omega-\Omega}\right)-\tilde{F}_{10}(5)=0{,}1\sqrt{\frac{35}{2}}\sqrt{\left|\frac{\omega-\Omega\,(1+A_2)}{(1+A_2)\,A_2\Omega}\right|}\times$$
$$\times\{\theta\,(A_2\,[\omega-\Omega\,(1+A_2)])+i\,\mathrm{sgn}\,A_2\theta\,(A_2\,[-\omega+\Omega\,(1+A_2)])\}. \qquad (4.92)$$

Near cyclotron-wave resonance incorporating $\sqrt{-\frac{3}{2}Y_2} = i\sqrt{5}$, we obtain (see (4.69))

$$f(\omega, \Omega) = \frac{A_2}{1+A_2}\left\{\frac{32}{9\pi^2}\left[\ln|\pi\xi_0| + 3\ln\left|\frac{1+A_2}{A_2}\right| + i\pi - \right.\right.$$
$$- 3i\pi\theta(-A_2) - 0{,}25 - 1{,}36\sqrt{\left|\frac{\omega - \Omega(1+A_2)}{(1+A_2)A_2\Omega}\right|}\Bigg\} \times$$
$$\times\left\{\theta\left(\frac{\omega-\Omega(1+A_2)}{A_2}\right) + i\,\mathrm{sgn}\,A_2\theta\left(\frac{-\omega+\Omega(1+A_2)}{A_2}\right)\right\}. \quad (4.93)$$

The principal singularity of the impedance that arises here corresponds to excitation of longwave cyclotron waves propagating in the direction of the constant d.c. magnetic field. The real and imaginary parts of the resonance addition to the impedance (4.91) are given in Fig. 11. In the case of a positive A_2 the real part of the addition to the impedance is antisymmetric with respect to the resonance point, while the imaginary part is symmetric. With a negative A_2 the situation is reversed where the real part is symmetric and appears as a peak at the cyclotron-wave resonance point with a relative magnitude of the order $0.5 \cdot 10^{-3}$.

We will now proceed to the vicinity of cyclotron resonance $\omega \to \Omega$ where we may take $|Y_2| \gg 1$. In order to investigate the frequency singularities of the impedance we will first obtain the asymptotic expansion of the function $F_0(Y_2)$ in the limit $|Y_2| \gg 1$. Then accounting consistent with (2.34) and (2.75) for

$$x_\varphi = \frac{2Y_2}{\pi} + \frac{4}{\pi} - \frac{3\pi}{4} + O\left(\frac{1}{Y_2}\right). \quad (4.94)$$

Then in accordance with (4.55) we have

$$\mathrm{Im}\,F_0(Y_2) = -\frac{2}{Y_2}\theta(Y_2)\,\mathrm{sgn}(\omega-\Omega)\left\{1 + \frac{\pi}{2Y_2}\left(\frac{4}{\pi} - \frac{3\pi}{4}\right)\right\}. \quad (4.95)$$

We will represent the real part of the function F_0 as the sum of the components $F_1 + F_2$, where

$$F_1(Y_2) = \int_0^1 \frac{dt\, t\varphi'(t)}{\varphi^2(t)[\varphi(t) - Y_2]} \simeq -\frac{1}{Y_2}\int_0^1 \frac{dt\, t\varphi'(t)}{\varphi^2(t)} -$$
$$-\frac{1}{Y_2^2}\int_0^1 \frac{dt\, t\varphi'(t)}{\varphi(t)} = -\frac{0{,}03}{Y_2} - \frac{0{,}11}{Y_2^2}, \quad (4.96)$$

$$F_2(Y_2) = \int_1^\infty \frac{dt\, t\varphi'(t)}{\varphi^2(t)[\varphi(t) - Y_2]}. \quad (4.97)$$

In order to obtain an asymptotic expansion of the last integral we will use the Mellin transform in the variable $|Y_2|$. Then for the Mellin transformant we have

$$M_2(z) = -2(\cos\pi z)^{\theta(Y_2)}\left(\frac{\pi}{2}\right)^{z-1}\frac{1}{\sin\pi z}\sum_{\mu=1}^{\infty}\frac{g_\mu(z)}{z-\mu-1}, \quad (4.98)$$

where the quantities $g_\mu(z)$ are coefficients of the expansion

$$t\varphi'(t)\,[\varphi(t)]^{z-3} = \left(\frac{\pi t}{2}\right)^{z-2} \sum_{\mu=0}^{\infty} g_{\mu}(z)\, t^{-\mu}, \quad t > 1. \tag{4.99}$$

For these coefficients we have

$$g_0 = 1, \qquad g_1' = 3\pi/4 - 4/\pi, \qquad g_1(2) = 4/\pi - 3\pi/4. \tag{4.100}$$

Using the inverse Mellin transform

$$F_2(Y_2) = \frac{1}{2\pi i} \int_{C-i\infty}^{C+i\infty} dz\, |Y_2|^{-z} M_2(z), \quad 0 < C < 1 \tag{4.101}$$

and shifting the contour to the right from Re z = C we obtain the following contributions from the poles of expression (4.98):

a) second order pole $z = 1$, $\mu = 0$:

$$\frac{4}{\pi^2} \frac{d}{dz} \left\{ \left| \frac{\pi}{2Y_2} \right|^z \frac{\pi(z-1)}{\sin \pi z} (\cos \pi z)^{\theta(Y_2)} \right\}_{z=1} = \frac{2}{\pi Y_2} \ln \left| \frac{\pi}{2Y_2} \right|, \tag{4.102}$$

b) simple pole $z = 1$, $\mu \neq 0$:

$$\frac{4}{\pi^2} \frac{\pi}{2Y_2} \sum_{\mu=1}^{\infty} \frac{g_{\mu}(1)}{-\mu} = -\frac{1}{Y_2} \int_0^{\infty} dt \left[\frac{t\varphi'(t)}{\varphi^2(t)} - \frac{2}{\pi t} \right] = \frac{0{,}90}{Y_2}, \tag{4.103}$$

c) second order pole $z = 2$, $\mu = 1$:

$$\frac{4}{\pi^2} \frac{d}{dz} \left\{ \frac{\pi(z-2)\, g_1(z)}{\sin \pi z} \left(\frac{\pi}{2|Y_2|} \right)^z (\cos \pi z)^{\theta(Y_2)} \right\}_{z=2} = \frac{1}{Y_2^2} \left[g_1'(2) + g_1(2) \ln \left| \frac{\pi}{2Y_2} \right| \right], \tag{4.104}$$

d) simple pole $z = 2$, $\mu \neq 1$:

$$\frac{1}{Y_2^2} \sum_{\mu \neq 1} \frac{g_{\mu}(2)}{1-\mu} = \frac{1}{Y_2^2} \left\{ 1 - \int_1^{\infty} dt \left[\frac{t\varphi'(t)}{\varphi(t)} - 1 - \left(\frac{4}{\pi} - \frac{3\pi}{4} \right) \frac{1}{t} \right] \right\} = \frac{0{,}74}{Y_2^2}. \tag{4.105}$$

Summing these contributions with y_2^{-2} accuracy we obtain

$$\operatorname{Re} F_0(Y_2) = \frac{2}{\pi Y_2} \ln \frac{\pi}{2|Y_2|} + \frac{0{,}87}{Y_2} + \frac{1}{Y_2^2} \left\{ \left(\frac{4}{\pi} - \frac{3\pi}{4} \right) \left[\ln \left| \frac{\pi}{2Y_2} \right| - 1 \right] + 0{,}63 \right\}, \tag{4.106}$$

which together with (4.95) yields the asymptotic expansion of the function $F_0(Y_2)$ for large absolute values of the argument.

We will now formulate the asymptotic expansion of the integral $F_{10}\left(\frac{3}{2}Y_2\right)$. In accordance with (4.63) and (4.94) we have

$$\operatorname{Im} F_{10}\left(\frac{3}{2} Y_2\right) = -\frac{\pi}{2Y_2} \theta(Y_2) \operatorname{sgn}(\omega - \Omega) + O\left(\frac{1}{Y_2^2}\right). \tag{4.107}$$

In order to obtain an asymptotic expansion of the real part

$$\mathrm{Re}\, F_{10}\left(\frac{3}{2} Y_2\right) \equiv I_F\left(\frac{3}{2} Y_2\right) = \frac{1}{\pi} \int_0^\infty \frac{dx}{x^2 + \frac{3}{2} Y_2} \ln\left|1 - \frac{3}{2} q(x)\left(1 + \frac{2x^2}{3Y_2}\right)\right| \tag{4.108}$$

We will use the Mellin transform:

$$M_F(z) = \int_0^\infty dy\, y^{z-1} I_F(y \operatorname{sgn} Y_2) = \frac{1}{\pi} \int_0^\infty dx \int_0^\infty d\chi\, \chi^{z-1} \frac{x^{2z-2}}{1 + \chi \operatorname{sgn} Y_2} \times$$
$$\times \ln\left|1 - \frac{3}{2} q(x)\left(1 + \frac{\operatorname{sgn} Y_2}{\chi}\right)\right|, \tag{4.109}$$

with the convention $y = x^2\chi$. We may integrate the right half of the last formula with respect to x by parts. Assuming $1/2 < \mathrm{Re}\, z < 1$ we obtain

$$M_F(z) = \frac{1}{2z-1} \frac{1}{\pi} \int_0^\infty d\chi\, \chi^{z-1} \int_0^\infty dx\, x^{2z-1} \frac{q'(x)}{q(x)} \frac{1}{1 - \psi \operatorname{sgn} Y_2}, \tag{4.110}$$

where $\psi = \chi\left[1 - \frac{3}{2} q(x)\right] / \left[\frac{3}{2} q(x)\right]$. Introducing the integration variable ψ and accounting for

$$\frac{1}{\pi} \int_0^\infty \frac{d\psi\, \psi^{z-1}}{\psi - \operatorname{sgn} Y_2} = -\operatorname{sgn} Y_2 \frac{(\cos \pi z)^{\theta(Y_2)}}{\sin \pi z}, \quad 0 < \mathrm{Re}\, z < 1, \tag{4.111}$$

we obtain $1/2 < \mathrm{Re}\, z < 1$)

$$M_F(z) = -\frac{(\cos \pi z)^{\theta(Y_2)}}{(2z-1) \sin \pi z} \int_0^\infty dx\, x^{2z-1} \frac{q'(x)}{q(x)} \left[\frac{\frac{3}{2} q(x)}{1 - \frac{3}{2} q(x)}\right]^z. \tag{4.1112}$$

Then we use the expansions

$$\frac{q'(x)}{q(x)} \left[\frac{\frac{3}{2} x^2 q(x)}{1 - \frac{3}{2} q(x)}\right]^z = \begin{cases} -5x5^{z-1} \sum\limits_{\nu=0}^\infty e_\nu(z)\, x^{2\nu}, & |x| < 1, \\ -\frac{1}{x}\left(\frac{3\pi}{4} x\right)^z \sum\limits_{\mu=0}^\infty d_\mu(z)\, x^{-\mu}, & x > 1, \end{cases} \tag{4.113}$$

where $e_0(z) = d_0(z) = 1$. In accordance with this

$$M_F(z) = -\frac{(\cos \pi z)^{\theta(Y_2)}}{(2z-1) \sin \pi z} \left\{\frac{2}{5} 5^z \sum_{\nu=0}^\infty \frac{e_\nu(z)}{2\nu + 1} + \left(\frac{3\pi}{4}\right)^z \sum_{\mu=0}^\infty \frac{d_\mu(z)}{\mu + 1 - z}\right\}. \tag{4.114}$$

We will substitute the derived expression into the inverse Mellin transform

$$I_F(Y_2) = \frac{1}{2\pi i} \int_{C-i\infty}^{C+i\infty} dz \left|\frac{3}{2} Y_2\right|^{-z} M_F(z), \quad \frac{1}{2} < C < 1, \tag{4.115}$$

after which in order to obtain the asymptotic expansion we will shift the integration contour to the right half-plane, accounting for the contributions of the poles that arise from this procedure. With sufficient accuracy for our purposes we account for the contribution of the nearest pole $z = 1$. Here the contribution of the double pole arises with $\mu = 1$:

$$\frac{d}{dz}\left\{\left|\frac{3Y_2}{2}\right|^{-z}\frac{(\cos\pi z)^{\theta(Y_2)}}{(2z-1)\sin\pi z}\left(\frac{3\pi}{4}\right)^z d_0(z)(z-1)\right\}_{z=1}=\frac{1}{Y_2}\left\{\ln\frac{\pi}{2|Y_2|}-2\right\}. \quad (4.116)$$

The contribution of the simple pole is given by the expression

$$-\frac{2}{3\pi Y_2}\left\{2\sum_{\nu=0}^{\infty}\frac{e_\nu(1)}{2\nu+1}+\frac{3\pi}{4}\sum_{\mu=1}^{\infty}\frac{d_\mu(1)}{\mu}\right\}=$$

$$=-\frac{2}{3\pi Y_2}\left\{2\int_0^1 dx\sum_{\nu=0}^{\infty}e_\nu(1)x^{2\nu}+\frac{3\pi}{4}\int_1^{\infty}\frac{dx}{x}\sum_{\mu=1}^{\infty}d_\mu(1)x^{-\mu}\right\}=$$

$$=\frac{1}{\pi|Y_2|}\left\{\int_0^1\frac{dx\,xq'(x)}{1-\frac{3}{2}q(x)}+\int_1^{\infty}\frac{dx}{x}\left[\frac{x^2q'(x)}{1-\frac{3}{2}q(x)}+\frac{\pi}{2}\right]\right\}. \quad (4.117)$$

Accounting for the results from tabulation of the integrals:

$$\int_0^1\frac{dx\,xq'(x)}{1-\frac{3}{2}q(x)}=-1{,}19,\quad \int_1^{\infty}dx\left[\frac{xq'(x)}{1-\frac{3}{2}q(x)}+\frac{\pi}{2x}\right]=0{,}51, \quad (4.118)$$

we obtain

$$\operatorname{Re}F_{10}\left(\frac{3}{2}Y_2\right)=\frac{1}{2Y_2}\left\{\ln\frac{\pi}{2|Y_2|}-2{,}43\right\}+O\left(\frac{1}{Y_2^2}\right). \quad (4.119)$$

Formulae (4.119) and (4.107) make it possible to write in these asymptotics:

$$\sqrt{-\frac{3}{2}Y_2}\,\mathrm{ctg}\left[\sqrt{-\frac{3}{2}Y_2}F_{10}\left(\frac{3}{2}Y_2\right)\right]\sim 1/F_{10}\left(\frac{3}{2}Y_2\right)=$$
$$=-2Y_2\left[\ln\left|\frac{2Y_2}{\pi}\right|+2{,}43+i\pi\theta(Y_2)\,\mathrm{sgn}(\omega-\Omega)\right]^{-1}. \quad (4.120)$$

Finally, substituting the derived asymptotic expansion of the function $F_0(Y_2)$ (4.95), (4.106) into (4.70) and using (4.120) we obtain

$$f(\omega,\Omega)=\frac{\omega-\Omega}{\omega}\left\{\frac{4}{3\pi^2}\ln|\pi\xi_0|-\frac{3}{4}\ln\left|\frac{\omega-\Omega}{\omega}\right|+\left(\frac{4}{\pi^2}-\frac{3}{\pi}\right)\ln\left|\frac{3\pi(1+A_2)}{20A_2}\right|+\right.$$
$$+\frac{3}{4}+\frac{1{,}03}{\pi}+i\left[\frac{4}{3\pi}-\frac{4}{\pi}\theta(A_2)+\frac{3\pi}{4}\,\mathrm{sgn}\,A_2\theta(A_2(\omega-\Omega))\right]\Big\}+$$
$$+\frac{20}{3\pi^2}\frac{A_2}{1+A_2}\left\{\frac{1}{3}\ln|\pi\xi_0|+1+\ln\left|\frac{3\pi(1+A_2)}{20A_2}\right|+i\pi\left(\frac{1}{3}-\theta(A_2)\right)-\right.$$
$$\left.-\pi^2\left[\ln\left|\frac{20A_2\omega}{3\pi(1+A_2)(\omega-\Omega)}\right|+2{,}43+i\pi\theta(A_2(\omega-\Omega))\,\mathrm{sgn}\,A_2\right]^{-1}\right\}. \quad (4.121)$$

It follows from formula (4.121) that due to interelectron interaction near cyclotron resonance $\omega = \Omega$ a logarithmic singularity in the impedance $\sim[\ln \times (\omega - \Omega)^{-1}$ arises. A graph of the real and imaginary parts of the addition to the impedance (4.121) is shown in Fig. 12.

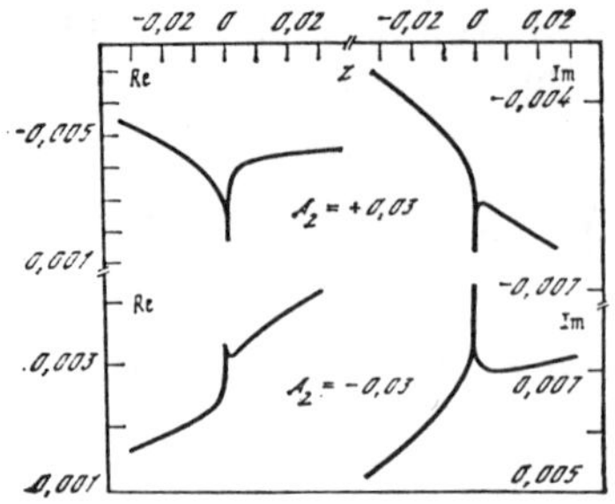

Fig. 12. The real and imaginary parts of the relative addition (4.121) when $z = 0$

$$\sim [\ln \times (\omega - \Omega)]^{-1}.$$

Both the real and imaginary parts of the addition to the impedance are represented as asymmetric peaks whose direction reverses with reversal of the sign of the constant A_2. In the case of a negative A_2 the peak of the real part has a relative order of approximately 10^{-3}.

4.4. Investigation of the resonance singularities of the impedance and the possibilities for experimental determination of the parameter A_2

In this section we will discuss results from the calculation of the impedance of a metal in the electron liquid model with diffuse scattering near the cyclotron-wave resonance frequency (4.93) and the cyclotron resonance frequency (4.121) by comparison to results from electron gas theory (4.78) and will compare these two corresponding results from a theory that posits specular electron scattering (2.78), (2.84), and (2.62). Since the strongest manifestation of interelectron interaction effects may be expected in resonance conditions, it may be useful to analyze these theoretical results for comparison to experimental data in order to obtain information on the parameters characterizing interelectron correlations in metals.

There will be a resonance dependence of the electron gas impedance on the magnetic field when the alternating frequency of the field coincides with the electron cyclotron frequency. In the case of specular reflection such a singularity is weaker $\sim(\omega - \Omega)^2 \ln (\omega - \Omega)$ than with diffuse scattering $\sim(\omega - \Omega) \ln (\omega - \Omega)$ (compare to (2.62), (4.78)). Moreover in the second case the singularity arises in a lower order of the expansion in the anomaly parameter $\sim|\xi_0|^{-2/3}$.

As demonstrated in section 2.3, incorporation of interelectron interaction will result in compensation of the most significant singular terms in cyclotron resonance, which eliminates the singularity existing in the electron gas. A principal singularity arises as a result at the cyclotron-wave resonance point attributable to Fermi-liquid cyclotron waves propagating along the d.c. magnetic field. The theoretical value of relative resonance correction (2.86) is 10^{-6} which lies beyond the measurement accuracy. On the other hand the first corrections attributable to Fermi-liquid interaction for the case of specular reflection $\sim 6\pi\xi_0 6^{-2/3}$ (2.60) in the same conditions has a value of $\sim 10^{-3}$, which lies within measurement accuracy.

Unlike the interelectron interaction-generated higher order non-resonance corrections with specular electron reflection in the case of diffuse scattering, higher corrections of the same order are, as indicated by (4.69), (4.93) resonance corrections. The root dependence for the impedance near the critical frequency of cyclotron waves was first discovered in the study by Baraff [25] where the theory of the impedance of an electron liquid with diffuse electron scattering was formulated using the approximate variational method. According to (4.69), (4.93) such a resonance addition to the impedance takes the form

$$\delta Z = 6{,}4\,\frac{v}{c^2}\,\frac{1+i\sqrt{3}}{|\pi\xi_0|^{1/3}}\,\frac{A_2}{1+A_2}\sqrt{\frac{\omega-\Omega(1+A_2)}{(1+A_2)A_2\Omega}}. \tag{4.122}$$

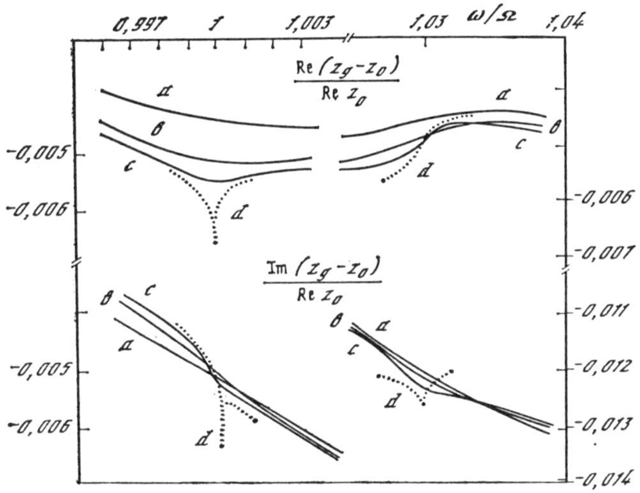

Fig. 13. The real and imaginary parts of the relative resonance addition to the impedance obtained numerically by formulae (4.69), (4.70) when $|\pi\xi_0|$ = 21600, A_2 = 0.03, ν/ω = 0.003 (a), 0.001 (b), 0.0003 (c), curves d - when ν = 0
The quantity $Z_0 = \pi v\sqrt{3}(1 - i\sqrt{3})|\pi\xi_0|^{-1/3}$ s^{-2} - the first term of asymptotic expansion (4.69)

The corresponding formula from study [25] has an undetermined coefficient which differentiates it from the one provided here. The impedance singularity in (4.122) is generated by cyclotron waves and is independent of the boundary reflection principle of the electrons although in the case of specular reflection (2.84) such a singularity will occur in the higher order of the expansion $\sim|\xi_0|^{-1}$. In the experimental conditions of study [23] the relative magnitude of the correction to the impedance near cyclotron-wave resonance (4.122) is 10^{-3}, which exceeds the corresponding addition in the case of specular reflection by three orders of magnitude, although it is less than the value of the peak discovered in the experiment at the critical frequency of the cyclotron wave [23].

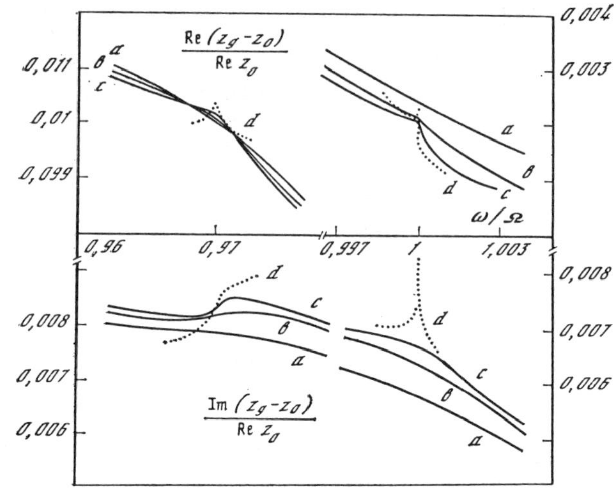

Fig. 14. The real and imaginary parts of the relative corrections to the impedance (4.69) when $|\pi\xi_0| = 21600$, $A_2 = 0.03$, $n/\omega = 0.003$ (a), 0.001 (b), 0.0003 (c); curves d - when $n = 0$

In the vicinity of cyclotron resonance the behavior of the impedance with diffuse electron scattering at the boundary of the metal is qualitatively different from the case of specular reflection. In the latter case, as noted above, the singularities compensate, which produces a regular dependence of the impedance on the magnetic field. On the other hand, in the case of diffuse scattering, not only does such compensation not occur, a singularity arises that is even stronger than in the gas model. In accordance with study [25] near cyclotron resonance the magnetic field-dependent addition to the impedance obeys the law $(\omega - \Omega + i\nu)^{-1}$. Our theory shows that there is no such impedance law. Consistent with (4.69), (4.121) the resonance dependence of the impedance takes the form

$$\delta Z = 10\pi \frac{v}{c^2} \frac{A_2}{1 + A_2} |\pi\xi_0|^{-2/3} (1 + i\sqrt{3}) \{\ln (A_2\omega/[(1 + A_2)(\omega - \Omega)]) + 3,18\}^{-1}. \tag{4.123}$$

We will provide results from the numerical calculation of the impedance of an electron liquid in diffuse scattering. Figs. 13 and 14 illustrate the magnetic field (or frequency) dependence, since ω/Ω: the ration of the resonance addition to the surface impedance is plotted on the X-axis. It is clear from a comparison of Figs. 13 and 14 that the relative position of the cyclotron and cyclotron-wave resonances on the curve plotting the impedance as a function of the magnetic field directly determines the sign of the Fermi-liquid constant A_2, as well as its absolute value.

Curves 1-3 obtained from numerical tabulation of formulae (4.69), (4.70) are drawn out for different values of the ratio of the collision frequency to the variable field frequency ν/ω = 0.003, 0.001, and 0.0003. Such values, obviously, satisfy inequality (2.85) which determines the conditions in which resonance dependencies may in fact arise. A comparison of the curves shows that in order to determine the Fermi-liquid interaction constants it is necessary to have a high metal purity that will provide a small yet quite realistic collision frequency.

CONCLUSIONS

1. Solution of the system of two integral equations for the electric field an electron liquid with diffuse scattering with a nonzero value of the parameter A_2 may be reduced to solving a single Wiener-Hopf equation.

2. The general equation for the impedance contains components that both have an analog in the case of the gas model and are entirely attributable to Fermi-liquid interaction.

3. The resonance singularity of the impedance at the cyclotron wave resonance point with diffuse electron scattering is manifest in a lower order of the expansion in the anomaly parameter than in the case of specular reflection. In the vicinity of cyclotron resonance a stronger singularity (compared to the gas model) occurs that is attributable to the existence of the A_2 constant.

4. The relative position of the singularities at the cyclotron and cyclotron-wave resonance points makes it possible to determine the sign and magnitude of the parameter A_2. The reason for this possibility is due to the small value of the collision frequency.

CONCLUSION

We will cover the primary results and conclusions obtained in our approach.

1. We have formulated a theory of the surface impedance of an isotropic metal in a magnetic field perpendicular to the surface that accounts for interelectron interaction characterized by the Fermi-liquid parameters A_1 and A_2.

2. With specular reflection or diffuse scattering of electrons on the metal surface in the anomalous skin-effect regime we have derived expansions of the impedance with term accuracy that largely determine the resonance dependence on the frequency and the magnetic field.

3. It is established that the Fermi-liquid interaction in the case of specular reflection with any sign of A_2 eliminates the logarithmic that existed at the cyclotron resonance point in the gas model.

4. It is demonstrated that due to interelectron interaction in the vicinity of cyclotron-wave resonance a principal singularity arises due to the excitation of longwave cyclotron waves. It is established that in the theory utilizing a diffuse boundary condition such a singularity arises in a lower order of the expansion in the anomalous parameter compared to the case of specular reflection.

5. Regardless of the nature of surface scattering, incorporation of the Fermi-liquid constant A_1 does not produce resonance dependencies of the impedance on frequency.

6. With diffuse scattering near cyclotron resonance it is discovered that unlike the case of specular reflection there is no total compensation of the singular terms in the impedance and a stronger resonance singularity arises, compared to the case of the gas model.

7. It is established that the relative position of the cyclotron-wave and cyclotron resonances on the curve plotting the impedance as a function of the magnetic field (or the frequency) in diffuse scattering directly determines the magnitude and sign of the constant A_2.

8. It is noted that high purity of the metal is required in order to assure clear manifestation of the resonance dependencies in the impedance; this purity will also generate a quite small yet quite realistic electron collision frequency. Corresponding conditions that impose constraints on the collision frequency are formulated.

9. Metals used in conjunction with the spherical Fermi surface model may be used for application of the developed theory; these include such metals as alkali metals.

10. The proposed method of calculating the impedance has a high degree of generality that makes possible its use for metal models that account for subsequent parameters of Fermi-liquid interaction (such as A_3).

It should be noted that the experimental investigation of phenomena to which this study is devoted is not yet on a level that would make it possible to perform a detailed comparison of theoretical predictions and experimental data. Specifically, the resolution for investigating the structure of the peak and the line shape of the signal near cyclotron-wave resonances is insufficient (Fig. 9); this process has been carried out only once to date in the experimental study by Baraff, Grimes, and Platzman [23] which was devoted to a consideration of the impedance of thick slabs. It would be desirable, in addition to enhancing measurement accuracy and sample purity, to record a signal of the magnetic field derivative of the impedance whose resonance singularities would be more clearly expressed.

In our view the further development of the theory would involve, first, investigation of the impedance of metallic samples of finite size such as slabs and films [56] with the assumptions discussed above and, second, a more detailed accounting of the conduction electron relaxation mechanisms on the surface, particularly the scattering diffusivity parameter falling outside the scope of the approximation (see, for example, [62]). One of the most important tasks would be a generalization of the theory that would employ concepts of the dependence of this parameter on the angle of incidence of the electron on the surface and would account for the actual electron scattering indicatrix on the surface. A comprehensive incorporation of this aspect within the scope of the electron Fermi-liquid model comprises a separate and rather complex task.

BIBLIOGRAPHY

1. Silin, V.P. "A theory of degenerate electron liquid" ZhETF, 1957, V. 33, p. 495-500.

2. Landau, L.D. "Fermi-liquid theory" ZhETF, 1956, V. 30, p. 1058-1064.

3. Silin, V.P. "Oscillations in a degenerate electron liquid" ZhETF, 1958, V. 35, p. 1243-1250.

4. Platzman, P.M., Jacobs, K.S. "Fermi liquid effects in cyclotron resonance" PHYS. REV., 1964, V. 134, p. 974-976.

5. Shults, S., Dunifer, G. "Observation of spin waves in sodium and potassium" PHYS. REV. LETT., 1967, V. 18, p. 283-287.

6. Silin, V.P. "Theory of degenerate electron liquid and electromagnetic waves" FMM, 1970, V. 29, p. 681-734.

7. Platzman, P.M., Wolff, P.A. "Spin-wave excitation in nonferromagnetic metals" PHYS. REV. LETT., 1967, V. 18, p. 280-283.

8. Walsh, W.M., Platzman, P.M. "Plasma wave propagation near cyclotron resonance in alkali metals" In: Tr. X Mezhdunar. konf. po fizike nizkikh temperatur [Papers of the 10th international conference on low temperature physics] Moscow VINITI, 1967, p. 214-218.

9. Walsh, W.M., Platzman, P.M. "Excitation of plasma waves near cyclotron resonance in potassium" PHYS. REV. LETT., 1965, V. 15, p. 784-786.

10. Platzman, P.M., Walsh, W. "Fermi liquid effects on plasma wave propagation in alkali metals" PHYS. REV. LETT., 1967, V. 19, p. 519-522.

11. Hamilton, D.C., McWhorter, A.L. "Plasma waves and Fermi liquid effects in alkali metals" BULL. AMER. PHYS. SOC. Ser. II, 1968, V. 13, p. 438.

12. Platzman, P.M., Waltsh, W.M., Jr. "Plasma waves and Fermi liquid effects in alkali metals" PHYS. REV. LETT., 1968, V. 20, Erratum, p. 89.

13. Mermin, N.D., Cheng, Y.C. "Fermi liquid effects in magnetoplasma modes in alkali metals" PHYS. REV. LETT., 1968, V. 20, p. 838-842.

14. Ying, S.C., Quinn, J.J. "Spin-independent oscillations of a degenerate electron liquid" PHYS. REV. LETT, 1968, V. 20, p. 1007-1010.

15. Ying, S.C., Quinn, J.J. "Spin independent oscillations of a degenerate electron liquid" PHYS. REV., 1969, V. 180, p. 193-217.

16. Silin, V.P. "Cyclotron waves in a degenerate electron liquid" ZhETF, 1968, V. 54, p. 1016-1021.

17. McIntyre, B.J., Ying, S.C., Quinn, J.J. "Angular dependence of spin-wave propagation in a degenerate electron liquid" PHYS. REV. LETT., 1968, V. 21, No. 17, p. 1244-1245.

18. Cheng, Y.C., Clarke, J.S., Mermin, N.D. "Magneto plasma modes in alkali metals" PHYS. REV. LETT., 1968, V. 20, p. 1486-1488.

19. Platzman, P.M., Walsh, W.M., Foo, K.N. "Fermi liquid effects on high frequency wave propagation in simple metals" PHYS. REV., 1968, v. 172, p. 689-692.

20. Frandsen, J.B., Gordon, R.A. "Effect of finite ωt on cyclotron wave propagation in metals in the longwave limit" PHYS. REV. B, 1976, V. 14, p. 4342-4344.

21. Saermark, K., Lebech, J. "Dispersion of cyclotron waves in metals and observability of many-body effects" PHYS. STATUS SOLIDI (B), 1977, V. 79, p. 723-730.

22. Gordon, R.A., Frandsen, J.B. "Comment on the determination of Fermi liquid parameters for m cyclotron wave measurements using empirical procedures" PHYS. REV., 1978, V. 17, p. 2785-2787.

23. Baraff, G.A., Grimes, C.C., Platzman, P.M. "Experimental evidence for the correlation produced magnetoplasma mode in potassium" PHYS. REV. LETT., 1969, V. 22, p. 590-592.

24. Baraff, G.A., "Surface impedance anomalies at perpendicular-field cyclotron resonance in the anomalous skin-effect regime" PHYS. REV., 1969, V. 187, p. 851-857.

25. Baraff, G.A. "Fermi liquid effects on surface impedance in anomalous skin effect regime" PHYS. REV. B, 1970, V. 1, p. 4307-4325.

26. Jones, H.C., Sondheimer, K.H. "Cyclotron resonance in the extreme anomalous relaxation region" PROC. ROY. SOC. LONDON A, 1964, V. 278, p. 256-273.

27. Baraff, G.A. "Enhancement of the anomalous skin effect fields beneath a rough surface and its effect on the correlation-produced magnetoplasma peak" PHYS. REV. B, 1970, V. 2, p. 637-644.

28. Kobelev, A.V. "The resonance singularities of the surface impedance of a metal generated by Fermi-liquid interaction" ZhETF, 1971, V. 61, p. 1202-1210.

29. Dingle, R.B. "The anomalous skin effect and the reflectivity of metals I" PHYSICS, 1953, V. 19, p. 311-347.

30. Gordon, A.M., Sondheimer, R.H. "Evaluation of the surface impedance in the theory of anomalous skin effect in metals" APPL. SCI. RES., 1953, V. B3, p. 297-305.

31. Reuter, T.G., Sondheimer, E.H. "The theory of the anomalous skin effect in metals" PROC. ROY. SOC. LONDON A, 1948, V. 195, p. 336-343.

32. Phillips, T.G., Baraff, G.A., Dunifer, G.L. "Fermi liquid effects in cyclotron phase resonance transmission through alkali metals" PHYS. REV. LETT., 1973, V. 30, p. 274-278.

33. Baraff, G.A. "Transmission of electromagnetic waves through a conducting slab. I. The two sided Wiener-Hopf solution" J. MATH. PHYS., 1968, V. 9, p. 372-384.

34. Baraff, G.A. "II. The peak at cyclotron resonance" PHYS. REV., 1968, V. 167, p. 625-636.

35. Baraff, G.A. "III. Effect of boundary scattering on helicon transmission near cut-off" PHYS. REV., 1969, V. 178, p. 1155-1165.

36. Baraff, G.A. "Transmission of electromagnetic waves through a conducting slab. IV. A simple multiple-reflection method" PHYS. REV. B, 1973, V. 7, p. 580-590.

37. Brinkman, W.F., Platzman, P.M., Rice, T.M. "A sum rule for the Landau-Fermi liquid parameters in metals" PHYS. REV., 1968, V. 174, p. 495-499.

38. Rice, T.M. "Landau-Fermi liquid parameters in Na and Ka" PHYS. REV., 1968, V. 175, p. 858-867.

39. Baraff, G.A. "Effect of Landau A_1 parameter on the microwave field within and electron gas" (I). PHYS. REV. B, 1973, V. 8, p. 5404-5412.

40. Baraff, G.A. "Cyclotron phase resonance in thin slab: Variational method. (II) PHYS. REV., 1974, V. 9, p. 1103-1114.

41. Dubovic, V.U., Fetisov, B.P. "On the possibility of determining the Landau coefficients in a charged Fermi liquid" PHYS. STATUS SOLIDI (b), 1973, V. 60, p. K95-K98.

42. Foc, E.N., Platzman, P.M. "Coupling of acoustic and cyclotron waves in an interacting electron gas" PHYS. REV. LETT., 1971, V. 27, p. 1568-1573.

43. Silin, V.P. "Theory of collective description of electron interaction in a solid" FMM, 1956, V. 3, p. 193-203.

44. Landau, L.D. "Theory of Fermi-liquid" ZhETF, 1958, V. 35, p. 97-105.

45. Gradshteyn, I.S., Ryzhik, I.M. "Tablitsy integralov, summ, ryadov i proizvedeniy" [Tables of integral, sums, series, and products] Moscow: Nauka, 1971, 1108 p.

46. McGroddy, J.C., Stanford, J.L., Stern, B. "Helicons and their effect on the impedance of metals" PHYS. REV., 1966, V. 141, p. 437-445.

47. Silin, V.P. "Zero-point sound and plasma oscillations in many particle systems" FIAN, No. 59, Moscow, 1972.

48. Kaner, E.A., Skobov, V.G. "Electromagnetic waves in metals in a magnetic field" ADV. PHYS., 1968, V. 17, p. 605-747.

49. Landau, L.D. "Oscillations in electron plasma" ZhETF, 1946, V. 16, p. 474-593.

50. Alodzhants, G.P., Silin, V.P. "Theory of cyclotron waves in metals" FMM, 1975, V. 39, p. 679-685.

51. Silin, V.P., Rukhadze, A.A. "Elektromagnitnye svoystva plazmy i plazmopodobnykh sred" [Electromagnetic properties of a plasma and plasma-like media] Moscow: Gosatomizdat" 1961, 244 p.

52. Mason, D.P. "Transmission and reflection of an electromagnetic wave incident normally on a plasma half-space" PHYSICA, 1972, V. 60, p. 41-46.

53. Dingle, R.B. "The evaluation of integrals appearing in the expression for the surface impedance" APPL. SCI. RES., 1953, V. B3, p. 69-99.

54. Kobelev, A.V., Silin, V.P. "The resonance properties of the surface impedance of a metal in a magnetic field normal to the metal surface" KVAT. SOOBSHCH PO FIZIKA, 1976, No. 9, p. 29-33.

55. Naberezhnykh, V.P., Dan'shin, N.K. "Diamagnetic resonance in a magnetic field normal to the metal surface" ZhETF, 1969, V. 56, p. 1223-1231.

56. Kondrat'ev, A.S., Kushma, A.Ye. "Elektronnaya zhidkost' normal'nykh metallov" [The electron liquid of normal metals] Leningrad: Izd-vo LGU, 1980, 200 p.

57. Nobl, V. "Metod Vinera-Khonfa" [The Wiener-Hopf method] Moscow: Izd-vo inostr. lit., 1962, 253 p.

58. Fok. V.A. "Solution of a single integral equation of mathematical physics" MAT. SB., 1944, V. 13, p. 3-15.

59. Kobelev, A.V., Silin, V.P. "Theory of surface impedance of a degenerate electron liquid of a metal in a magnetic field" FMM, 1975, V. 39, p. 231-240.

60. Kobelev, A.V., Silin, V.P. "A theory of the surface impedance of a metal in a magnetic field" FMM, 1977, V. 44, p. 499-510.

61. Kobelev, A.V., Silin, V.P. "The surface impedance of a metal in the cyclotron-wave and cyclotron resonance region" FMM, 1981, V. 52, p. 903-916.

62. Okulov, V.I., Ustinov, V.V. "Surface scattering of conduction electrons and the kinetic phenomena in metals" FNT, 1979, V. 9, p. 213-252.

SUBJECT INDEX